STRUCTURE AND BONDING

Volume 37

Editors:
J. D. Dunitz, Zürich · J. B. Goodenough, Oxford
P. Hemmerich, Konstanz · J. A. Ibers, Evanston
C. K. Jørgensen, Genève · J. B. Neilands, Berkeley
D. Reinen, Marburg · R. J. P. Williams, Oxford

With 40 Figures and 43 Tables

Springer-Verlag
Berlin Heidelberg GmbH 1979

ISBN 978-3-662-15420-5 ISBN 978-3-540-35230-3 (eBook)
DOI 10.1007/978-3-540-35230-3

Library of Congress Catalog Card Number 67-11280

Originally published by Springer-Verlag Berlin Heidelberg New York in 1979
Softcover reprint of the hardcover 1st edition 1979

Typesetting: R. & J. Blank, München. Printing and bookbinding: Brühlsche Universitätsdruckerei, Lahn-Gießen
2152/3140-543210

Contents

Local and Cooperative Jahn-Teller Interactions in Model Structures. Spectroscopic and Structural Evidence
D. Reinen, C. Friebel 1

The Frameworks (Bauverbände) of the Cubic Structure Types
E. E. Hellner 61

Polyhalogen Cations
J. Shamir 141

SPRINGER-VERLAG

D-6900 Heidelberg 1
P. O. Box 105280
Telephone (06221) 487·1
Telex 04-61723

D-1000 Berlin 33
Heidelberger Platz 3
Telephone (030) 822001
Telex 01-83319

SPRINGER-VERLAG
NEW YORK INC.

175, Fifth Avenue
New York, N.Y. 10010
Telephone (212) 477-8200

Local and Cooperative Jahn-Teller Interactions in Model Structures
Spectroscopic and Structural Evidence*

Dirk Reinen and Claus Friebel

Fachbereich Chemie der Universität Marburg und Sonderforschungsbereich 127 (Kristallstruktur und chemische Bindung), Lahnberge, 3550 Marburg/Lahn, Germany.

Table of Contents

I. Introduction . . . 2

II. Local Jahn-Teller Effects . . . 4

III. The Cooperative Jahn-Teller Effect . . . 14

IV. Local and Cooperative Jahn-Teller Distortions in Nitrocomplexes $A_2^IM^{II}T^{II}(NO_2)_6$ . . . 26
A. Alkaline Earth Complexes [M^{II}: Ca, Sr, Ba] with Cu^{2+} and Co^{2+} . . . 27
B. Lead Complexes [M^{II}: Pb] with Cu^{2+} . . . 30

V. Local and Cooperative Jahn-Teller Interactions in Host Lattice Structures with Distorted Octahedra . . . 41

VI. Various Examples . . . 52

VII. References . . . 57

The structural chemistry of solid transition metal compounds is quite often controlled by the Jahn-Teller effect. In particular d^9-, d^4- and low-spin d^7-configurated cations in octahedral coordination with σ-antibonding E_g ground states are strongly affected and may induce structural deformations and symmetry reductions. After a short introduction into the essential features of the *local* Jahn-Teller effect the phenomena of *cooperative* Jahn-Teller interactions are discussed and reviewed. The cooperative Jahn-Teller order patterns in different types of host lattice structures with Cu^{2+}, Ag^{2+}, Cr^{2+}, Mn^{3+} and low-spin Co^{2+}, Ni^{3+} in octahedral coordination are considered. Emphasis is laid upon the analysis of the multistage phase transitions from static to partially dynamic and finally to completely dynamic structures which occur with increasing temperature. Nitrocomplexes $A_2^IM^{II}Cu(NO_2)_6$ [M^{II}: Sr^{2+}, Pb^{2+}] are extensively discussed as model compounds in which cooperative Jahn-Teller interactions induce different structural modifications and order patterns. EPR spectroscopy proves to be the most powerful method for analysing the symmetry aspects and static to dynamic phase transitions of Jahn-Teller unstable solid compounds and is treated in some detail.

---

* Dedicated to Professor Otto Schmitz-DuMont on the occasion of his 80th birthday on 13th February 1979.

"It is a great merit of the Jahn-Teller effect that it disappears when not needed."
J.H. van Vleck (1939)

It is also true, however, that quite a number of phenomena in the chemistry of transition metal compounds can hardly be understood without the Jahn-Teller theorem.

I. Introduction

The Jahn-Teller theorem states that degenerate ground states are not possible [1,2]. There will always be a normal coordinate in the point group of the complex or molecule which provides a mechanism for lifting this degeneracy. While the theorem is valid in all symmetries with the exception of linear molecules for *orbitally* degenerate states [1], there is a restriction in the case of spin degeneracy in so far as Kramers doublets are not split by the Jahn-Teller effect [2]. In this article we will be concerned with orbitally degenerate ground states of transition metal ions in an octahedral coordination of equal ligands. Furthermore, we will restrict ourselves to σ-antibonding ground stes E_g (compare, however, Chap. VI) for which especially large energetic effects are expected. The d^n configurations leading to twofold degenerate ground states of this symmetry are those with:

$$n = 4\ [t_{2g}^3\, e_g^1 - {}^5E_g : Cr^{2+}, Mn^{3+}],\ n = 9\ [t_{2g}^6\, e_g^3 - {}^2E_g : Ni^+, Cu^{2+}, Ag^{2+}]$$

and

$$n = 7\ \text{(low-spin)}\ [t_{2g}^6\, e_g^1 - {}^2E_g : Co^{2+}, Ni^{3+}]$$

The only vibrational mode in O_h symmetry which can be active in removing the degeneracy of an electronic E_g state is also of E_g symmetry, and may lead to distortions corresponding to a tetragonal elongation or compression [D_{4h}] of the octahedra or to deformations with orthorhombic symmetries [D_{2h}] (Fig. 2).
Two questions will be of dominant interest in the following:

1. To which distortion geometries does the Jahn-Teller active E_g vibrational mode lead and which distortion geometry is energetically favored, the tetragonal elongation or compression?
2. How will the distorted octahedra arrange with respect to each other in a crystal structure of a certain symmetry, and which are the most common geometric patterns of cooperative Jahn-Teller ordering?

We will mainly study structures in which the octahedra are either isolated from each other or connected via common corners in two or three dimensions. Lattices with chains of octahedra bridged by common faces will also be discussed shortly. The most regular structure with a 3-dimensional array of linearly connected octahedra is the perovskite lattice corresponding to the stoichiometry ABX_3 (Fig. 1). If there is a 1:1 cation ordering in the octahedral sites, the elpasolite structure – $A_2BB'X_6$ – results (Fig. 1). While in the perovskite lattice each octahedrally coordinated site is occupied by a Jahn-Teller cation ($KB^{II}F_3$: B^{II} = Ni, Zn → Cr, Cu), only half of these sites may be occupied by Jahn-Teller unstable cations in the elpasolite lattice (Ba_2ZnWO_6 → Ba_2CuWO_6). The ordered cation distribution in the octahedral sites leads to quite large B^{II}–B^{II} distances and reduces the magnetic interactions between the paramagnetic d^n cations. Indeed EPR spectroscopy which is the most powerful method for analyzing the Jahn-Teller effect of d^9 and low-spin d^7 cations yields remarkably sharp signals.

In order to be sure that a distortion of a Cu^{2+} compound, for example, is Jahn-Teller induced, a comparison with the isomorphous compounds with Zn^{2+} or Ni^{2+} is useful. These ions possess orbitally nondegenerate ground states in octahedral coordination and have nearly the same ionic radii as Cu^{2+}. If the latter compounds contain Ni^{2+} and Zn^{2+} in regular octahedral coordination, while the ligand environment is strongly distorted in the case of Cu^{2+}, the Jahn-Teller effect must be responsible for the symmetry reduction.

The local Jahn-Teller effect in isolated octahedra is the subject of the next chapter. Experimental examples of Cu^{2+} doped into compounds of perovskite and elpasolite structure will be given. In Chap. III the cooperative Jahn-Teller effect as the consequence of high concentrations of Jahn-Teller centers in a solid matrix will be considered. In Chap. IV we discuss as illustrative model examples the highly interest-

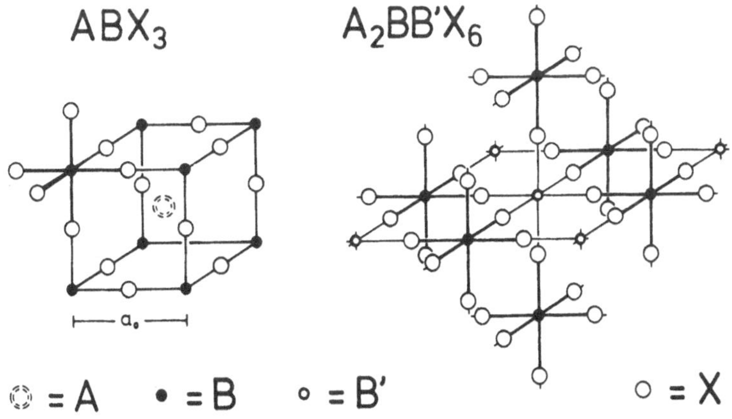

Fig. 1. The perovskite and elpasolite structures [left: perovskite unit cell (if A site empty: ReO_3 structure); right: 3-dimensional frame-work of corner-connected octahedra in the elpasolite structure]

ing spectroscopic and structural results which are obtained for nitrocomplexes with Cu^{2+} and low-spin Co^{2+}. Complexes of this type have become attractive for many research groups within the last decade, mainly because of the multistage phase transitions which they undergo as the consequence of dynamic and static Jahn-Teller effects. They will be treated rather extensively. The subjects of Chap. V are d^4 and d^9 cations in solid compounds in which the octahedral sites are already distorted along a 4-fold or 3-fold axis in the host lattice. Explicitly layer structures of K_2NiF_4 and related types as well as linear chain compounds are considered. Various further examples of Jahn-Teller instability which illustrate the importance of the Jahn-Teller effect in the solid state chemistry of transition metal compounds are finally discussed in the last chapter. The interesting case of d^9 and d^4 cations in cuboctahedral ligand fields with strongly σ-antibonding T_{2g} ground states is treated in detail. Some examples which demonstrate the influence of the Jahn-Teller effect on the high-spin ↔ low-spin transition of d^7 cations in octahedral coordination will be given.

II. Local Jahn-Teller Effects

Because there are many excellent reviews covering the theory of electronic E_g ground states, the degeneracy of which is lifted by the Jahn-Teller active vibrational E_g mode (Fig. 2)[3–6], we will present only a short outline of those aspects which are needed in the following chapters.

The linear coupling between electronic and vibrational motions leads to a potential surface which is called the "mexican hat" (Fig. 2). The upper and lower potential surfaces have energies:

$$E_{\pm} = E_0 \pm V\rho + \frac{M\omega^2}{2}\rho^2 \tag{1}$$

E_0 is the electronic energy in the absence of the Jahn-Teller effect. The second term measures the strength of the linear Jahn-Teller coupling, and the third term represents the potential energy associated with the E_g vibrations. ρ is the radial coordinate measuring the extent of distortion, M is the mass of one of the six ligands and ω the frequency of the radial vibration. The configurations of minimum energy in the lower surface are points on a circle with $\rho_{min} = \frac{|V|}{M\omega^2}$. The Jahn-Teller stabilization energy with respect to E_0 is:

$$E_{JT} \equiv E_{min} - E_0 = \frac{M\omega^2}{2}\rho^2_{min} = \frac{V^2}{2M\omega^2} \tag{2}$$

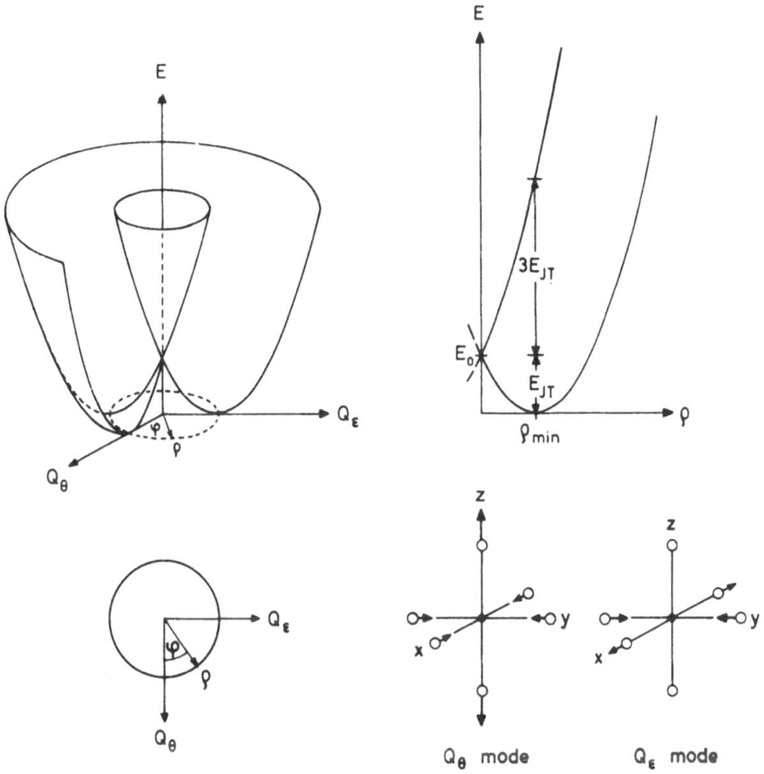

Fig. 2. Linear coupling of electronic E_g states and vibrational E_g modes [components $Q_\theta(\varphi = 0)$, $Q_\epsilon(\varphi = 90°)$]; "mexican hat" potential surfaces

The frequency ω of the radial vibration can be calculated from the following expression:

$$\omega = \frac{\hbar}{MA^2} = \left(\frac{2E_{JT}}{M\rho^2_{min}}\right)^{1/2} \tag{3}$$

in which A is the zero-point amplitude. The vertical distance between the minimum positions and the upper potential surface is $4E_{JT}$ (Fig. 2). The ground state and excited state eigenfunctions are:

$$\psi_+ = \sin\frac{\varphi}{2} d_{z^2} + \cos\frac{\varphi}{2} d_{x^2-y^2}$$
$$\psi_- = \cos\frac{\varphi}{2} d_{z^2} - \sin\frac{\varphi}{2} d_{x^2-y^2} \tag{4}$$

For d^9 configurated cations ($V<0$) ψ_+ is to be correlated with the ground state, while in the case of cations with a d^4 or low-spin d^7 configuration ($V>0$) ψ_- is associated with the lower potential surface. The angular coordinate φ in Eq.(4) describes the (tetragonal or o-rhombic) symmetry of the octahedra and their orientation in space, as described below. Explicitly the displacements of the ligands from their original octahedral positions on the x-, y- and z-axes, respectively, by the action of the Q_θ and Q_ϵ components of the E_g vibrations (Fig. 2) are:

$$\Delta x = \frac{\rho}{\sqrt{3}} \cos(\varphi + 240°), \qquad \Delta y = \frac{\rho}{\sqrt{3}} \cos(\varphi + 120°)$$

$$\Delta z = \frac{\rho}{\sqrt{3}} \cos\varphi, \qquad \rho = (\sum_i 2\,\Delta i^2)^{1/2}, \quad [i = x, y, z] \tag{5}$$

In the linear coupling approximation φ is undetermined. Equation (1) shows that there is an infinite number of energetically equivalent configurations of minimum energy. The introduction of higher order anharmonicity and coupling terms yields a more realistic description. Explicitly the third-order coupling induces the following depencende of the energy on the angular coordinate φ:

$$E_\pm = E_0 \pm V\rho + \frac{M\omega^2}{2}\rho^2 + V_3\rho^3 \cos 3\varphi \tag{6}$$

A warping of the lower potential surface results. Corresponding to the invarience of $E_\pm$ with respect to a change of φ by $\frac{2\pi}{3}$ three additional minima develop, the positions of which depend on the sign of the third order coupling constant V_3. With $V_3 < 0$ these minima are calculated from Eq.(6) to have positions at $\varphi = 0°$, $120°$, $240°$ with saddle-points at $\varphi = 60°$, $180°$, $300°$ separating them (Fig. 3). They are lowered with respect to the minimum positions of the mexican hat by the additional small energy:

$$\delta E = |\beta| \cong 2\left|\frac{V_3}{V}\right| \cdot \rho_{min}^2 \cdot E_{JT} \tag{7}$$

The energy barrier separating the minima is $2\,|\beta|$. As can easily be verified by the use of Eq. (5) the situation just described is equivalent with an energetic stabilization of the *tetragonal elongation*. $\varphi = 0°$, $120°$, $240°$ correspond to distortion geometries with the long cation – ligand bond lengths directed along the octahedral z-, x- and y-axes, respectively. In the alternative situation of $V_3 > 0$ the positions of the minima and the saddle-points have to be interchanged. The octahedra are tetragonally compressed along z($\varphi = 180°$), x($\varphi = 300°$) and y($\varphi = 60°$), respectively, in that case. With the exception of the mentioned six points with D_{4h} symmetry all other positions on the circle in the $\rho\varphi$-plane correspond to o-rhombic symmetries (D_{2h}).

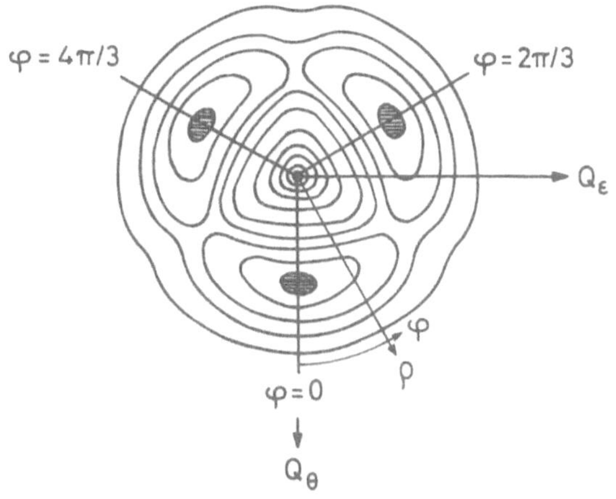

Fig. 3. Warping in the lower potential surface by nonlinear Jahn-Teller coupling

After the treatment of the static properties of local Jahn-Teller centers the dynamic aspects have to be considered by the inclusion of kinetic energy contributions. The vibrational energy $[E_{\upsilon,j}]_{\beta=0}$ which adds to the electronic energy of Eq.(1) in the case of strong linear and vanishing nonlinear Jahn-Teller coupling is:

$$[E_{\upsilon,j}]_{\beta=0} = \left(\upsilon + \frac{1}{2}\right)\hbar\omega + j^2\alpha, \quad \left[\alpha = \frac{(\hbar\omega)^2}{4E_{JT}}\right] \tag{8}$$

The first term represents the radial vibration energy (quantum number υ), the second energy contribution $[E_j]_{\beta=0}$ is associated with the angular oscillations (quantum number j). While υ is an integer, j may adopt the values 1/2, 3/2, 5/2, ... Each j defines a doubly degenerate vibronic E state. The energy variations and splittings of these angular E levels which are associated with the lower potential surface of the mexican hat (Fig. 2) are given as a function of the nonlinear coupling strength in Fig. 4[7]. The dotted line corresponds to the anisotropy energy $|\beta|$. The ratio of the angular frequency ω' and the radial frequency ω, defined in Eq. (3), has the estimated value[4]:

$$\frac{\omega'}{\omega} \approx 3\left(\frac{2|\beta|}{4E_{JT}}\right)^{1/2} \tag{9}$$

For strong nonlinear Jahn-Teller coupling there is a near-degeneracy of the lowest vibronic E level and one of the A states originating from the first excited E level; the energetic separation between the two mentioned states is called the tunneling splitting 3Γ and very small in this case. The near-degeneracy occurs between E and A_2 for V, $V_3 < 0$ [d^9: D_{4h}, elongated] (Fig. 4). If V, $V_3 > 0$ [d^4, low-spin d^7: D_{4h}, com-

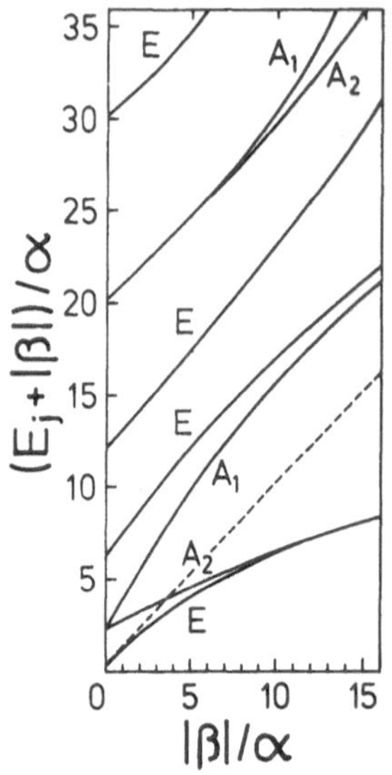

Fig. 4. Angular energies of lowest vibronic levels (strong linear coupling limit) as a function of nonlinear coupling strength (for V, $V_3 < 0$: d^9, D_{4h} elongated) [α as defined in Eq. (8)]

pressed] the notations of A_1, A_2 have to be interchanged in Fig. 4. In the following the d^9 case will be discussed in some more detail. Because the two lowest vibronic states E and A_2 are much lower in energy than the next excited A_1 level for strong nonlinear coupling, the calculation can be restricted to the vibronic triplet (E + A_2) ("three-state model"). The corresponding perturbation matrix is[4]:

$$\begin{array}{llll} \Phi_{A_2} & 3\Gamma + G_1 & -\frac{1}{\sqrt{2}} G_\epsilon & \frac{1}{\sqrt{2}} G_\theta \\ \Phi_{E\theta} & & G_1 - \frac{1}{2} G_\theta & \frac{1}{2} G_\epsilon \\ \Phi_{E\epsilon} & & & G_1 + \frac{1}{2} G_\theta \end{array} \qquad (10)$$

The parameters G_1 and G_θ, G_ϵ may represent the isotropic and anisotropic Zeeman interactions for example. They have the values [$2u \ll 1$]:

$$\begin{aligned} G_1 &= (g_0 + 4u - 5u^2)\beta S_z H \\ G_\theta &= (2u - u^2)(2n^2 - l^2 - m^2)\beta S_z H \\ G_\epsilon &= (2u - u^2)\sqrt{3}(l^2 - m^2)\beta S_z H \end{aligned} \qquad (11)$$

[n, l, m: directional cosines of magnetic field with respect to the octahedral z-, x-, y-axes; $g_0 = 2.00_2$; u: orbital contribution to the spectroscopic g factor]

If $3\,\Gamma$ is considerably larger than the anisotropic Zeeman interactions G_θ and G_ϵ, the following g factors of the lowest vibronic doublet are calculated from matrix (10):

$$g = g_0 + 4\,u - 5\,u^2 \pm (2\,u - u^2)\,[1 - 3\,(n^2\,l^2 + l^2\,m^2 + m^2\,n^2)]^{1/2} \quad (12)^1$$

EPR spectra with the cubic anistropy described by Eq. (12) have been observed for Cu^{2+}-doped MgO single crystals[8]. The angular dependence of the g tensor in the {100} planes is determined by $2\,\vartheta$:

$$g = g_0 + 4\,u - 5\,u^2 \pm (u - 1/2\,u^2)\,(1 + 3\,\cos^2 2\,\vartheta)^{1/2} \quad (13)$$

yielding two EPR signals with g values: $g_0 + 6\,u - 6\,u^2$ and $g_0 + 2\,u - 4\,u^2$ parallel to the cubic ⟨100⟩ axes and $g_0 + 5\,u - 5.5\,u^2$ and $g_0 + 3\,u - 4.5\,u^2$ in the ⟨110⟩ directions. This kind of angular dependence is typical for the dynamic behaviour which is associated with the lowest vibronic doublet.

If the tunneling splitting $3\,\Gamma$ is considerably smaller than G_θ and G_ϵ, the ground state wave-functions are completely localized in the three minima at $\varphi = 0°$, $120°$ and $240°$, respectively. The Jahn-Teller effect is static, and the following matrix is more appropriate than the equivalent matrix (10) for calculating energies in the three-state region[4]:

$$\begin{matrix} \Phi_z & \Gamma + G_1 + G_\theta & \Gamma & \Gamma \\ \Phi_x & & \Gamma + G_1 - \dfrac{G_\theta - \sqrt{3}\,G_\epsilon}{2} & \Gamma \\ \Phi_y & & & \Gamma + G_1 - \dfrac{G_\theta + \sqrt{3}\,G_\epsilon}{2} \end{matrix} \quad (14)$$

The vibronic eigenfunctions $\Phi_i (i = x, y, z)$ connected with the diagonal energies of matrix (14) are Born-Oppenheimer products of electronic and vibrational wavefunctions χ_i:

$$\begin{aligned} \Phi_z &= d_{x^2-y^2}\,\chi_z \\ \Phi_x &= d_{y^2-z^2}\,\chi_x = -\tfrac{1}{2}\,(d_{x^2-y^2} + \sqrt{3}\,d_{z^2})\,\chi_x \\ \Phi_y &= d_{z^2-x^2}\,\chi_y = -\tfrac{1}{2}\,(d_{x^2-y^2} - \sqrt{3}\,d_{z^2})\,\chi_y \end{aligned} \quad (15)$$

The g parameters are easily calculated from the diagonal energies of matrix (14):

$$\begin{aligned} g_\parallel &= g_0 + 8\,u - 7\,u^2 \\ g_\perp &= g_0 + 2\,u - 4\,u^2 \end{aligned} \quad (16)$$

1 This expression is to be modified in the presence of strain. The magnitude of g is not changed, however, if the strain is completely random.[4]

If ϑ is the angle between the preferred axis of the distorted octahedron and the magnetic field direction, the angular dependence of the g factor is given by:

$$g^2 = g_{\|}^2 \cos^2\vartheta + g_{\perp}^2 \sin^2\vartheta \tag{16a}$$

While the magnitude of the g parameters is identical for the three minima, the orientation of the CuL_6 polyhedra is different, namely the long Cu^{2+} – ligand bonds are directed along the octahedral x-, y- and z-axes, respectively.

If $3\,\Gamma$ becomes much larger than G_θ and G_ϵ, the diagonal energies of matrix (10) represent a better approximation. The wave-functions are not localized in the minima anymore because the energetic (near-)degeneracy of the vibronic triplet is lifted. The finite overlap of the nuclear wave-functions χ_i demands that the electronic parts of the functions [Eq. (15)] have to be orthogonal. The new eigenfunctions are the following linear combinations of Φ_z, Φ_x, Φ_y:

$$\begin{aligned} \Phi_{A_2} &= \frac{1}{\sqrt{3}}(\Phi_z + \Phi_x + \Phi_y) \\ \Phi_{E\theta} &= \frac{1}{\sqrt{2}}(\Phi_y - \Phi_x) \\ \Phi_{E\epsilon} &= \frac{1}{\sqrt{6}}(2\,\Phi_z - \Phi_x - \Phi_y) \end{aligned} \tag{17}$$

We have considered D_{4h} symmetries only so far. For the more general case of o-rhombic symmetries the following g parameters substitute those of Eq. (16) [static Jahn-Teller effect]:

$$g_x = g_0 + 4\,u - 5\,u^2 - (2\,u - u^2)(\cos\varphi - \sqrt{3}\sin\varphi) \tag{18}$$

or, if possible anisotropies in the orbital contributions u are allowed for in addition [9,10]:

$$\begin{aligned} g_x = g_0 &+ 4\,u_x - 2\,(u_y^2 + u_z^2) - (u_x u_y - u_y u_z + u_z u_x) \\ &- [2\,u_x + (2\,u_z^2 - u_y^2) - (2\,u_x u_y + u_y u_z - u_z u_x)]\cos\varphi \\ &+ \sqrt{3}\,[2\,u_x + u_y^2 - (u_y u_z + u_z u_x)]\sin\varphi \end{aligned} \tag{19}$$

$$u_i = \frac{k_i^2}{E_i}\xi_0, \quad [i = x, y, z]$$

g_y and g_z are generated by cyclic permutation of u_x, u_y, u_z and by changing φ by $\frac{4\pi}{3}$ every time. The k_i factors are covalency parameters and ξ_0 is the free ion LS-coupling parameter, with a value of 830 cm^{-1} for Cu^{2+}. The E_i represent electronic

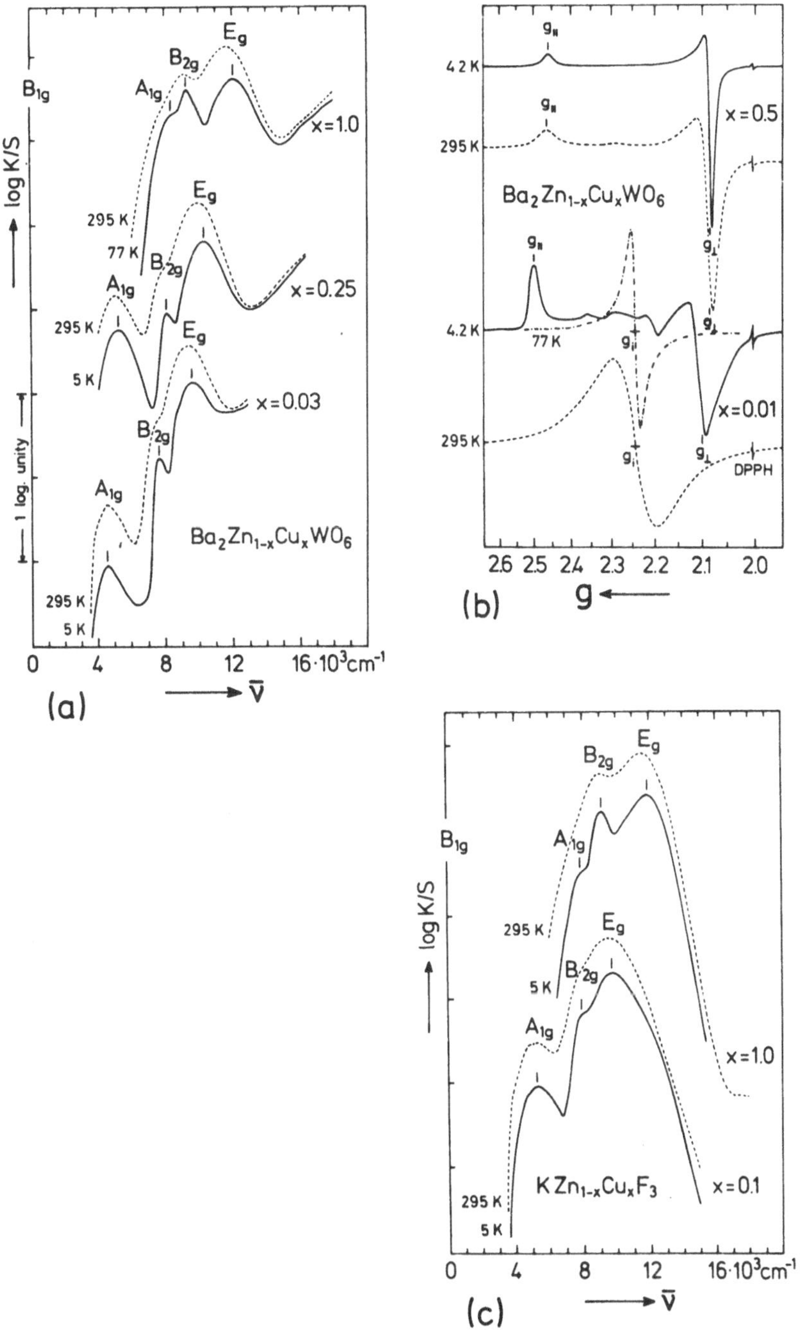

Fig. 5. Ligand-field (a) and EPR spectra (b) of elpasolite-type mixed crystals $Ba_2Zn_{1-x}Cu_xWO_6$ and ligand-field spectra of perovskite-type mixed crystals $KZn_{1-x}Cu_xF_3$ (c)

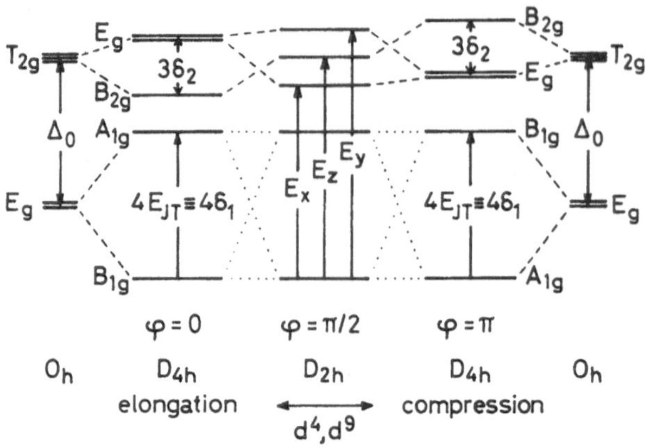

Fig. 6. Energy level diagram of d^4 and d^9 cations in ligand fields of D_{4h} and D_{2h} symmetries

energies which are obtained from the ligand-field spectra (Figs. 5,6). For $\varphi = 0$ and $u_x = u_y = u_\perp$, $u_z = u_\parallel$ Eq. (19) simplifies to:

$$\begin{aligned} g_\parallel &= g_0 + 8\,u_\parallel - 3\,u_\perp^2 - 4\,u_\parallel u_\perp \\ g_\perp &= g_0 + 2\,u_\perp - 4\,u_\parallel^2 \end{aligned} \qquad (20)$$

Figure 5 shows typical EPR powder spectra of Cu^{2+} in octahedral coordination, the host lattice being of the elpasolite type (Fig. 1). At very low temperatures the signal is anisotropic with g values (Table 1) which are in agreement with Eq. (20). They indicate a static distortion with the symmetry of a tetragonal elongation. By raising the temperature the EPR signal becomes isotropic, which can be accounted for by the assumption of a dynamic Jahn-Teller deformation of the CuL_6 octahedra. Obviously there is a rapid temperature-induced change of the preferred axis between the octahedral x-, y- and z-directions. This change has the symmetry of a pseudorotation around a threefold axis by $\frac{2\pi}{3}$ and $\frac{4\pi}{3}$ and can be followed quantitatively by corresponding variations of the angular parameter φ in the $\rho\varphi$-plane which characterizes the minimum positions of the lower potential surface in the mexican hat (Fig. 3). This pseudorotation with the angular frequency ω' [Eq. (9)] is fast with respect to the time scale of EPR spectroscopy and leads to motional narrowing in the EPR signal[3]. kT has become comparable to the height of the potential barrier $2\,|\beta|$ separating the minima. Excited angular levels (Fig. 4) which are not localized in one of the three minima anymore become thermally populated, and rapid vibrational relaxation occurs. The squares of the trigonometric functions which determine the

Table 1. Structural (Cu–L bond lengths [pm]) and spectroscopic data (ligand-field energies [10^3 cm^{-1}] and EPR parameters) of Cu^{2+} in various mixed crystals

	$Ba_2Zn_{1-x}Cu_xWO_6$		$KZn_{1-x}Cu_xF_3$		$Zn_{1-x}Cu_xSb_2O_6$
x	0.01	1.0	0.1	1.0	$0 < x \leq 1.0$
d_l	≈ 226[a] (2x)	242 (2x)	≈ 218[a] (2x)	225 (2x)	
d_m	≈ 206[a] (4x)	198 (4x)	≈ 196[a] (4x)	196 (2x)	
d_s				189 (2x)	
$\bar{d}$[b]	≈ 213	213	≈ 203	203	
ρ	≈ 23	≈ 51	≈ 25	≈ 38	
φ	–	0°	–	109°, 251°	
$E_{JT} \equiv \delta_1$	1.10	2.15	1.35	≅ 2.0	≅ 1.0
δ_2	0.55	0.95	0.55	0.95	0.8
Δ_0[c]	6.75	6.90	6.50	≅ 7.0	≅ 8.6
$g_\parallel$	2.50_2	2.44_2	2.56_7	–	
$g_\perp$	2.10_0	2.07_2	2.11_5	–	
$k_\parallel$[d]	0.78	0.80	0.85_5	–	
$k_\perp$[d]	0.82	0.77_5	0.88_5	–	
Ref.	11)	11,14)	13,21)	12,21)	18)

[a] Estimated from ligand-field data by AOM calculations.
[b] Average Cu–L bond lengths defined by $\bar{d} = (d_l + d_s + d_m)/3$.
[c] $\Delta_0 = E_{x,y} - \delta_2 - 2\,\delta_1$ (Fig. 6).
[d] Calculated from Eqs. (20) and (19) with the assumption of D_{4h} symmetry.

electronic ground state [Eq. (4)] have to be replaced by the corresponding expectation values:

$$\langle\cos^2\frac{\varphi}{2}\rangle + \langle\sin^2\frac{\varphi}{2}\rangle = 1\ , \quad \langle\sin\varphi/2 \cdot \cos\varphi/2\rangle = 0 \tag{21}$$

The dynamic limit is characterized by:

$$\langle\cos^2\varphi/2\rangle = \langle\sin^2\varphi/2\rangle = \tfrac{1}{2}\,,$$

and one obtains from Eqs. (18) and (19):

$$g_{iso} = g_0 + 4\,u - 5\,u^2 = \tfrac{1}{3}\,(g_x + g_y + g_z) \tag{22}$$

The ligand-field spectra exhibit three d–d bands, corresponding to the splitting of the octahedral ^2E_g ground state ($4\,E_{JT} = 4\,\delta_1$) and to the transitions to the split levels of the excited $^2T_{2g}$ state [15,16]. They reflect the symmetry of tetragonally distorted CuO_6 octahedra, however, irrespective whether the Jahn-Teller effect is of a static or dynamic nature (Figs. 5, 6; Table 1).

III. The Cooperative Jahn-Teller Effect

Before discussing cooperative Jahn-Teller interactions in general some experimental examples will be presented. If the Cu^{2+} concentration in the elpasolite host lattice Ba_2ZnWO_6 is continuously raised, a phase transition from the cubic $[a_c]$ to a tetragonal structure $[a_0, c_0]$ with $c_0/a_0 > 1$ $[a_0 < a_c < c_0]$ is observed at x = 0.23 (Fig. 7). The corresponding phase line has an unusual curvature. At concentrations below the inflection point at x = 0.35 the phase transition appears to be continuous, while it is of first order for x > 0.35 [19]. The structural and spectroscopic[2] results reveal that the CuO_6 octahedra are tetragonally elongated, with the long axes orientated parallel to the c-axis in the tetragonal modification $[\varphi = 0]$[11,14]. This symmetry of cooperative Jahn-Teller ordering is called *ferrodistortive* (Fig. 8).

The ligand-field bands of the mixed crystals $Ba_2Zn_{1-x}Cu_xWO_6$ – in particular the low-energy transition – show a pronounced shift to higher wave-numbers with increasing x above the phase transition (Figs. 5 a, 7). This result indicates a more dis-

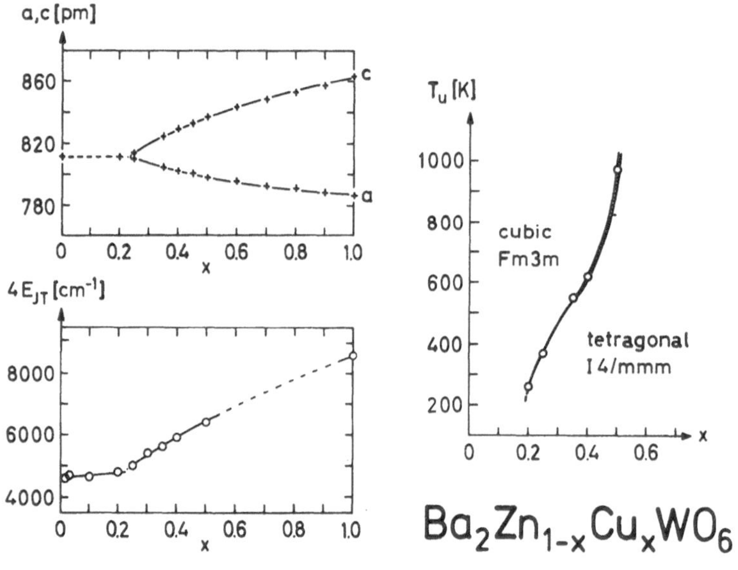

Fig. 7. Unit cell dimensions, phase line and ^2E_g ground level splittings ($4\,E_{JT}$) for mixed crystals $Ba_2Zn_{1-x}Cu_xWO_6$

2 Nicely resolved EPR powder spectra could be observed even for Ba_2CuWO_6. The nearest neighbor Cu^{2+}–Cu^{2+} distances in the structure (≈ 560 pm) are large enough as a consequence of the ordered distribution of W and Cu over the octahedral sites to avoid extreme dipolar broadening. The EPR spectra of $KZn_{1-x}Cu_xF_3$, however, are too broad at higher x-values to yield any reliable information.

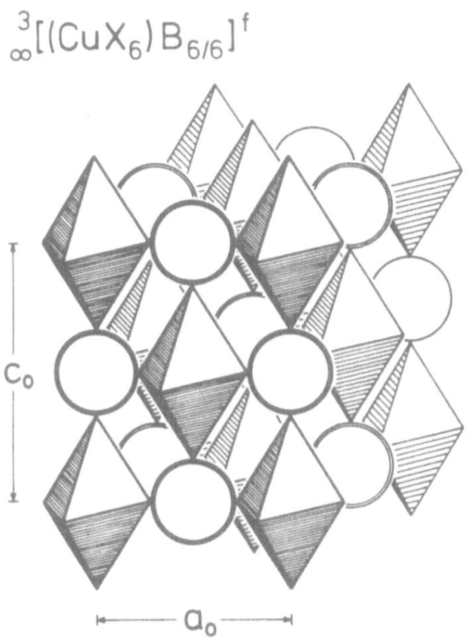

Fig. 8. Ferrodistortive order of elongated octahedra in elpasolite-type compounds $A_2BB'X_6$ [B' (*open circles*): non-Jahn-Teller cations]

tinct tetragonal distortion of the CuO_6 octahedra at higher than at low Cu^{2+} concentrations. We were able to estimate the bond lengths within the CuO_6 polyhedra from the ligand-field energies [14] using the angular overlap model (AOM) [20]. Obviously the minimum at $\varphi = 0$ in the lower potential surface of the mexican hat (Fig. 3) is appreciably lowered relative to those at $\varphi = 120^\circ$ and 240° with increasing x, because the cooperative-elastic interactions between the CuO_6 polyhedra within the solid matrix are steadily becoming more pronounced. Thus, two cooperative phenomena are observed at and above the phase transition in this structure type which is extremely favorable with respect to large Jahn-Teller deformations of the CuO_6 polyhedra:

(a) A symmetry reduction from cubic to tetragonal ($c_0/a_0 > 1$) which is induced by orbital ordering. The pattern of polyhedra orientation is ferrodistortive (Figs. 8, 11b).
(b) A considerable increase in the extent of the Jahn-Teller deformation of the single CuO_6 polyhedra with increasing Cu^{2+} concentration (Fig. 7).

The second result is not always obtained, but depends strongly on the specific connection pattern of the polyhedra in the host lattice structure and on the pattern of the cation distribution as well. For example no shift of the ligand-field bands is observed within the mixed crystal series $Zn_{1-x}Cu_xSb_2O_6$ with increasing x (Table 1), though a continuous phase transition from the tetragonal trirutile lattice to a monoclinic variant of this structure type occurs at $x \approx 0.7$. This and further examples of Jahn-Teller induced phase transitions with weak and strong effects corresponding to (b) are investigated and discussed elsewhere[18]. Geometric features which particularly favor strong cooperative phenomena of kind (b) are the following:

(1) host lattice structures with isolated or corner-connected octahedra;
(2) structures with close-packed anionic layers;
(3) cation distributions which occur perpendicular to close-packed anionic layers (relevant for compounds in which non-Jahn-Teller cations are also present in the octahedral sites; example: elpasolite structure).

About 50 mole % Cu^{2+} can be substituted into the Zn^{2+} site of $KZnF_3$ without a symmetry reduction of the cubic unit cell $[a_c]$[21]. At higher Cu^{2+} concentrations the structure becomes tetragonal [a, c] with $c_0/a_0 < 1$ $[a/\sqrt{2} \equiv a_0 > a_c > c/2 \equiv c_0]$ (Fig. 9). The structural analysis of $KCuF_3$ reveals tetragonally elongated octahedra which posses a distinct o-rhombic component with three different Cu–F bond lengths d_l, $d_m \approx d_s$ [12] (Table 1). While the d_m bonds order parallel to the tetragonal c-axis, the $d_l (d_s)$ bonds of neighbored CuF_6 polyhedra in the (001) planes are orientated perpendicular to each other. This kind of cooperative order is called *antiferrodistortive* (Fig. 9).

A unit cell with an antiferrodistortive order of tetragonally elongated octahedra can be described by consisting of two ferrodistortive sublattices which are rotated with respect to each other by $\frac{2\pi}{3}$ around the threefold axis of the cubic unit cell and which may be represented by two vectors in the directions $\varphi = \frac{2\pi}{3}$ and $\frac{4\pi}{3}$ in the $\rho\varphi$-plane (Figs. 2, 10a). The resultant vector describes the distortion geometry of the unit cell ($\varphi = 180° \rightarrow$ tetragonal with $c_0/a_0 < 1$). If the sublattices contain distorted octa-

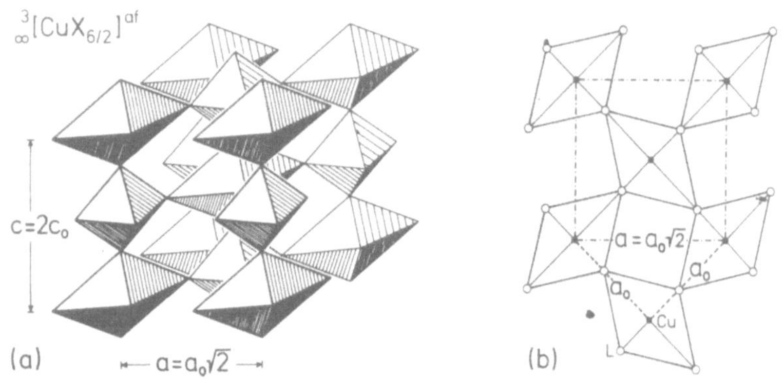

Fig. 9. Antiferrodistortive order of elongated octahedra in perovskite-type compounds ABX_3 [3-dimensional sketch (a) and projection into (001) plane (b)]

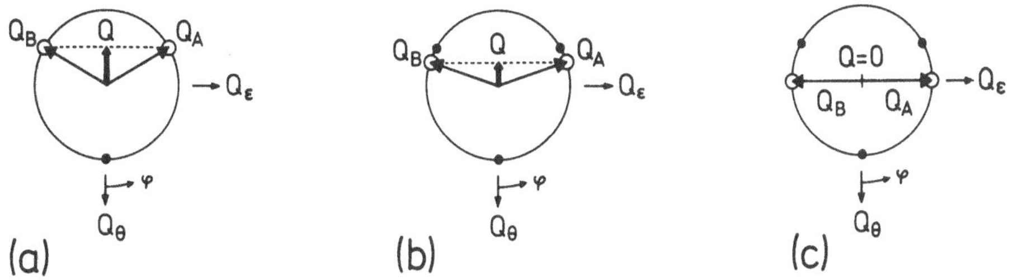

Fig. 10. Antiferrodistortive order with sublattices A and B at $\varphi = 120°, 240°$ (a), at $\varphi = 90°, 270°$ [$\pm Q_\epsilon$-symmetries] (c) and at $\varphi = 109.4°, 250.6°$ [$KCuF_3$] (b)

hedra with the geometry of the $\pm Q_\epsilon$ mode without contributions from Q_θ($\varphi = 90°$, $270°$), the antiferrodistortive order retains the cubic symmetry of the unit cell (Fig. 10c). The compounds $KCuF_3$ and MnF_3 (distorted ReO_3 structure, Fig. 1) represent examples between the two extreme cases with $\varphi = 109.4°, 250.6°$ and $\varphi = 96.6°, 263.4°$, respectively (Tables 1, 2; Fig. 10b). The sublattice angular parameters and the M–F bond lengths d_x, d_y and d_z are combined by the relation:

$$\operatorname{tg}\varphi = \sqrt{3}\,\frac{d_x - d_y}{2\,d_z - d_x - d_y} \quad [3] \tag{23}$$

Depending on the topology of the crystal structure different symmetries of cooperative Jahn-Teller order have been predicted. The results are derived on the basis of the theory of pair interactions between cations with strong Jahn-Teller coupling and of molecular field theory[22–29]. In the case of the spinel lattice with octahedra which are connected by common edges the ferrodistortive order is predicted to be the most stable one. The molecular field of all neighbored polyhedra acting on any polyhedron leads to an alignment of the preferred octahedral axes parallel to the c-direction of the unit cell [$\varphi = 0$]. For the 3-dimensional array of linearly connected octahedra, as it is realized in the perovskite structure (Fig. 1), the antiferrodistortive order comes out to be the energetically favored geometric pattern, however. In the NaCl structure finally both kinds of connections via common corners and edges occur, rendering this structure type as most interesting for realizing different Jahn-Teller orderings. The elpasolite lattice of Ba_2ZnWO_6 for example (Fig. 1) can be looked at as being of the NaCl type in a simplified picture by considering the octahedral WO_6 groups as single ligands (and disregarding the 12-coordinated Ba^{2+} ions). The same consideration applies to the elpasolite-related nitrocomplexes $A^{I}_{2}B^{II}Cu(NO_2)_6$ which are discussed in the next chapter. In the latter compounds the ferrodistortive order – as in Ba_2CuWO_6 – *and* the antiferrodistortive order can indeed be realized just by varying the chemical constitution slightly.

3 Relation (23) can be derived from Eq.(5) with the equalities: $d_x(d_y, d_z) = \bar{d} + \Delta x(y,z)$. $\bar{d}$ is the equilibrium distance within the undistorted octahedron $[\bar{d} \cong \frac{1}{3}(d_x + d_y + d_z)]$.

The molecular field can be formally treated as a tetragonal strain [26], because these concepts are symmetry-equivalent. The strain energies – with $\gamma < 0$ for d^9 cations – are of the magnitude [4]:

$$G_\theta^s = \gamma \rho_s \cos\alpha\,, \qquad G_\epsilon^s = \gamma \rho_s \sin\alpha \tag{24}$$

A ferrodistortive order with the long octahedral z-axes parallel to the crystallographic c-direction $[\alpha = 0]$ corresponds to a strain of G_θ symmetry:

$$G_\theta^s = -|\gamma|\,\rho_s\,, \qquad G_\epsilon^s = 0 \tag{24a}$$

Under the condition of strong linear and nonlinear Jahn-Teller coupling the nearly degenerate vibronic triplet of Eq. (15) and in Fig. 4 is split into a low-lying singlet A_1^I and an excited doublet $(A_1^{II} + B_1^I)$ by the action of the strain term in Eq. (24a). The energies of A_1^I and $(A_1^{II} + B_1^I)$ [26] have the values $-|\gamma|\,\rho_s$ and $\frac{1}{2}|\gamma|\,\rho_s$, respectively, as is easily derived from matrix (14). If the Zeeman interaction is also included, the g values which characterize the minimum at $\varphi = 0$ (A_1^I ground state) should be observed. The experimental EPR data of ferrodistortive elpasolite compounds [11] are indeed in agreement with Eqs. (16), (16a). A phase transition of first order is predicted for the ferrodistortive order and found for the mixed crystals $Ba_2Zn_{1-x}Cu_xWO_6$ with $x > 0.35$ (Fig. 7). Only if the anisotropy energy $|\beta|$ is considerably smaller than the molecular field strength, the latent heat will be very small, and the transition may appear experimentally as being of second order [26]. This is obviously the case for the just mentioned mixed crystals in the region $0.35 > x > 0$.

The model calculation in the molecular field approximation is similar for antiferrodistortive and ferrodistortive ordering in the first step. The energies for either of the two ferrodistortive sublattices can be readily obtained from matrix (10) for strong linear Jahn-Teller coupling. The elastic interaction between the two sublattices is taken into account by adding an antiferrodistortive coupling term to the diagonal energies, which is of the same magnitude for each of the three states [27]. The molecular field in each sublattice can again be described by the strain components G_θ^s and G_ϵ^s and induces the energetic contributions:

$$\alpha = 120°, 240°: \quad G_\theta^s = \frac{1}{2}|\gamma|\,\rho_s\,, \qquad G_\epsilon^s = \mp\frac{\sqrt{3}}{2}|\gamma|\,\rho_s \tag{24b}$$

$$\alpha = 90°, 270°: \quad G_\theta^s = 0\,, \qquad G_\epsilon^s = \mp|\gamma|\,\rho_s \tag{24c}$$

While the molecular field tends to stabilize sublattices with $\pm Q_\epsilon$ symmetries [Fig. 10c, Eq. (24c)], the anisotropy contributions from the nonlinear Jahn-Teller coupling shift the sublattices towards the positions with tetragonal symmetry [Fig. 10a, Eq. (24b)]. Depending on the ratio of molecular field strength and anisotropy energy situations between the extremes (Fig. 10b) may be stabilized, as already discussed. If one approaches the phase transition, dynamic contributions become of major importance and shift the sublattices towards $\varphi = 90°$ and $270°$ [27]. The phase transition, transforming the tetragonal unit cell with $c_0/a_0 < 1$ into the cubic high-temperature modification, is predicted to be of second order.

The single crystal EPR spectrum of a Cu^{2+} compound with an antiferrodistortive order is expected to consist of *two* signals in the (001) planes, each corresponding to one of the two sublattices. The g values for each of these sublattices – with the assumption of strong Jahn-Teller coupling and large molecular fields ($\Gamma \ll G_\theta^s, G_\epsilon^s$) – are those of Eq. (16). If an electronic overlap between the two inequivalent Cu^{2+} sites is present (via superexchange for example), the two signals may eventually collapse into one. This situation arises if the exchange integral J is large with respect to the anisotropy contribution to the Zeeman energy $(g_\parallel - g_\perp)\beta H$ [3]. A fast electronic exchange between two sublattices resembles the mechanism leading to the dynamic Jahn-Teller effect. Correspondingly the g value of the unified EPR signal along an arbitrary magnetic field direction in the (001) plane is obtained by taking the *linear* average of the molecular g values $g_\parallel$ and $g_\perp$ which characterize the sublattices A and B [3,98], and using Eq. (16):

$$\begin{aligned} g_{ex}(\perp [001]) &= \tfrac{1}{2}(g_\parallel + g_\perp) = g_0 + 5u - 5.5u^2 \\ g_{ex}(\parallel [001]) &= g_\perp = g_0 + 2u - 4u^2 \end{aligned} \qquad (25)$$

Equation (25) bases on the simplified model of an isotropic exchange interaction within a pair of dissimilar ions [identical ions which differ in orientation] with $s = 1/2$ [3] and results for $x\ [\equiv J/(g_\parallel - g_\perp)\beta H] \gtrsim 40$ [98]. Interesting situations with more signals in the (001) plane, which show a pronounced angular dependence, are only expected at lower x-values. If one introduces 2γ as the canting angle between the long axes (or between the planes of the short Cu–L bond lengths) of the CuL_6 polyhedra in the sublattices A and B, a modified g tensor results [98]:

$$\begin{aligned} g_{ex}(\parallel [100]) &= \cos^2\gamma\, g_\parallel + \sin^2\gamma\, g_\perp \\ g_{ex}(\parallel [010]) &= \sin^2\gamma\, g_\parallel + \cos^2\gamma\, g_\perp \\ g_{ex}(\parallel [001]) &= g_\perp \end{aligned} \qquad (26)$$

Equation (26) is again valid only for large values of x. $2\gamma = 90°$ describes the undisturbed antiferrodistortive order (Fig. 9b) and changes Eq. (26) into Eq. (25). An extension to o-rhombic sublattices A and B with $\pm\varphi \mp 120°$ is easily possible.

The g tensor of Eq. (25) with $g_{ex}(\perp [001]) > g_{ex}(\parallel [001]) > g_0$ reflects the *tetragonal compression of the unit cell* (Fig. 10). It differs essentially from the molecular g values, however, which characterize *tetragonally compressed octahedra* [Eq. (18) with $\varphi = 180°$] and for which a sequence $g_\perp > g_0 \gtrapprox g_\parallel$ is valid:

$$\begin{aligned} g_\perp &= g_0 + 6u - 6u^2 \\ g_\parallel &= g_0 - 3u^2 \end{aligned} \qquad (27)$$

Basing on a paper of Abe and Ono several authors propose to calculate molecular g values from the crystal g parameters by taking *quadratic* instead of *linear* averages[98a)]:

$$g_{ex}(\| [100]) = [g_\|^2 \cos^2\gamma + g_\perp^2 \sin^2\gamma]^{1/2}$$
$$g_{ex}(\| [010]) = [g_\|^2 \sin^2\gamma + g_\perp^2 \cos^2\gamma]^{1/2} \quad (28)$$
$$g_{ex}(\| [001]) = g_\perp$$

The model calculations[3,98)] mentioned above suggest that this procedure is not correct. The numerical differences when calculating molecular g values by applying Eq. (25) and (28) [with $2\gamma = 90°$], respectively, are only very small in most cases, however. We have applied Eq. (25) or (28) to explain the EPR spectra of nitrocomplexes $A_2^I PbCu(NO_2)_6$[30,31,33)] (Chap. IV). Equation (28), extended to the case of o-rhombic sublattices, proved to be useful for analyzing the disturbed antiferrodistortive pattern in Ba_2CuF_6[48)] (Chap. V). Several further examples will be discussed in the following which demonstrate that electronic coupling is quite common in Cu^{2+} compounds.

The orbital ordering which characterizes the different cooperative orientation patterns of Jahn-Teller distorted octahedra is also responsible for definite magnetic structures. The interesting correlation between the symmetry of Jahn-Teller ordering and the magnetic structure will be considered now and illustrated by simple examples. The *simplest model structure* for studying cooperative Jahn-Teller interactions would be a cubic *primitive unit cell* with octahedra which are isolated from each other. Compounds $T^{II}(ONC_5H_5)_6X_2$ [T^{II}: $Fe^{2+} \rightarrow Zn^{2+}$; X: ClO_4^-, BF_4^-] with pyridine N-oxide as ligands crystallize in a tetragonally distorted version of such a lattice type at 298 K (Fig. 11a)[34,35)]. They exhibit interesting magnetic properties, and it is especially surprising that the two Cu^{2+} compounds behave quite differently. It could be deduced from heat capacity data that the perchlorate complex is a one-dimensional antiferromagnetic system below 1.0K, while the fluoborate compound becomes a planar antiferromagnet at $T < 1.4$ K[34,36)]. We have performed single crystal EPR measurements with these two complexes which have rhombohedral unit cells at 298 K as the other compounds, but show phase transitions to unit cells with lower symmetries at temperatures of $\cong$ 90 K (BF_4^-) and below 70 K (ClO_4^-). From the angular dependence of the g tensor at 77 K and 4.2 K, respectively, a ferrodistortive order of elongated CuO_6 octahedra could be deduced for the fluoborate (Fig. 11b), while an antiferrodistortive order is found for the perchlorate (Fig. 11c)[37)]. In the latter case the cooperative g_{ex} values of Eq. (25) are observed, in contradiction to those of Eq. (16) which are found for the fluoborate complex. These results are in complete agreement with the reported magnetic structures, as follows from the inspection of the superexchange patterns in Fig. 12. A ferrodistortive order of elongated octahedra should indeed lead to planar antiferromagnetism in the case of magnetic exchange between d^9 cations [interacting half-filled $d_{x^2-y^2}$ orbitals in (001) planes], in close analogy to the case with corner-connected polyhedra in Fig. 12a.

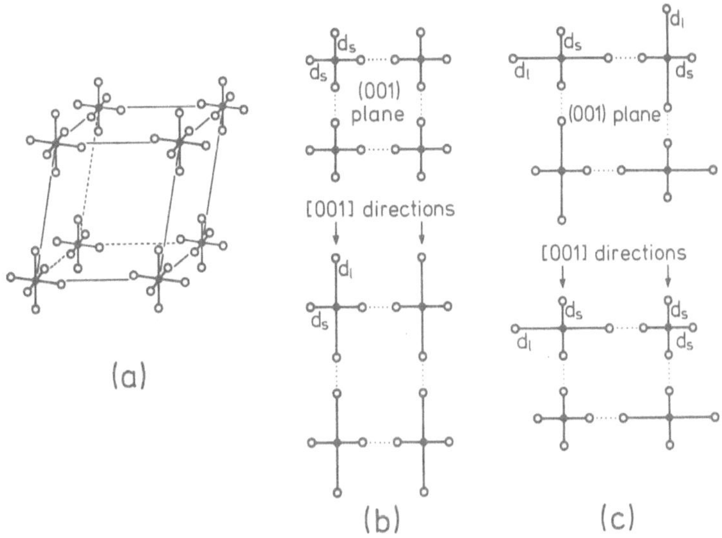

Fig. 11. CuO_6 octahedra (dynamic) in the unit cells of complexes $Cu(ONC_5H_5)_2X_2$ at 298 K [from [34]] (**a**), ferrodistortive [$X = BF_4^-$] (**b**) and antiferrodistortive order [$X = ClO_4^-$] (**c**) of elongated CuO_6 octahedra (static) in low-temperature modifications

For the antiferrodistortive alternative an antiparallel spin-spin coupling along the [001] direction is expected, while the interaction between filled d_{z^2} and half-filled $d_{x^2-y^2}$ orbitals in the (001) planes is negligible in first approximation (compare Fig. 12b). In both cases the long-range elastic Jahn-Teller forces are much stronger

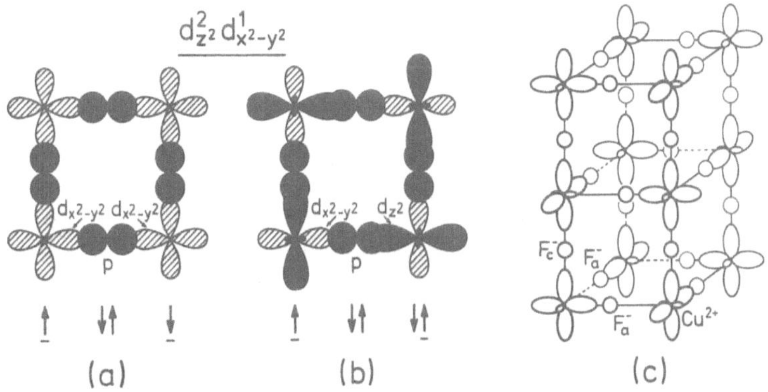

Fig. 12. Orbital ordering in (001) planes of structures with corner-connected elongated CuL_6 octahedra for ferrodistortive (**a**) and antiferrodistortive coupling (**b**); the 3-dimensional orbital order in the perovskite structure ($KCuF_3$) is also shown (**c**)

than the long-range magnetic interactions. This is clearly reflected by the considerably higher temperature of the Jahn-Teller induced structural phase transitions. The low-temperature unit cell distortions should be of pseudotetragonal symmetry with $c_0/a_0 > 1$ and < 1 for the ferro- and antiferrodistortive order[37], respectively (Fig. 11). This example demonstrates again that the elongated octahedral coordination represents the most stable distortion geometry. The antiferrodistortive order pattern seems to be slightly favored with respect to the ferrodistortive alternative, because the g tensor of newly prepared complexes with X: ClO_3^-, NO_3^- gives evidence for antiferrodistortively coupled CuO_6 polyhedra in the latter compounds also[37a]. In any case the energy difference between the two cooperative patterns is very small, because the tiny perturbation induced by a change of the anionic groups is sufficient to alter the ordering symmetry. A similarly interesting case is represented by the complex $[Cu(en)_3]SO_4$. Again the isolated CuN_6 octahedra exhibit a dynamical Jahn-Teller distortion at 298 K, but undergo a phase transition to statically elongated octahedra in disturbed antiferrodistortive order at 180 K[124].

If we proceed from isolated complex units to octahedra which are corner-connected in three dimensions we have as a model example the perovskite $KCuF_3$ with an antiferrodistortive order of elongated CuF_6 octahedra (small o-rhombic component, Table 1). Again in agreement with the discussed symmetry correlation (Fig. 12b,c) $KCuF_3$ is a linear antiferromagnet [exchange constants $J/k = -190$ K for the intrachain coupling]. The very weak interchain interaction within the (001) planes [$|J'|/k \approx 0.2$ K] provides a mechanism for the observed magnetic long-range order with $T_N \approx 30$ K[38], however. Similar considerations are possible for d^4 cations and will be discussed later.

A further model lattice type with isolated octahedra is the cubic K_2PtCl_6 structure (Fig. 13a). It can be derived from the elpasolite lattice $A_2BB'X_6$ simply by omitting the B′ cations. Compounds $A_2^{II}Cu(OH)_6$ [A^{II}: Sr, Ba] adopt a Jahn-Teller modified structure of this type[39] which is characterized by an antiferrodistortive order of elongated $Cu(OH)_6^{4-}$ octahedra. This order is different from those considered above, however. The octahedra in z = 1/4 and z = 3/4 (Fig. 13a) constitute the two ferrodistortive sublattices in which the long Cu–O bond lengths are located in the planes perpendicular to [001]. The order pattern is "disturbed", however, in the sense that the angle 2γ between the long axes in successive planes along [001] deviates appreciably from 90°: $2\gamma = 122°$ and 119° for the Sr^{2+} and Ba^{2+} compounds, respectively. It is interesting to note that the introduction of either additional Jahn-Teller ions or non-Jahn-Teller ions into the empty B′ sites of the K_2PtCl_6 lattice [perovskite and elpasolite structures, respectively] induces strong elastic correlations between the CuL_6 polyhedra via anionic bridges (Figs. 1,13a). The energetic and geometric situation is quite different from that just discussed.

Experimental data with respect to cooperative Jahn-Teller ordering in perovskite and elpasolite compounds as well as in lattices with isolated octahedra are collected in Table 2 for transition metal ions with electronic E_g ground states. The most inter-Chaps. II and III can be summarized as follows:

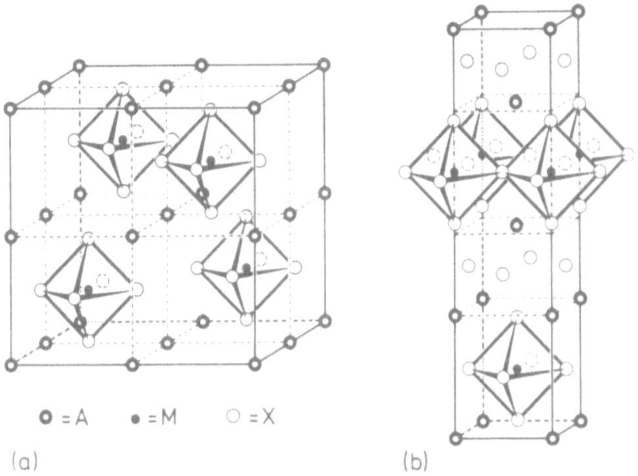

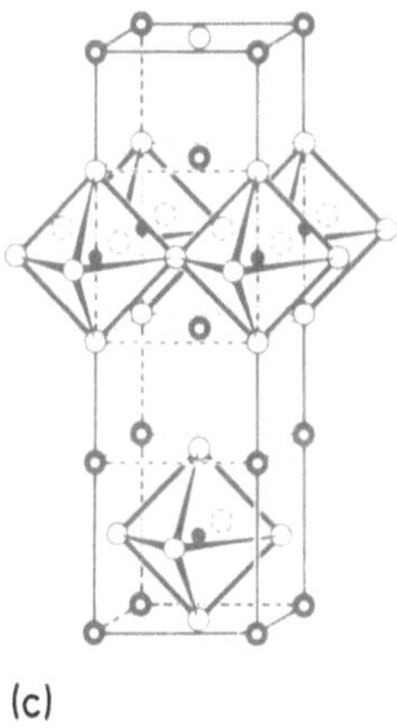

Fig. 13. Unit cells of compounds $A_2MX_{6(4)}$: cubic K_2PtCl_6 (a), tetragonal layer structures Ba_2NiF_6 (b) and K_2NiF_4 (c)

In all cases in which Cu^{2+} is isomorphously substituted into a host lattice site of *regular octahedral symmetry*, the Jahn-Teller effect induces a tetragonal *elongation* – sometimes with an o-rhombic component superimposed (Tables 1, 2). The same seems to be true for d^4 and low-spin d^7 cations. A statement concerning the relative stabilities of cooperative Jahn-Teller ordering patterns is also possible. In general the antiferrodistortive order of elongated octahedra is the slightly preferred Jahn-Teller pattern in compounds with d^4 and d^9 cations (Table 2). An apparent exception is the elpasolite lattice which seems to stabilize the ferrodistortive order. But even in this structure type the antiferrodistortive pattern can be observed, as in $(NH_4)_3MnF_6$ which crystallizes with $c_0/a_0 < 1$[41] and in the lead nitrocomplexes of Cu^{2+} for example. These cases will be treated in the next chapter. Included in Table 2 are also structural data of layer structures in which corner-connected octahedra extend only in two dimensions. Compounds of this type are the subject of Chap. V.

Table 2. Structural and spectroscopic data of representative examples with different types of Jahn-Teller ordering [abbreviations: f = ferrodistortive, af = antiferrodistortive, af* = disturbed antiferrodistortive, d = dynamic, d* = partially or planar dynamic].

Compound	Parent structure		T–L bond lengths [pm][c] d_l	d_m	d_s	$\bar{d}$
(a) Isolated octahedra						
$Tl_2Cu(OH_2)_6(SO_4)_2$	Tutton salt [$P2_1/a$]		231.7	201.7	195.7	209.7
β-$Cu(ONC_5H_5)_6(BF_4)_2$ (< 90 K)	cubic primitive		≅ 230[d]	≅ 198[d]	≅ 198[d]	209[e]
β-$Cu(ONC_5H_5)_6(ClO_4)_2$ (5 K)	cubic primitive		≅ 229[d]	≅ 199[d]	≅ 199[d]	209[e]
$Cu(terpy)_2(NO_3)_2$[a]	–		≅ 229	≅ 208.5	≅ 199	212
$Ba_2Cu(OH)_6$	K_2PtCl_6		280.5	197.2	195.8	224.5
(b) 1-dimensionally connected octahedra[a]						
Na_2CuF_4	–		237	192	190	206
α-$CsCuCl_3$ (> 423 K)	$CsNiCl_3$ [113]		251 (3×)		239 (3×)	245
β-$CsCuCl_3$	$CsNiCl_3$		277.5	≈ 232[f]	≈ 232[f]	247
α-$CsCrCl_3$	$CsNiCl_3$		262 (3×)		242 (3×)	252
β-$RbCrCl_3$	$CsNiCl_3$	$\overline{X',Y'}$:	(≅ 259)	(≅ 259)	≅ 240	253
		Z:	≅ 269	≅ 243	≅ 243	252
γ-$RbCrCl_3$ (100 K)	$CsNiCl_3$	X': Y':	≈ 275[f]	≈ 245[f]	≈ 242[f]	254
		Z:	272	≈ 240[f]	≈ 240[f]	251
(c) 2-dimensionally connected octahedra[a]						
Ba_2CuF_6	Ba_2ZnF_6 [100]		232.5	193.5	185	204
K_2CuF_4	K_2NiF_4 [132]		222	192	192	202
			228	195	186	203
$(NH_4)_2CuCl_4$	K_2NiF_4		279	233	230	247
Rb_2CrCl_4	K_2NiF_4		274	243	240	252
La_2CuO_4	K_2NiF_4		246	191	191	209
$CsMnF_4$	$TlAlF_4$		(201)	(201)	181	194
(d) 3-dimensionally connected octahedra						
$KCrF_3$	$KZnF_3$		232	200	195	209
K_2NaMnF_6	Ba_2ZnWO_6		206	186	186	193
Cs_2KNiF_6 (5K)[b]	Ba_2ZnWO_6		≈ 191	≈ 179	≈ 179	≈ 183[g]
$KCuF_3$, Ba_2CuWO_6: see Table 1						
$A^I_2M^{II}T^{II}(NO_2)_6$ [T^{II}: Cu, Co]: see Table 3						
(e) Further examples[a]						
Ba_2CuTeO_6	Ba_2NiTeO_6 [110]		≈ 218[f]	≈ 205[f]	≈ 203[f]	209
$Ba_3CuSb_2O_9$	$Ba_3NiSb_2O_9$ [112]		208.0 (3×)		206.5 (3×)	207

[a] Compare Chap. V.
[b] Compare Chap. VI.
[c] Values in parentheses: apparent spacings and distortions by partial dynamic averaging.
[d] Estimated from AOM calculations.
[e] From dynamic α-phase (295 K).
[f] Averaged over acentric displacements of T^{II}.
[g] Estimated from ionic radii [Shannon and Prewitt [74]].
[h] From bond lengths and rms displacements, respectively.

ρ [pm][h,c]		φ[j,c]		JT order	Ligand-field data [10^3 cm^{-1}]			Coval. parameters		Ref.
					E_{JT}	δ_2	Δ_0	k_z	$k_{x,y}$	
39;	–	≈ ± 120°[k];	≈ ± 120°[k]	af*	2.0	0.8	7.7_5	0.82	0.88	40)
– ;	37	– ;	1°	f	1.85	0.9	8.5_5	0.78	0.82	51,35)
– ;	35	– ;	≈ ± 117°	af	1.75	0.9	8.5_5	0.78	0.82	
31;	–	± 138° ;	± 136°	af	1.85	≅ 0	11.2	0.78	0.78	42)
97;	–	± 121° ;	± 121°	af*	≈ 3.7	≈ 0.7	≈ 7.8	0.79	0.74	39)
53;	–	2° ;	≅ 0°	f	≈ 2.2	0.8	≈ 7.0	0.85	0.89	43,13)
– ;	–	– ;	–	af*(d)	–	–	–	–	–	44)
≅ 53;	–	0°, ± 120° ;	–	af*	1.45	≈ 0	≅ 6.5	–	–	45)
– ;	–	– ;	–	af*(d)	≅ 1.55	≅ 0.3	≅ 7.4	–	–	46)
(22);	–	(180°) ;	–	af*(d*)	≅ 1.65	≈ 0	≅ 7.7	–	–	66,55)
30 ;	–	0° ;								
≅ 37;	–	≈ + 120° ;	–							
		≈ – 120° ;	–	af*	1.65	≅ 0.5	7.8	–	–	65,55)
≅ 37;	–	0° ;	–							
51[i];	–	± 130° ;	± 129°	af*	≅ 2.5	0.9	≅ 6.9	0.84	0.84	47,48,49)
35 ;	–	± 120° ;	–	af	≅ 2.0	0.8	≅ 6.9	≈ 0.82	≈ 0.83	50,13)
44[i];	–	± 108° ;	–							51)
55;	–	± 117° ;	–	af	≈ 2.7	≈ 0.7	≈ 6.6	≈ 0.76	≈ 0.72	52)
37;	–	≈ ± 115° ;	–	af	1.9	≅ 0.4	7.3	–	–	53,54,55,136)
64[l];	–	0° ;	–	f	–	–	–	–	–	56)
(23)[i];	37	(180°) ;	–	af(d*)?	3.8	1.3	13.3	–	–	57,58)
40;	–	± 113° ;	–	af	≅ 2.3	1.0	8.5	–	–	59)
23;	–	0° ;	–	f	2.25	0.7	14.4	–	–	58,59)
≅ 14;	–	0° ;	–	f	1.5	≈ 0.2	≈ 13.0	–	–	61,62)
≅ 16;	–	≈ 6° ;	–	f	1.9	0.7	6.8_5	–	–	63)
– ;	≈ 40	– ;	–	d	≅ 1.55	≅ 1.0	≅ 7.1	–	–	64)

[i] ρ Values [pm] from bond lengths of Ba_2ZnF_6: 10 [100)], K_2NiF_4: 3 [132)], La_2NiO_4: 36 [107)], $CsFeF_4$: 12 [106)].

[j] From bond lengths and EPR data, respectively.

[k] Two sublattices which are not exchange-coupled; o-rhombic distortion of $Cu(OH_2)_6^{2+}$ octahedra corresponding to $\varphi = 9°; 8°$.

[l] Calculated from EPR data [third order terms included, Eq. 19)].

Concluding this chapter some statements concerning the experimental access to the Jahn-Teller parameters may be important:

1. E_{JT} can be taken from the ligand-field spectra (Figs. 5, 6). They are of considerable magnitude in the considered cases, in agreement with the assumption of strong Jahn-Teller coupling. Similar energies and energy variations are found for Cu^{2+} in various oxidic mixed crystals different from those discussed [17,18].
2. $\rho_{min}(\equiv \rho)$ is easily calculated from structural data by applying Eq. (5) if the Jahn-Teller effect is static. In the dynamic case ρ_{min} can be estimated from the rms amplitudes [95] [compare Chap. IV].
3. The angular coordinate φ is accessible from the g tensor or the structural data [Eqs. (18), (19), (5)].
4. Information about the potential barrier height $2\,|\beta|$ [Eq. (7)] may readily be obtained from the temperature dependence of the ligand-field and EPR spectra.

IV. Local and Cooperative Jahn-Teller Distortions in Nitrocomplexes $A_2^IM^{II}T^{II}(NO_2)_6$

Ni^{2+} compounds $A_2^IM^{II}Ni(NO_2)_6$ [A^I: K, Rb, Tl, Cs; M^{II}: Ca, Sr, Ba or Pb] crystallize with cubic unit cells (Fm3). The structure is equivalent to the elpasolite type if the NO_2^- groups are regarded as one-atom ligands (Fig. 1). The NO_2^- groups are located symmetrically with respect to the octahedral x-, y- and z-axes (Fig. 14) and in-

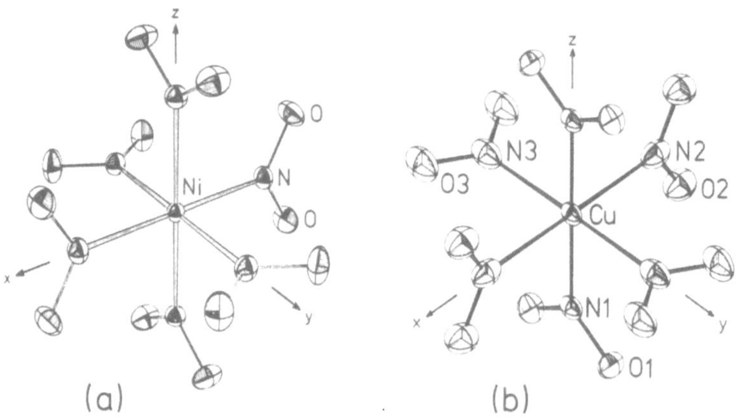

Fig. 14. $T(NO_2)_6^{4-}$ polyhedra in $K_2SrNi(NO_2)_6$ **(a)** [from [69]] and β'-$Cs_2PbCu(NO_2)_6$ **(b)** [from [89]]

duce a 12-coordination of M^{II} by oxygen. The N-atoms are ligated to the transition metal ion, as can be concluded from the ligand-field parameters[67,30,68] and has been confirmed by structural analysis[69] (Table 3). The Ni–N bond lengths seem to be significantly larger in the Cs than in K(Tl, Rb) complexes, which is indicated by much smaller Δ values in the former compounds[30]. Similarly the increasing Fe–N bond length in nitrocomplexes $A_2^IPbFe(NO_2)_6$ along the series A^I = K, Tl, Rb, Cs is very sensitively reflected by Mössbauer spectroscopy for example, where a linear dependence of the isomer shift from the unit cell dimension is found[70].

A. Alkaline Earth Complexes [M^{II}: Ca, Sr, Ba] with Cu^{2+} and Co^{2+}

The substitution of Ni^{2+} by Cu^{2+} in complexes with alkaline earth ions in the M^{II} positions leads to o-rhombic structures (Fmmm), the symmetry of which is pseudotetragonal with $c_0/a_0 > 1$, however[67,30]. One can easily deduce from the EPR and ligand – field data that the unit cell distortion is caused by tetragonally elongated CuN_6 octahedra (with a tiny o-rhombic component) in ferrodistortive order[30] (Fig. 8). The g values of the stoichiometric Cu^{2+} compounds agree closely with those of Cu^{2+}-doped $Cs_2SrCd(NO_2)_6$ in which possible interactions between the CuN_6 polyhedra are minimized. The ligand-field spectra show only two d–d bands (Fig. 15a). The excited $^2T_{2g}$ level is not split within the band-width limits, an indication of only weak Cu–N π-bonding[67,30,16]. In agreement with this result the g values are well described by Eq. (16), which implies an isotropic covalency parameter k[30,31]. The conclusions from spectroscopic results are confirmed by structural data of three K complexes[71] (Table 3). The Cs compounds exhibit the smallest Δ and the largest E_{JT} values as well as the largest deviations of the c_0/a_0 ratio from 1.00 in the series of Cu^{2+} nitrocomplexes[30]. Hence they contain CuN_6 polyhedra with the strongest Jahn-Teller distortions and the largest average Cu–N bond lengths. Bond length data which have been estimated from the ligand-field energies on the basis of the Angular Overlap Model are also given in Table 3. The tiny o-rhombic distortion component of the unit cell and of the $Cu(NO_2)_6^{4-}$ polyhedra is probably induced by the orientations of the NO_2^- groups which are not consistent with a tetragonal symmetry (Fig. 14). This symmetry effect is reflected by the EPR spectra as well (Fig. 15b)[30]. ρ_{min} is calculated from Eq. (5) to be $\cong$ 32 pm for the K, Tl and Rb complexes and estimated considerably larger for the Cs compounds ($\approx$ 38 pm). This implies energies $\hbar\omega$ of the radial E_g vibration of about 150 cm^{-1} for the Cs complexes [Eq. (3)] and roughly 15% higher values for the $Cu(NO_2)_6^{4-}$ polyhedra in the K, Tl and Rb compounds. The linear Jahn-Teller coupling constant V [Eq. (2)] is slightly lower and the energy barrier 2 $|\beta|$ about 70% higher in the case of the Cs complexes. The latter value was estimated under the assumption that the nonlinear coupling constant V_3 changes by the same ratio as V [Eq. (5)].

In the corresponding alkaline earth compounds with Co^{2+} the transition metal ion possesses a low-spin ground state, as can be suggested from the ligand-field spectra

Table 3. Structural and spectroscopic data of nitrocomplexes $A_2^IM^{II}T^{II}(NO_2)_6$ (abbreviations for the type of Jahn-Teller order as in Table 2)

T^{II}	M^{II}	A^I	m, T[K][a]	T^{II}–N bond lengths [pm][b] d_l	d_m	d_s	$\bar{d}$	ρ [pm][b,f]	JT order	Ligand-field data [10^3 cm^{-1}] E_{JT}	Δ_0	B	g (∥c)	g(⊥ c)[g]	Covalency parameters $k_\parallel$	$k_\perp$	Ref.
Ni	Sr	K		–	–	–	207.8	0 ; –	–	–	13.55	0.70	2.14_3		0.86		30,68,69)
	Pb	K		–	–	–	208.0	0 ; –	–	–	13.55	0.70	2.13_9		0.85		
	Sr	Cs		–	–	–	≅ 213[c]	0 ; –	–	–	12.5	0.70	2.16_4		0.88		30)
	Pb	Cs		–	–	–	≅ 213[c]	0 ; –	–	–	12.5	0.70	2.16_0		0.87		
Cu	Sr	K		231.0	204.1	202.9	212.7	32 ; –	f	1.90	12.7	–	2.25_0	2.07_0 2.05_7	0.80	0.81	30,71)
	Sr	Cs		≅ 239[d]	≅ 206[d]	≅ 206[d]	≅ 217[e]	≈ 38 ; –	f	2.20	11.6	–	2.26_5	2.07_0 2.05_8	0.81	0.80	
Co	Ba	K	α, 230		–		≈ 198[c]	(0); –	d	≅ 2.00	≅ 12.6	≅ 0.65	2.11				72,73)
			β, 140	210	193	191	198	21 ; –	f	≅ 2.00	≅ 12.6	≅ 0.65	2.02	2.15			
Cu	Pb	K	α, 298	–	–	–	211.8	(0); 31	d	1.75	≅ 12.7	–	2.12_4		k values [$k_\parallel \cong k_\perp$] ≅ 0.79 ± 0.02		30,33,84)
			β, 276	(216.6)	(215.3)	205.8	212.6	(12); 28	af(d*)	–	–	–	2.07_1	≅ 2.14_7			33,90)
			γ, 160	≅ 223	≅ 205	≅ 204	210.7	≅ 21 ; –	af	1.85	≅ 12.7	–	2.06_0	2.14_3 2.15_2			30,33,88)
Cu	Pb	Rb	β, 294	(217.7)	(216.8)	206.3	213.6	(13); 31	af(d*)	1.90	≅ 12.4	–	2.07_2	≅ 2.15_2			30,33,90)
Cu	Pb	Cs	α, ≅ 400	–	–	–	217.4	(0); –	d	–	–	–	2.12_6				30,33,81)
			β, 323	(222.5)	(222.5)	207.5	217.5	(17); –	af(d*)	–	–	–	2.07_0	≅ 2.15_0			33,86)
			β′, 293	(223)	(223)	207	217.7	(18); 33		2.20	≅ 11.6	–	2.06_8	2.14_8 2.15_2			30,33,89)
			γ, 160	≅ 230	≅ 211.5	≅ 207	216.2	≅ 24 ; –	af	2.20	≅ 11.6	–	2.06_2	2.14_7 2.15_3			30,33,87)

[a] Modification, temperature.
[b] Values in parentheses: apparent spacings and distortions by dynamic or partial dynamic averaging.
[c] Estimated from X-ray powder data.
[d] Estimated from X-ray powder data and AOM calculations.
[e] From the analogous lead complex.
[f] From bond lengths and rms displacements, respectively.
[g] Frequently split by a small o-rhombic symmetry component.

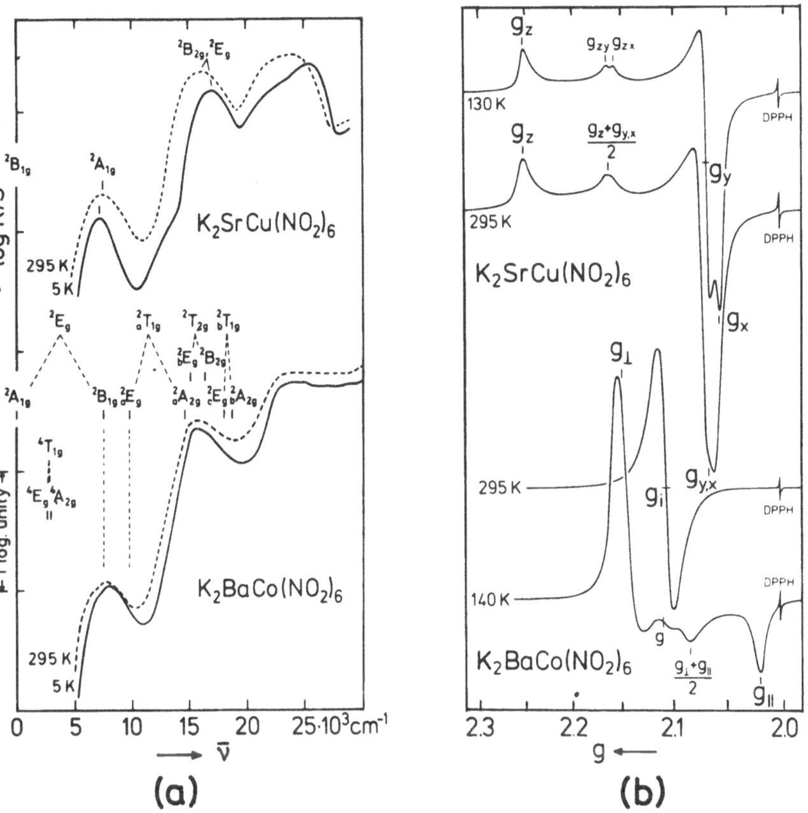

Fig. 15. Ligand-field (a) and EPR spectra [35 Gc] (b) of nitrocomplexes $K_2SrCu(NO_2)_6$ and $K_2BaCo(NO_2)_6$. Co^{2+} spectrum (a): band positions fitted with $\Delta_0 = 12600\ cm^{-1}$, $B = 650\ cm^{-1}$; $C/B = 4.4$, $E_{JT} = 2000\ cm^{-1}$; octahedral parent terms are also given.

already (Fig. 15a)[67,72]. The EPR spectra confirm this suggestion and give evidence that – at lower temperatures – the CoN_6 octahedra are tetragonally elongated (obviously without an o-rhombic component) and ordered in the ferrodistortive pattern (Fig. 15b)[72]. The Δ/B ratio of $\cong 20$ is somewhat below the theoretical value which characterizes the cross-over from the high-spin to the low-spin d^7 configuration (Tanabe-Sugano diagram) for a regular octahedral coordination. Nevertheless the low-spin configuration is stabilized by a strong Jahn-Teller splitting of the ^2E_g state ($\approx 8000\ cm^{-1}$). The energetic separation between the $^2A_{1g}(^2E_g)$ ground state and the first excited quartet level is about $3000\ cm^{-1}$. Though the ground state splittings of the Co^{2+} complexes are comparable to those of the Cu^{2+} analogues, the extent of distortion measured by the radial parameter ρ is much smaller in the CoN_6 polyhedra (Fig. 15, Table 3). This follows from the single crystal structure determination of $K_2BaCo(NO_2)_6$ (Fmmm, pseudotetragonal)[73] and from the smaller c_0/a_0 ratios compared to Cu^{2+} [72]. The reason is the considerably smaller ionic radius of Co^{2+} (low-spin) compared to Cu^{2+} ($r_{Co^{2+}}/r_{Cu^{2+}} \cong 0.93$) [74], by which the elastic energy necessary for distorting the $T^{II}(NO_2)_6^{4-}$ octahedra is enhanced. The same observation was made for the K, Tl and Rb complexes of Cu^{2+} with respect to the Cs compounds with larger average Cu–N bond lengths. With the structural data of $K_2BaCo(NO_2)_6$ and using Eqs. (5) and (3) ρ_{min} is calculated to be 21 pm, while ω has a value of about 250 cm^{-1}. Compared with the stochiometrically equivalent Cu^{2+} complexes the ratios of the radial frequencies and of the linear coupling constants are: $\omega_{Co}/\omega_{Cu} = |V_{Co}|/|V_{Cu}| \cong 1.5$, and the anisotropy energy $|\beta|$ turns out to be only half as large for Co^{2+} compared to Cu^{2+}. Here it again was assumed that V_3 is lowered by the same ratio as V when substituting Co^{2+} by Cu^{2+}.

While the alkaline earth nitrocomplexes of Cu^{2+} do not show phase transitions up to the decomposition temperatures at about 500 K–600 K, the Co^{2+} compounds transform into cubic structures. The phase transitions are continuous and take place at $260\,K \leqslant T \leqslant 370\,K$ for the compounds with A^I: K, Rb and M^{II}: Sr, Ba [72,73]. The apparent second order nature of the transitions implies that the molecular field strength is considerably larger than the anisotropy energy. The much smaller value of $|\beta|$ for Co^{2+} compared to Cu^{2+} is in essential agreement with this consideration and is reflected by the lower transition temperatures of the Co^{2+} compounds also. The EPR spectra of the cubic high-temperature modifications are isotropic above the transition temperature (Fig. 15) and indicate that the phase transition is caused by a change from the static to the dynamic Jahn-Teller effect. The same conclusion is possible from the IR spectra at 298 K. While in $K_2SrNi(NO_2)_6$ and $K_2BaCo(NO_2)_6$ [cubic phase with dynamically distorted $Co(NO_2)_6^{4-}$ octahedra] the δ-NO_2 bending mode is found at 825 and 823 cm^{-1}, respectively, this band is split in the o-rhombic complexes $K_2SrCu(NO_2)_6$ [815 and 823, 826 cm^{-1}] and $Rb_2BaCo(NO_2)_6$ [815 and 824 cm^{-1} (more intensive component)] [72].

B. *Lead Complexes [M^{II}: Pb] with Cu^{2+}*

In the lead nitrocomplexes of Ni^{2+} the bonding properties within the $Ni(NO_2)_6^{4-}$ polyhedra are not essentially different from those in the corresponding alkaline earth compounds. The ligand-field energies and g parameters of the Pb complexes are only slightly smaller than those of the Sr compounds (Sr^{2+} and Pb^{2+} are nearly identical in size), and the same is true for the covalency parameters derived from the EPR data (Table 3)[30]. We would not confirm the large differences in g found by Hathaway and Slade[75]. While the statement of comparable T^{II}–N bonding in Pb^{2+} and Sr^{2+} complexes is true for Cu^{2+} also (Table 3), there is a significant change in the symmetry of the elastic interactions between the $T^{II}(NO_2)_6^{4-}$ polyhedra when substituting Sr^{2+}(Ca^{2+}, Ba^{2+}) by Pb^{2+}. The Jahn-Teller distortion of the elpasolite unit cell is again pseudotetragonal, but with $c_0/a_0 < 1$ now. Multistage phase transitions of first order transform the pseudotetragonal γ- and β-modifications into the

Table 4. Jahn-Teller induced phase transitions in nitrocomplexes $A^{I}_{2}PbCu(NO_2)_6$

A^I	Phase	Transition temperature T_u [K][a]			Unit cell parameters [pm]					Space group	
		X-ray; N[b]	DSC[c]	EPR[d]	a	b	c	γ	T[K]		
K	α				1069	–	–	–	300	Fm3	30,88)
	↕	≈ 277[30]	281 ± 1 [0.56]	280 ± 1							
	β				1075	–	1050	–	275	(Fmmm)[e]	88)
	↕	–	274 [0.52]	273							
	γ				2 × 1075	2 × 1075	2 × 1053	90.2°[f]	250	$C\bar{1}$	88)
Rb	α				1077	–	–	–	306	Fm3	30)
	↕	≈ 306[30]	315 ± 2 [0.72]	313 ± 2							
	β				1083	1082	1061	–	295	Fmmm	30,90)
	↕	–	276 [0.21]	276 ± 1							
	γ				–	–	–	–	–	–	
Cs	α				1097	–	–	–	420	Fm3	30)
	↕	393 ± 5[81]	391 ± 3 [1.50]	386 ± 5							
	β				1103	1100	1075	–	323	Fmmm	86)
	↕	–	–	302 ± 8							
	β'				1104	1101	1074	–	293	Fmmm[g]	89)
	↕	–	–	282 ± 7							
	γ				2 × 1097	2 × 1095	2 × 1069	90.1°[f]	160	B2/b	87)

[a] $T_u \pm \Delta T$: temperature range of hysteresis.
[b] Neutron diffraction.
[c] Differential scanning calorimetry; numbers in square brackets: transition enthalpies in kJ $mole^{-1}$; data from [80] and [77] [K only].
[d] Data from [33] and [78] [K only].
[e] Incommensurate structure.
[f] Dimensions of 8-fold pseudocell with Z = 32 (Figs. 1, 17).
[g] Monoclinic phase with very small deviation from Fmmm [87].

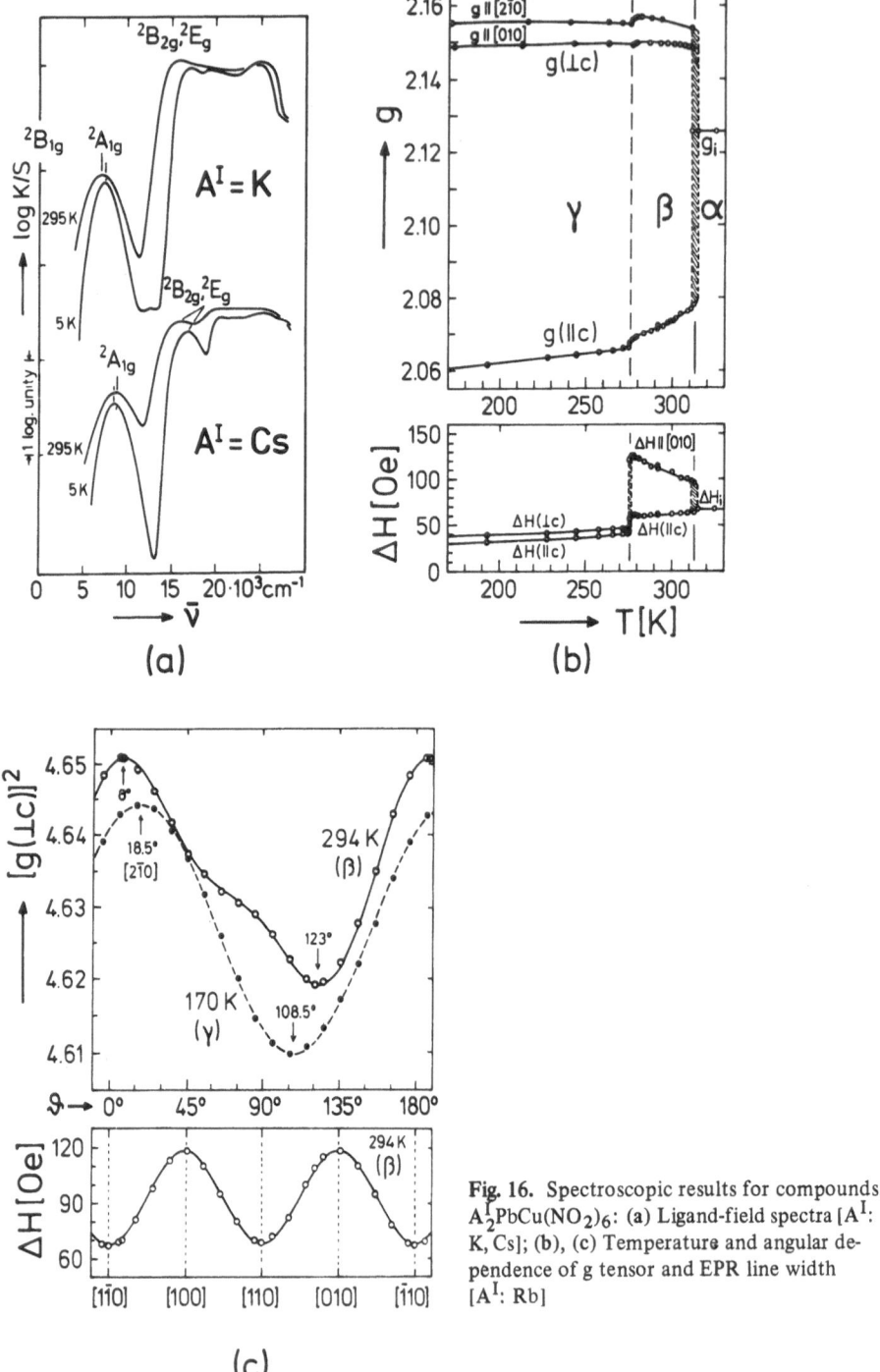

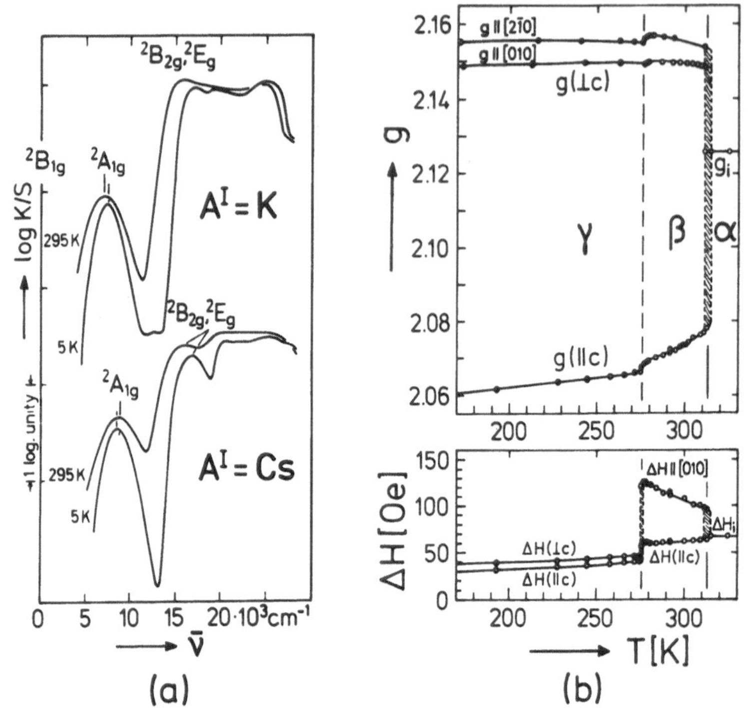

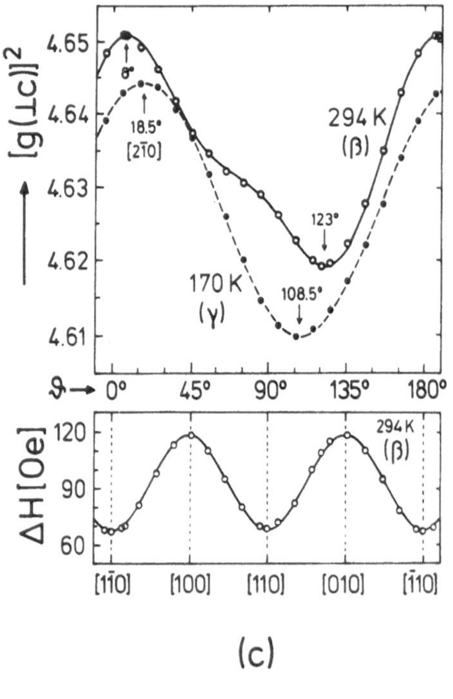

Fig. 16. Spectroscopic results for compounds $A^I_2PbCu(NO_2)_6$: (a) Ligand-field spectra [A^I: K, Cs]; (b), (c) Temperature and angular dependence of g tensor and EPR line width [A^I: Rb]

cubic α-phase. The transition temperatures have been determined by different methods (Table 4)[33,76–81]. The high-temperature α-modification (space group Fm3) is characterized by a dynamical Jahn-Teller effect, as can be concluded from the isotropic EPR signals (Fig. 16b)[30]. The CuN_6 octahedra are regular, but the temperature ellipsoids of the nitrogen (and oxygen) atoms are indicative of a 3-dimensional dynamics[81–84]. There is a distinct increase in the rms displacements in particular in the Cu–N bond directions compared with those in the Ni^{2+} complexes. Together with the EPR results this finding leads to the conclusion that the CuN_6 coordination is regular only in time average.

We will now discuss the structures of the γ- and β-modifications in greater detail. There has been considerable controversy in literature, whether the β- and γ-modifications contain tetragonally elongated CuN_6 octahedra in antiferrodistortive order [30,31,33] (Fig. 17) or compressed octahedra in ferrodistortive order[77–79,85]. We have proposed the former alternative and gave experimental and theoretical evidence, which is specified below.

The EPR data of the γ-phase obtained from single crystal and powder spectroscopy[30,33,76–79] yield g tensors which can be quantitatively accounted for by Eq. (25). If one averages over the small o-rhombic component imposed on the unit cell by the orientations of the NO_2^- ligands, the experimental g values are reproduced by these equations with nearly the same values for the orbital contributions which are found for the alkaline earth complexes with $c_0/a_0 > 1$[30,31] (Table 3):

$$\begin{aligned}&K_2SrCu(NO_2)_6:\quad g(\parallel c) = g_\parallel = 2.25_0\,,\quad g(\perp c) = g_\perp \cong 2.06_3\,,\\ &u = 0.032,\ [\text{elongated octahedra, ferrodistortive; Eq. (16)}];\end{aligned} \tag{29}$$

$$\begin{aligned}&\gamma\text{-}K_2PbCu(NO_2)_6:\quad g(\parallel c) = g_\perp = 2.06_0\,,\quad g(\perp c) = (g_\parallel + g_\perp)/2 \cong 2.14_8\,, \rightarrow g_\parallel = 2.23_6\,,\\ &u = 0.030_5,\ [\text{elongated octahedra, antiferrodistortive}].\end{aligned}$$

A possible o-rhombic component of the ferrodistortive sublattices has not been considered in this approximate calculation. Alternatively Harrowfield and Pilbrow [77–79] discussed the observed g values as originating from tetragonally compressed CuN_6 octahedra [Eq. (27)]. In this description an unreasonably large angular delocalization of the ground state wave-function has to be assumed, however, in order to account for the large deviation of g(||c) from g_0. The angular parameter is calculated from the experimental g factors to extend approximately between the limits $\varphi \approx \pm \frac{2\pi}{3}$ [with u = 0.03 (Eq. (29))]. The same considerable delocalization of the ground state wave-function between the minima at $\varphi \approx \pm 120°$ is also the basic requirement of Eq. (26). The model of exchange-coupled sublattices with elongated octahedra in antiferrodistortive order proves to be the correct alternative, however. This is indicated by the structural results which will be given now for the *γ-phases*. It is very difficult indeed to grow untwinned single crystals of the γ-modifications, because always a domain structure develops when passing the transition temperatures[33,78,86]. Nevertheless a neutron diffraction powder experiment on γ-Cs_2PbCu

$(NO_2)_6$ [87] and a single crystal analysis of γ-$K_2PbCu(NO_2)_6$ [88] have been performed. Neglecting the monoclinic angle, which deviates only very little from 90°, and the small difference between a and b, the structure of the Cs complex can be based on a pseudotetragonal 8-fold elpasolite cell (Fig. 1, Table 4). It contains eight ferrodistortive sublattices of elongated CuN_6 octahedra, from which always two are coupled corresponding to four antiferrodistortive pairs. There are eight symmetry-equivalent orientations of these pairs with respect to each other in the unit cell, from which one is illustrated in Fig. 17a. The doubling along the c-direction of the pseudotetragonal lattice occurs because successive equivalent Cu^{2+} layers along [001] are rotated by 90° with respect to each other. The smallest unit cell [monoclinic space group: B2/b] is also indicated in Fig. 17a[87]. γ-$K_2PbCu(NO_2)_6$ seems to have an analogous structure[88]. Though the Cu–N bond length data are not very accurate, the Jahn-Teller distortion of the CuN_6 octahedra seems to be appreciably smaller in the Pb [K: $\rho \cong 21$ pm; Cs: $\rho \cong 24$ pm] compared to the Sr complexes [K: $\rho \cong 32$ pm; Cs: $\rho \approx 38$ pm] (Table 3). One may conclude that it is energetically more difficult to induce tetragonally elongated octahedra within an antiferrodistortive order pattern than within the ferrodistortive alternative, if the quasi-elpasolite structure of the nitrocomplexes is considered. The realization of an antiferrodistortive order is accompanied by a strong deformation of

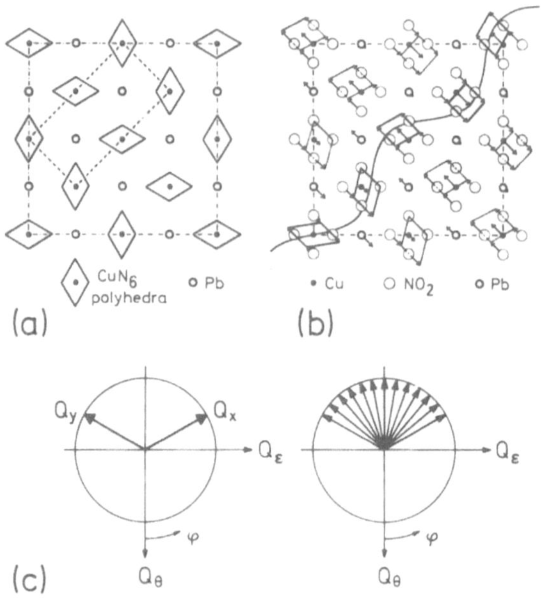

Fig. 17. Cooperative Jahn-Teller ordering in complexes $A^I_2PbCu(NO_2)_6$: **(a)** Antiferrodistortive order of elongated CuN_6 octahedra (small o-rhombic component) in γ-modifications [(001) planes, z = 0; z = 1/2: rotation of (z = 0) plane around [001] by 90° [87]; **(b)** Incommensurate β-phase of K complex [from[88]]; (c) Antiferrodistortive order in γ-phase and "spatial disorder" in β-modification of K complex[88].

the MO_{12} coordination with quite different Pb—O bond lengths, as is easily seen by inspecting Fig. 17a. While the highly polarizable Pb^{2+} ion with its lone pair tolerates this distortion, the ionic alkaline earth ions cannot be stabilized in these sites. Obviously the ferrodistortive order which does not imply any distortion of the MO_{12} sites is energetically preferred in the case of rigid M^{2+} ions. Similar arguments may be used for the NH_4^+ ion in the B′ site of the elpasolite $(NH_4)_2NH_4MnF_6$ which was mentioned before. In contrast to analogous compounds with alkaline ions in this site (Table 2) $(NH_4)_3MnF_6$ crystallizes with $c_0/a_0 < 1$ as the copper lead nitrocomplexes[41,57].

In general the antiferrodistortive pattern is more stable than the ferrodistortive order in a 3-dimensional frame-work of corner-connected octahedra (perovskites, lead nitrocomplexes). As discussed just before the antiferrodistortive order leads to appreciable deformations of those octahedral sites which are occupied by non-Jahn-Teller ions in elpasolite-type structures. Thus the ferrodistortive order may be stabilized also (oxidic elpasolites, alkaline earth nitrocomplexes). The lower transition temperatures of the lead compared to the alkaline earth compounds indicate weaker cooperative-elastic interactions and are possibly caused by the less distinct distortion of the CuN_6 polyhedra (Table 3). Finally it should be mentioned that the g tensor in the (001) planes of the γ-modifications has an orientation [K: 15°; Rb: 18.5° (Fig. 16c); Cs: ≈ 23°][33) which, at least for the Rb complex, correlates with the [120] and [210] directions of the large pseudocell [18.4°] (Fig. 17a). These directions correspond to the diagonals of the true monoclinic unit cell in the ab-plane of the lattice and indicate an electronic coupling between sublattices with the same orientation. Though the g values themselves do not change significantly during the phase transition from the γ- to the β-modifications (Fig. 16b), the angular dependence of the g tensor in the (001) planes is markedly different (Fig. 16c). This behavior which is similar for the K, Rb and Cs compounds is complex and can be described as a superposition of the angular dependence which characterizes the γ-phases and of a component which shows a 2 ϑ dependence[33]:

$$g(\perp c)^2 = A + b_1 \cos^2(\vartheta - \Phi) + b_2 \cos^2 2\vartheta \qquad [\text{Rb: } A = 4.616;\ \ b_1 = 3\,b_2 = 0.028;\ \ \Phi = 18.5°] \tag{30}$$

The second term in Eq. (30) is caused by the small symmetry deviation of the β-phases from tetragonal, which these complexes show in close analogy to the γ-modifications (Table 4). The third term resembles the angular dependence of Eq. (13), but is restricted to the (001) plane. Eq. (13) refers to a vibronic E state (Fig. 4), with the electronic components listed in Eq. (17), and follows from matrix (10) for $3\Gamma \gg G_\theta, G_\epsilon$. The energetic situation for the β-phases differs from that one just considered in so far as the molecular field which stabilizes the two ferrodistortive sublattices and the antiferrodistortive coupling between them introduces an additional strain component [D_{4h} symmetry] corresponding to $\alpha \approx 180°$ [Eq. (24) → $G_\theta^s \approx |\gamma|\rho_s$; $G_\epsilon^s \approx 0$]. This strain component stabilizes a Φ_{E_θ} wave-function (B_1 level) with only a slight admixture of Φ_{E_ϵ} (A_1 level) [Eq. (17)]. This admixture is introduced by the small deviation of α from π and hence of G_ϵ^s from 0. If the small G_ϵ^s contribution which induces the 2ϑ dependence in the third term of Eq. (30) is not taken into account, the following wave-function and g tensor [(001) plane] is calculated from matrix (10):

$$\Phi_{E_\theta} = \frac{1}{\sqrt{2}} (\Phi_y - \Phi_x)$$
$$g(\perp c) = g_0 + 5\,u - 5.5\,u^2 \qquad (31)$$

We interpret this behavior as caused by *excited* vibronic levels which are partly occupied at higher temperatures. After all the angular dependence of the g tensor in the β-phases [Eq. (30)] still indicates the presence of a static antiferrodistortive pattern as in the γ-modifications which is, however, superimposed by a "planar dynamic" Jahn-Teller effect. The $g(\perp c)$ parameters, characterizing the planar delocalization in the excited state ("planar dynamics") [Eq. (31)] and the static antiferrodistortive order [Eq. (25)], respectively, are identical. Only the small lower symmetry component of the unit cells which induces the $\cos^2 2\vartheta$ term in Eq. (30) allows to suggest the superimposed planar dynamic behavior as discussed. Further evidence for a delocalization of this type is obtained from the EPR line width which exhibits a pure $\cos^2 2\vartheta$ dependence (Fig. 16c)[33]. At the $\gamma \rightarrow \beta$ phase transition excited vibronic states with the properties given in Eq. (31) are sufficiently populated to induce a partial equilibration of the alternating long and short Cu–N bond lengths in the (001) planes of the β-phases.

In connection with the concept of a planar dynamics some EPR spectroscopic results we obtained for mixed crystals $TlCd_{1-x}Cu_x(NO_2)_3$ should also be mentioned. These compounds crystallize with slightly hexagonally distorted unit cells (compare Fig. 1) below $x \approx 0.1$ [$x = 0$: $a = 762\,pm \approx \frac{1}{\sqrt{2}}\, a_{cub.}$; $c = 1845\,pm \approx \sqrt{3}\, a_{cub.}$] and have pseudocubic lattices between $0.4 > x > 0.1$. This type of hexagonal distortion with $c/a < \sqrt{6}$ is quite common in perovskite and elpasolite structures and fills the gap between the cubic and the o-thombic lattices with decreasing tolerance factor. At very low x-values the EPR signals are isotropic at 298 K and anisotropic at 4.2 K with g values indicating tetragonally elongated CuN_6 octahedra. With increasing x this type of spectrum was superimposed by an isotropic signal which could not be frozen in even at 4.2 K. A single crystal spectrum of $TlCd_{0.82}Cu_{0.18}(NO_2)_3$ showing only this signal gave the following results at 4.2 K for the g tensor and the line width ΔH:

$$g\perp\{100\} = 2.11_6\,, \qquad g\perp\{110\} = 2.12_0$$
$$\Delta H\perp\{100\} = 190\,Oe\,, \qquad \Delta H\perp\{110\} = 120\,Oe \qquad (32)$$

This behavior of "cubic anisotropy" is equivalent to the corresponding component for β-$A_2^IPbCu(NO_2)_6$ in Eq. (31), but not restricted to one plane in this case. Similar isotropic signals which also could not be frozen in at 4.2 K are found for mixed crystals $Sr_2Zn_{1-x}Cu_xW(Te)O_6$ in the range $0.05 < x < 0.5$ as well[11].

The results of a recent single crystal neutron diffraction study of β'-$Cs_2PbCu(NO_2)_6$ [4] are not in contradiction with the assumption of a (partial or complete) planar dynamics[89]. The $Cu(NO_2)_6^{4-}$ octahedra are found to be tetragonally compressed, but the thermal ellipsoids of the N- and O-atoms in the (001) plane exhibit anomalous behavior. It is striking that the long axes of these ellipsoids are directed along or (nearly) parallel to the Cu–N bonds (Fig. 14). This is different from the situation in the [001] direction and in the alkaline earth complexes, where the longest axes of N and O do not orientate with respect to the Cu–N bond directions. Also the thermal ellipsoids have considerably longer axes along or parallel to the Cu–N bonds in the (001) plane than is observed in the [001] direction and in the nitrocomplexes with $c_0/a_0 > 1$. These results are in agreement with the model of a time-averaging process in which the NO_2^- groups along the c-axis are not involved. For β-$Rb_2PbCu(NO_2)_6$ similar rms displacements are observed[90].

Y. Yamada et al. have given structural evidence for a somewhat different interpretation in the case of β-$K_2PbCu(NO_2)_6$. A careful investigation and analysis of weak satellite reflections by means of X-ray and neutron diffraction showed that this phase has an incommensurate structure with an angular disorder of elongated octahedra approximately along the [110] directions of the elpasolite-like pseudocell (Fig. 17b)[88]. Again the average picture corresponds to compressed CuN_6 octahedra (Table 3). Yamada et al. interpret their results as a spatial disorder in close analogy to a "fan spin" ordering (Fig. 17c). This mechanism substitutes the time-averaging process in our model. If one accepts, however, that Fig. 17b is the instantaneous picture of a cooperative vibrational mode which induces a planar dynamic structure, there may be no essential difference between the two concepts. It should only depend on the physical method and on the temperature within the stability range of the β-phase then, if one sees a static, a partially or completely planar-dynamic behavior. In the same paper an analogous incommensurate structure is suggested for β-$Rb_2PbCu(NO_2)_6$ also from preliminary X-ray results. In a very recent investigation finally the same authors studied the satellite reflections which appear in the β'-, β- and γ-phases of $Cs_2PbCu(NO_2)_2$[91]. They confirm the existence of four modifications instead of three in the case of the Cs complex found by single crystal EPR spectroscopy[33] (Table 4). It is further stated in this paper that the β- and β'-phases differ from the γ-modification only by a partial (β', incommensurate structure) and complete disorder (β) along the [001] direction, respectively, while the (001) planes are still considered to be ordered in an antiferrodistortive pattern[4] (Fig. 17a). This description and the picture of an at least partial planar dynamics in the β- and β'-modifications deduced from EPR spectroscopy[33] are not necessarily contradictory. Further spectroscopic and structural measurements on copper lead nitrocomplexes in dependence on temperature within the stability ranges of their β-modifications seem to be necessary, however, to improve the present models.

4 The β-modification crystallizes with the space group Fmmm[86] as the β-phases of the other lead nitrocomplexes. The structural difference between β'- and β-$Cs_2PbCu(NO_2)_6$ is a very small monoclinic symmetry component in the former modification[87] (Table 4).

The mechanism of the successive phase transitions in the structure of $K_2PbCu(NO_2)_6$ which corresponds to a distorted cubic face-centred arrangement of CuN_6 polyhedra is studied on the basis of an order-disorder model by Kashida[92]. One interesting result of these calculations is that the antiferrodistortive order is stabilized with respect to the ferrodistortive alternative if the elastic next-nearest neighbor interactions [along the edges of the pseudocell (Fig. 17a)] are stronger than the nearest neighbor interactions. We do not agree with the statement, however, that the greater covalency in the lead complexes is responsible for the realization of the antiferrodistortive order in these cases. We think that the coupling of the CuN_6 polyhedra to the lattice deformations which are not included into Kashida's model are the deciding energetic factor which determines the Jahn-Teller ordering symmetry. This concept has been extensively discussed before in connection with the high polarizability of Pb^{2+} in comparison to the rigid alkaline earth ions. Superexchange interactions are found in the lead *and* alkaline earth complexes, which is not in agreement with essentially differing covalency in the two classes of compounds. The symmetries of superexchange interactions in elpasolite compounds may be readily compared to those in the perovskite structure, if the $B'X_6$ polyhedra in $A_2BB'X_6$ (Fig. 1) are regarded as single ligands with strongly extended exchange path-ways. The resulting face-centered arrangement of Cu^{2+} ions is composed of four interpenetrating cubic primitive lattices. Hence planar and linear antiferromagnetism is expected for a ferrodistortive and antiferrodistortive order, respectively, also in elpasolite-type structures (Fig. 12). Indeed $K_2BaCu(NO_2)_6$ is a planar antiferromagnet with an exchange constant of $J/k = -1.0$ K, while $K_2PbCu(NO_2)$ is a linear Heisenberg antiferromagnet [$J/k \cong -2.8$ K][93]. In the case of the lead complex the heat capacity curve with a broad maximum at 2.8 K exhibits a second much weaker peak at even lower temperatures, which may indicate a transition to 3-dimensional magnetic order. Again the observed Jahn-Teller order patterns correlate with the magnetic structures. Coming back to the order-disorder model of Kashida the phase transition from the static antiferrodistortive γ-phase to the "disordered" β-modification could be explained by the greater entropy in the latter case. In a similar theoretical study of antiferrodistortive interactions between isolated octahedra, but in cubic primitive arrangement, Schröder and Thomas have also demonstrated that the transition from the tetragonal to the cubic phase may occur in two steps[29]. Besides the modification with an antiferrodistortive order as shown in Fig. 11c ($c_0/a_0 < 1$) a second antiferrodistortive phase is predicted to exist which exhibits disorder in one sublattice and is entropy-favored ($c_0/a_0 > 1$; Fig. 18). We will come back to this specific model when we discuss linear-chain structures in the next chapter.

In one respect the antiferrodistortive order in the lead nitrocomplexes is different from the one found in the perovskite lattice. While the molecular field in the perovskite compounds is determined by a large Q_ϵ component which shifts the sublattices appreciably from $\varphi = 120°, 240°$ towards $\varphi = 90°, 270°$ (Fig. 10b), the sublattice distortions in the lead nitrocomplexes obviously correspond to φ-values which slightly deviate from 120° and 240° towards 180° (Table 3). This observation may possibly be related with the first order nature of the phase transitions (Table 4).

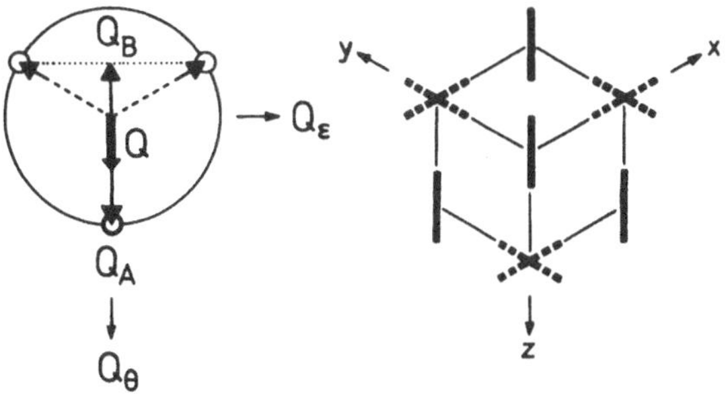

Fig. 18. Antiferrodistortive Jahn-Teller ordering with disorder in one sublattice [from [29]]

We will now consider the problem how to get information about the extent of the underlying Jahn-Teller deformation (ρ_{min}) if the Jahn-Teller effect is dynamic. Using ideas of Dunitz and Orgel [94] Ammeter et al. have proposed an interesting procedure to calculate approximate values for the radial distortion parameter ρ_{min} from the rms displacements in the case of dynamically distorted CuL_6 octahedra [95]. They define the Jahn-Teller contribution to the mean square amplitude U of the Cu–L stretching vibrations by [compare Eq. (5)]:

$$\rho_{min} \equiv \sqrt{6}\,[\Delta U(Cu{-}L)_{dyn.} - \Delta U(M{-}L)_{stat.}]^{1/2} = \sqrt{6}\,\Delta U_{JT}^{1/2} \qquad (33)$$

M is a non-Jahn-Teller ion of comparable size and electronic configuration to Cu^{2+} (e.g.: Ni^{2+}, Zn^{2+}) or Cu^{2+} in a statically distorted coordination. The ΔU's are the differences between the mean square amplitudes of the metal ion and the ligand atom along the Cu(M)–L bond directions. The $\Delta U(M{-}L)_{stat.}$ amplitudes for the alkaline earth nitrocomplexes of Ni^{2+} and Cu^{2+} are calculated from the structural data [69,71] to have values around 20 pm^2 by this procedure, while the $\Delta U(Cu{-}L)_{dyn.}$ amplitudes in the case of α-$A^I_2PbCu(NO_2)_6$ [A^I: K, Tl, Cs] are about ten times larger [81]. The ρ_{min} values as calculated from Eq. (33) are collected in Table 3. In the case of planar dynamic Jahn-Teller effects ρ_{min} may be estimated from the U amplitudes in the (001) plane by the modified equation:

$$\rho_{min} \equiv \sqrt{2}\,(2\,\Delta U_{JT}^{xy} + \Delta z^2)^{1/2} \qquad (34)$$

The ρ_{min} values for the totally dynamic α- and planar dynamic β-modifications of lead copper nitrocomplexes, calculated from Eqs. (33) and (34), are higher than those for the static γ-phases, but comparable to those of the alkaline earth compounds (Table 3). Though the ground state splittings $4\,E_{JT}$ are about the same in the compounds with $c_0/a_0 < 1$ and > 1, the static Jahn-Teller deformations of the CuN_6 octahedra are less distinct in the γ-phases of the lead complexes. This is not unexpected because the elastic forces which oppose a cooperative Jahn-Teller ordering are much larger for an antiferrodistortive than a ferrodistortive pattern in elpasolite-type structures, as has been discussed before.

Using an all-valence electron SCFMO method [96] total energies have been computed for $T^{II}(NO_2)_6^{4-}$ complexes (T^{II}: Cu^{2+}, Co^{2+}). This method had been quite successfully applied to $Ni^{III}F_6^{3-}$ polyhedra [97]. The calculated bond lengths, Jahn-Teller parameters E_{JT}, $|\beta|$, ρ and φ as well as the high-spin $\leftrightarrow$ low-spin separations showed quite reasonable agreement with experimental data of elpasolite compounds $A_2^IB^INi^{III}F_6$, obtained independently [61]. This example will be discussed in detail in Chap. VI. In the case of $Cu(NO_2)_6^{4-}$ polyhedra the computational results are in essential agreement with the spectroscopic and structural evidence also, but quantitatively less satisfactory than for the NiF_6^{3-} polyhedra [99]. The value of the energy barrier $2\,|\beta|$ separating the minima from the saddle-points in the lower potential surface [Eq. (7), Fig. 3] is quite reasonable (400 cm^{-1}), however. From the temperature dependence of the $^2B_{2g} \rightarrow {}^2A_{1g}$ transition ($\equiv 4\,E_{JT}$) in the ligand-field spectra of $A_2^IPbCu(NO_2)_6$ [A^I: K, Rb, Tl] (Fig. 16a) a barrier height of about 700 cm^{-1} is derived which has to be overcome during the transitions from the γ- to the α-modifications [16]. Obviously the model of isolated complex units which seems to be a satisfactory approximation for NiF_6^{3-} in elpasolites $A_2^IB^INi^{III}F_6$ is not valid for $Cu(NO_2)_6^{4-}$ polyhedra in nitrocomplexes without reservation. The results in [99] further suggest that the t_{2g} molecular orbitals are essentially nonbonding with a node close to the nitrogen atoms – in accord with lacking splitting effects of the excited T_{2g} level in the ligand-field spectra [30,16] (Table 3).

The last section of this chapter is devoted to some interesting results obtained by applying high pressure techniques to lead nitrocomplexes. As was mentioned already the phase transitions of this class of compounds can also be followed by IR spectroscopy. Especially the δ-NO_2 bending mode around 800 cm^{-1} registrates the symmetry changes very sensitively and confirms the given interpretations in most respects [67,72,86]. Applying high pressures up to 35 kbar removes continuously the symmetry splittings of the δ-NO_2 vibration as well as the splittings of the symmetric and asymmetric Cu–N stretching vibrations in the case of β'-$Cs_2PbCu(NO_2)_6$, and the unit cell becomes cubic [86]. It is rather unlikely, however, that the distortion of the single CuN_6 polyhedra has been removed. The lowering of the ground state when going from O_h to D_{4h} symmetry as the consequence of the Jahn-Teller effect corresponds to an energy barrier $2\,E_{JT}$ (= 4450 cm^{-1}) which seems too high to be met by a pressure of 35 kbar. An alternative, but somewhat speculative explanation may be given: Under high pressure the lead ions become less polarizable, so that possibly a ferrodistortive Jahn-Teller order as in the alkaline earth complexes becomes successively more stable. The sublattices at $\varphi \cong 120°$ and 240° are forced to move towards $\varphi = 0°$, until at about 35 kbar a sublattice structure with 90° and 270° is reached (Fig. 10c) which is planar dynamic (or incommensurate) as the $\beta'(\beta)$-phase under normal pressure.

V. Local and Cooperative Jahn-Teller Interactions in Host Lattice Structures with Distorted Octahedra

It has been demonstrated that the preferred Jahn-Teller distortion symmetry of Cu^{2+} ions in octahedral coordination is the tetragonal elongation (sometimes with an o-rhombic component). This statement is valid if the octahedral host lattice site is of regular O_h symmetry. Furthermore it has been shown that in structural frameworks which consist of corner-connected octahedra in three dimensions the antiferrodistortive order is energetically stabilized with respect to the ferrodistortive alternative. This statement is also true for d^4 cations ($KCrF_3$, Table 2). Elpasolite-type compounds $A_2BB'X_6$, in which B is a Jahn-Teller unstable and B' a nonpolarizable and rigid cation with an orbitally nondegenerate ground state, crystallize with a ferrodistortive order of elongated $B'X_6$ octahedra, however. This is again found to be true for d^4 ($A^I_2B^IMn^{III}F_6$ with A^I: Cs, Rb, NH_4; B^I: Na, K) and low-spin d^7 cations as well (Tables 2, 3). In $(NH_4)_3MnF_6$ and γ-$A^I_2PbCu(NO_2)_6$ complexes which have pseudotetragonal structures with $c_0/a_0 < 1$ and contain the polarizable NH_4^+ and Pb^{2+} ions in the B' site the cooperative order pattern is different, however.

We will now study local and cooperative Jahn-Teller effects in structures in which the octahedra are connected by common corners only in two dimensions (Fig. 13b, c). As the consequence of this layer structure the terminal ligands are somewhat stronger bonded to the Ni^{2+} ions in the tetragonal K_2NiF_4 [132] and Ba_2NiF_6 [100] lattice types, leading to slightly tetragonally compressed NiF_6 octahedra along [001] already in the host structure. The isomorphous substitution of Zn^{2+} by Cu^{2+} in Ba_2ZnF_6 results in anisotropic EPR signals with $g(\parallel c) = 1.99_0$ and $g(\perp c) = 2.36_1$ (4.2 K) up to 30 mole % of Cu^{2+} [10] (Fig. 19, I). These g values correspond to those of Eq. (27), if a possible anisotropy in u is taken into account, and can be derived from Eq. (19) with $\varphi = 180°$ and $u_x = u_y = u_\perp$, $u_z = u_\parallel$:

$$g_\parallel = g_0 - 3u_\perp^2 , \qquad g_\perp = g_0 + 6u_\perp - 6u_\perp^2 \tag{35}$$

The well resolved copper hyperfine structure in the $g(\parallel c)$ signal at lower Cu^{2+} concentrations gives evidence that the CuF_6 polyhedra are largely isolated from each other in the host lattice structure. It is obviously possible to stabilize Cu^{2+} in tetragonally compressed octahedral coordination if the host lattice site itself has already this symmetry. This host lattice effect can be treated as a strain with the components [Eq. (24) with $\alpha = 180°$]: $G^s_\theta = |\gamma|\rho_s$, $G_\epsilon = 0$. It is introduced into the calculation as a perturbation comparable in magnitude to the nonlinear coupling. The resulting energy equation for the lower potential surface is then [compare Eq. (6)] [10]:

$$\frac{E}{E_{JT}} = \rho'^2 - 2\rho' - \frac{|\beta|}{E_{JT}} \cos 3\varphi \cdot \rho'^3 + \frac{|\gamma|\rho_s}{E_{JT}} \cos\varphi \tag{36}$$
$$[\rho' \equiv \rho/\rho_{min} \; ; \; E = E_- - E_0]$$

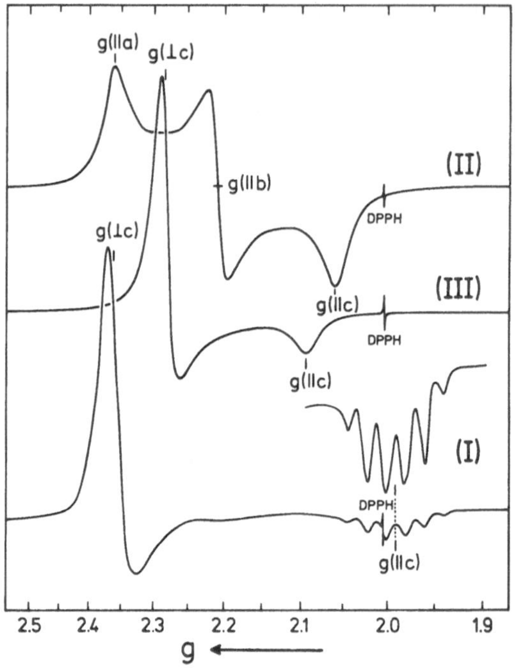

Fig. 19. EPR spectra [35 Gc, 295 K] of $Ba_2Zn_{0.9}Cu_{0.1}F_6$ (I), Ba_2CuF_6 (II) and K_2CuF_4 (III)

The EPR data for mixed crystals $Ba_2Zn_{1-x}Cu_xF_6$ [$x \leqslant 0.3$] correspond to a lower Jahn-Teller surface with only one minimum at $\varphi = \pi$ [$\rho' = 1$] (Fig. 20a). This situation results if $|\gamma| \rho_s$ becomes larger than $9\,|\beta|$. With increasing nonlinear Jahn-Teller coupling two minima symmetrically to $\varphi = \pi$ develop, separated by a saddle-point at $\varphi = \pi$ (Fig. 20b). If the hieght of this barrier is sufficiently large, an antiferrodistortive order of two o-rhombic sublattices results. This description is in agreement with the structural and spectroscopic evidence for Cu^{2+} concentrations of $x > 0.6$. In the region $0.3 < x < 0.6$ the energy barrier at $\varphi = \pi$ is small [$\cong 0.2\,|\beta|$ for $|\gamma| \rho_s / |\beta| = 6$ (Fig. 20b)], and interesting delocalization effects are observed which resemble those of the lead nitrocomplexes in the β-modifications. The extent of the tetragonal compression at low Cu^{2+} concentrations is much larger than that of the ZnF_6 host lattice site, as we have estimated from the ligand-field energies on the basis of AOM calculations [$x = 0$: $\rho_{min} = 10$ pm ; $0 < x \leqslant 0.3$: $\rho_{min} \cong 28$ pm]. With increasing x the cooperative interactions between the CuF_6 polyhedra become stronger and induce an enhanced linear and nonlinear Jahn-Teller coupling [$x = 0.63$: $\rho_{min} \cong 43$ pm ; $x = 1.0$: $\rho_{min} = 51$ pm]. The compressed coordination changes into an antiferrodistortive order of elongated octahedra (with an o-rhombic component) (Fig. 21). The antiferrodistortive order is "disturbed" by nonlinear Cu–F–Cu bond angles in the (001) planes, however (Fig. 21, Table 2), in correspondence with the o-rhom-

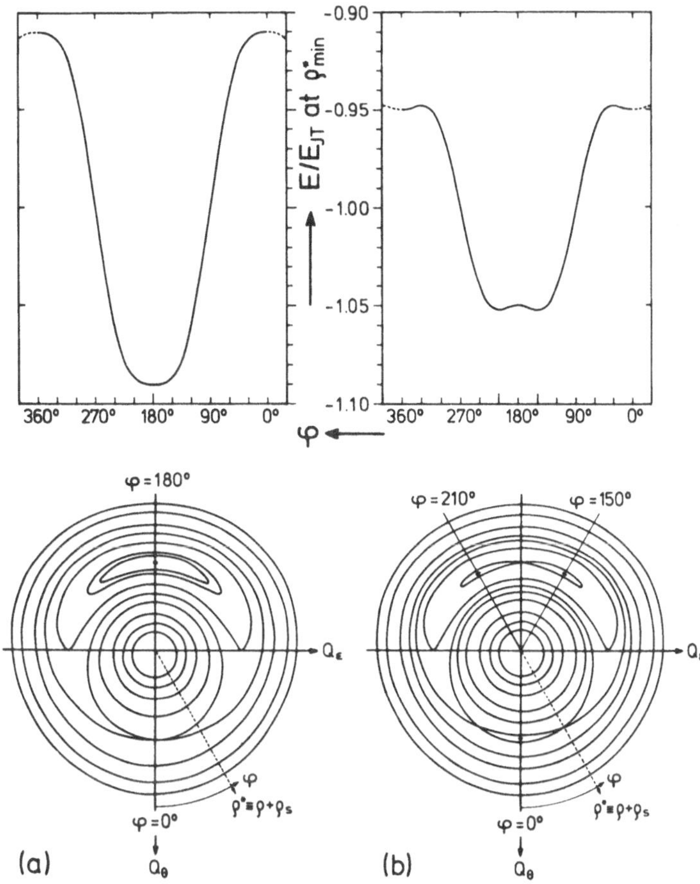

Fig. 20. Energy contour diagrams of lower potential surface in the presence of strain ($\alpha = 180°$) and nonlinear Jahn-Teller coupling for $|\gamma|\,\rho_s/|\beta| = 10$ (**a**) and 6 (**b**), respectively [d^9 cations; $|\beta| = 0.01\ E_{JT}$]

bic symmetry of the EPR spectra at higher Cu^{2+} concentrations (Fig. 19, II). This structural description[47)] corrected the result of compressed CuF_6 octahedra, originally reported for Ba_2CuF_6 [100)]. It is interesting to note that a refinement of the X-ray data with respect to the Cu–F bond lengths and Cu–F–Cu bond angles in the (001) planes was possible by means of EPR spectroscopy[48)]. These corrected results were confirmed by a neutron diffraction study, performed later[49)] (Table 2). As the consequence of the host lattice effect the shortest Cu–F bond lengths of the o-rhombically distorted CuF_6 octahedra are observed along [001], while the long and intermediate bond distances alternate in the (001) planes. This finding is different from the situation in $KCuF_3$ where the Cu–F spacings of intermediate length extend along [001] (Table 1).

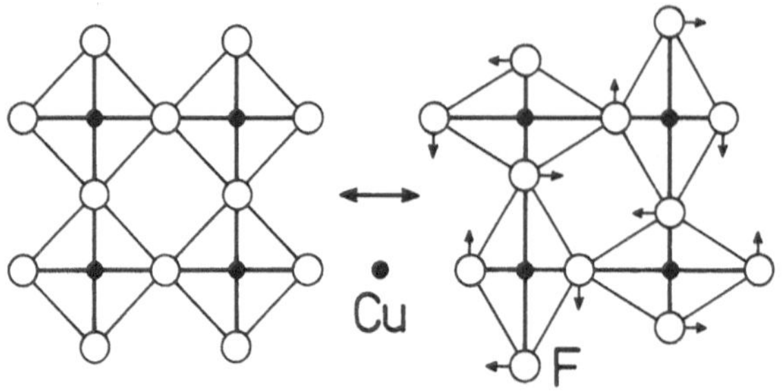

Fig. 21. The change from compressed CuF_6 octahedra (ferrodistortive) to elongated octahedra (antiferrodistortive) in mixed crystals $Ba_2Zn_{1-x}Cu_xF_6$ [view into (001) plane of Ba_2ZnF_6 structure (Fig. 13b); the arrows indicate additional rotations of the CuF_6 polyhedra around [001] which characterize the structure at high x-values]

The structural and spectroscopic investigation of mixed crystals $K_2Zn_{1-x}CuF_4$ yielded results[101] similar to those of compounds $Ba_2Zn_{1-x}Cu_xF_6$ [10]. Because of the less distinct host lattice effect (compression of metal – ligand bond lengths along [001]) a delocalization between two minima symmetrically to $\varphi = \pi$, as in Fig. 20b, is already found at very small x-values, however. The CuF_6 octahedra in K_2CuF_4 were originally described as being tetragonally compressed also[102]. The observed planar ferromagnetism [$J/k \cong 10$ K; $T_c = 6.5$ K][38] is not in accord with this finding and has lead Khomskii and Kugel to propose an antiferrodistortive order of elongated octahedra[103]. For this order pattern a weak parallel spin-spin interaction can indeed be derived from simple superexchange considerations (Fig. 12b). The conclusion of elongated CuF_6 octahedra in antiferrodistortive order (with linear Cu–F–Cu bonds) was possible from EPR powder data also[13]. The observed g parameters (Fig. 19, III) can be quantitatively accounted for by Eq. (25), if anisotropic u values are introduced in addition:

$$
\begin{aligned}
g(\parallel c) &= g_0 + 2u_\parallel - 3u_\perp^2 - u_\parallel u_\perp \\
g(\perp c) &= g_0 + 5u_\perp - 4.5u_\perp^2 - u_\parallel^2
\end{aligned}
\tag{37}
$$

but are not in agreement with Eq. (35) which corresponds to compressed CuF_6 octahedra (Fig. 19, I). A recent structure determination has confirmed the antiferrodistortive model[50], deduced from spectroscopic and magnetic results. A further refinement of the F^- positions by neutron diffraction[51] yielded a distinct o-rhombic component, with the Cu–F bond lengths orientated similar to those in $KCuF_3$ (Tables 1 and 2). The stochiometrically analogous Rb, Tl and Cs compounds are isomorphous with K_2CuF_4 [104]. $Cs_2Ag^{II}F_4$ and $Rb_2Ag^{II}F_4$ could be prepared as well[121]. The g values of the EPR powder spectra obey Eq. (37) as those of compounds $A^I_2CuF_4$, and the ligand-field data are also consistent with corresponding values of the stochiometrically equivalent Cu^{2+} compounds[122] (Table 5). Obviously the mentioned Ag^{2+} fluorides have the K_2CuF_4 structure and hence an antiferrodistortive order of essentially elongated AgF_6 octahedra which may be disturbed slightly as in Ba_2CuF_6, however. The values Δ_0 and E_{JT} of Ag^{2+} are increased by a factor of about 1.6 ± 0.1 compared to Cu^{2+}, while the covalency parameters k [$\approx k_\perp \approx k_\parallel$] [Eq. (19)] are smaller for Ag^{2+} than for Cu^{2+} by 15–20% (Table 5). The k parameters of $A^I_2Ag(Cu)F_4$ have been evaluated from the EPR data[13,122] with the assumption of D_{4h} symmetry and by using Eq. (37) [with $\lambda_0(Ag^{2+}) = 1850\,cm^{-1}$].

Table 5. Ligand-field data [$10^3\,cm^{-1}$] and covalency parameters for $K_2(Rb_2)CuF_4$ and $Rb_2(Cs_2)AgF_4$ (I) and the compounds $CuPbF_6$ and $AgSnF_6$ (II), respectively

	(I)		(II)	
	Cu^{2+}	Ag^{2+}	Cu^{2+}	Ag^{2+}
Δ_0	7.0	11.4_5	≈ 6.0	10.1
E_{JT}	2.0_5	3.2	$\cong 1.5$	2.1
k	0.85	0.68	≈ 0.90	0.76

Cr^{2+} ions may be stabilized in the K_2NiF_4 structure type also: $A^I_2CrCl_4$ (A^I: K, NH_4, Rb, Cs). For the Rb and Cs compounds tetragonally compressed $CrCl_6^{4-}$ octahedra are reported as the result of powder neutron diffraction studies with the rather approximate values: $Rb_2(Cs_2)CrCl_4$: ≈ 243 (240) pm (2 ×), ≈ 257 (261) pm (4 ×)[53]. The observed low-temperature ferromagnetism is not consistent with this

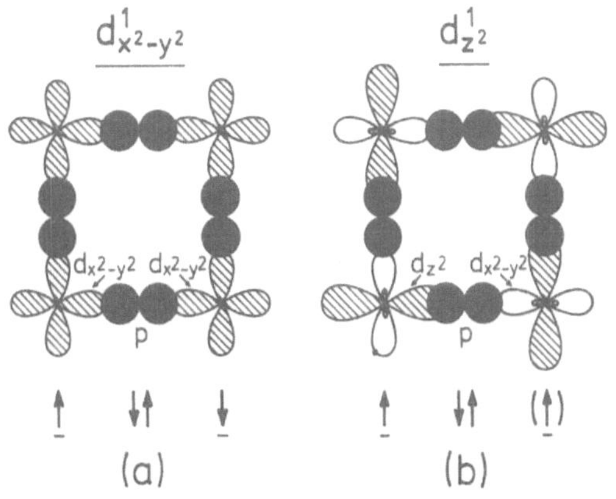

Fig. 22. Orbital ordering in (001) planes of structures with corner-connected TL_6 polyhedra [T: d^4 cations]: (**a**) compressed octahedra in ferrodistortive order; (**b**) elongated octahedra in antiferrodistortive order

ordering pattern, for which antiferromagnetism is expected (Fig. 22a). An antiferrodistortive order of elongated octahedra, however, would lead to alternating empty $d_{x^2-y^2}$ and half-filled d_{z^2} orbitals along the Cr–Cl–Cr bonds in the (001) planes and induce ferromagnetic interactions (Fig. 22b), in agreement with the experiment[55]. A very recent single crystal neutron diffraction study of Mn^{2+}-doped Rb_2CrCl_4 confirms the conclusions from the magnetic data[136]. The $CrCl_6$ octahedra are slightly o-rhombic with Cr–Cl bond lengths of: 243 pm (2x), 240 pm (2x), 274 pm (2x) (Table 2). As in K_2CuF_4 the spacings of medium lengths are orientated ∥ [001]. In a review on this highly interesting class of "transparent ferromagnets"[105] a preliminary structural result for Rb_2CrCl_4 is cited which is also in accord with two orientations of elongated $CrCl_6$ octahedra in the (001) planes (compare Fig. 9b).

A layer structure very similar to those considered before is the $TlAlF_4$ type, from which $CsFeF_4$ [space group P4/nmm] is a variant with slightly nonlinear Fe–F–Fe bridging[106]. In the latter compound the FeF_6^{3-} octahedra are considerably compressed in the directions of the terminal F^- ligands along [001] [ρ = 12 pm] (Fig. 23a). The extent of compression becomes much larger if Fe^{3+} is substituted by Mn^{3+} (Table 2)[57]. The temperature ellipsoids of the bridging F^- ligands in the (001) plane along the Mn–F–Mn bond directions are anomalously long compared to the geometric situation in $CsFeF_4$, however (Fig. 23), and resemble closely those of the NO_2^- ligands in the β-phases of $A_2^IPbCu(NO_2)_6$ (Fig. 14). This result may be caused either by an antiferrodistortive Jahn-Teller pattern of elongated MnF_6 octahedra, which is disordered along [001] or averaged by twinning with respect to the (110) plane, or by a "planar dynamic" Jahn-Teller effect. If one accepts the latter explanation and compares the mean square amplitudes of $CsMnF_4$ and $CsFeF_4$, a radial distortion parameter $\rho \cong 37$ is estimated from Eq. (34). This value leads to an underlying static (slightly o-rhombic) elongation with Mn–F bond lengths of about 181 pm (∥ [001]), 186 pm and 216 pm, if Eq. (5) is applied. As in Ba_2CuF_6 but different from K_2CuF_4 the shortest transition metal–ligand bond length seems to appear in

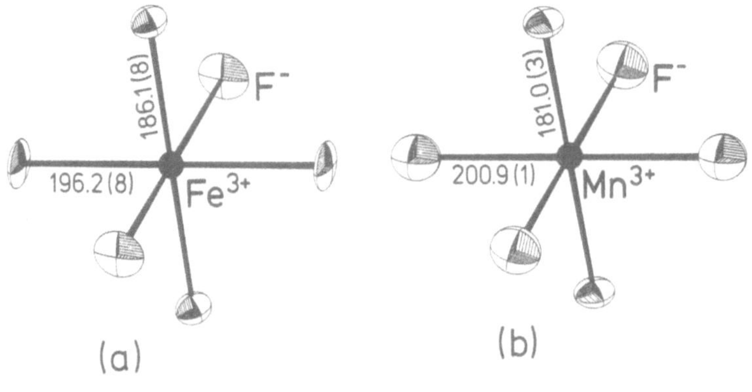

Fig. 23. Bond lengths [pm] and fluorine thermal ellipsoids of FeF_6^{3-} and MnF_6^{3-} polyhedra in $CsFeF_4$ and $CsMnF_4$, respectively

the [001] direction. The magnetic measurements indicate ferromagnetism below T_c = 21 K[58], again in agreement with elongated octahedra in antiferrodistortive order. From neutron diffraction diagrams the magnetic moments were found to lie in the (001) planes[57]. The compounds A^IMnF_4 [A^I: Rb, NH_4, K] with smaller A^I cations have similar properties and structures which are lower symmetry versions of the $CsFeF_4$ or $CsMnF_4$ type. From the spectroscopic and magnetic results analogous or closely related cooperative Jahn-Teller ordering symmetries can, however, be suggested[58].

The given examples demonstrate that even in the presence of strain with (α = 180°)-symmetry and of different magnitude the elongated configuration is adopted by d^4 and d^9 cations in octahedral coordination. Only at low concentrations of Jahn-Teller cations, i.e. if the Jahn-Teller effect is essentially local, the compressed geometry may eventually be stabilized. While in the compounds with K_2NiF_4 structure discussed so far the octahedral host lattice sites were tetragonally compressed, they are considerably elongated in the isostructural La_2NiO_4 [193 pm (4 x); 224 pm (2 x); ρ = 36 pm][107]. This is obviously caused by the influence of the La^{3+} ions, which – in contrast to K^+ with only one positive charge – strongly attract the axial oxygen ligands (compare Fig. 13c). It is expected that Cu^{2+} ions adopt this geometry when substituted into these sites. The distortion of the CuO_6 octahedra in La_2CuO_4 exceeds that one of NiO_6 considerably, as expected [191 pm (4 x) 246 pm (2 x); ρ = 64 pm][56] (Table 2). While the La^{3+} ions are coordinated by 9 oxygen atoms in La_2CuO_4 (Fig. 13c), the coordination number is only 8 for the smaller rare earth ions in compounds Ln_2CuO_4 [Ln = Nd, Sm, Eu, Gd][56]. In this structural variant the layers are differently packed, with the Cu^{2+} ions in a square planar oxygen coordination without axial ligands now. While it is difficult to prepare La_2NiO_4 without a certain amount of Ni^{3+} [107], compounds $SrLnNiO_4$ (Ln: La, Nd, Sm, Eu, Gd) with only Ni^{3+} in the octahedral sites can be synthesized under an oxygen pressure of 2 kb[108]. The distortion of the NiO_6 octahedra in $SrLaNiO_4$ [191 pm (4 x), 223 pm (2 x) ; ρ = 37 pm] is appreciably larger than the elongation of the AlO_6 octahedra in $CaSmAlO_4$ [184 pm (4 x), 204 pm (2 x) ; ρ = 23 pm ; also K_2NiF_4 structure type][109] and may be induced by the Jahn-Teller instability of Ni^{3+} in the low-spin configuration. Magnetic and susceptibility measurements on the black compounds $SrLnNiO_4$ indicate, however, that the σ-antibonding electrons of Ni^{3+} are semiconducting[108]. Thus a description in the bonding scheme assumed in this paper for the discussion of the Jahn-Teller effect is possibly only correct in first approximation.

The tendency to avoid a compressed octahedral coordination and to choose an elongated configuration within an antiferrodistortive order pattern is observed for *isolated* octahedra with Cu^{2+} and low-spin Co^{2+} also. Compounds $T^{II}(terpy)_2X_2 \cdot nH_2O$ with T = Ni^{2+} contain TN_6 octahedra which are compressed along the [001] direction of the (pseudo-)tetragonal unit cells as the result of the rigid structure of the terpyridine ligands ("host lattice effect"). Substituting Ni^{2+} by Cu^{2+} (Co^{2+}) leads to a "splitting" of the T–N spacings in the (001) planes into a long and an intermediate bond length as the consequence of the Jahn-Teller effect[42]. The CuN_6 octahedra are elongated, but with a distinct o-rhombic component (Table 2). While the short Cu–N distances are orientated along [001], the intermediate and long Cu–N spacings are bound into an antiferrodistortive order in the (001) planes. It is remarkable, however, that the EPR signals are not exchange-narrowed. Obviously the electronic coupling between neighbored Cu^{2+} ions in these planes is extremely weak or vanishing, so that the g tensor of elongated CuN_6 polyhedra can be observed. Phase transitions to dynamic modifications with averaged Cu–N bond lengths within the (001) planes are observed with increasing temperature[42]. After all there is a close analogy between these complexes and the layer structures discussed in the beginning of this chapter, if the distortion geometry of the CuL_6 octahedra and the cooperative Jahn-Teller order are compared.

We will now discuss host lattices in which the octahedral sites are distorted along a 3-fold axis. Deformations of this kind cannot lift the orbital degeneracy of an E_g ground state, but may nevertheless influence the geometry and extent of Jahn-Teller distortions. They are found in structures in which the octahedra are partly or entirely connected with each other via common faces. In rhombohedral Ba_2NiTeO_6 for example there are groups of three octahedra with common faces which are bridged by single octahedra via common corners[110] (Fig. 24a). The Ni^{2+} ions occupy the outer sites in the groups of three octahedra and possess a trigonally distorted coordination [201 pm (3 ×), 217 pm (3 ×)]. Cs_2NaCrF_6 crystallizes in the same structure

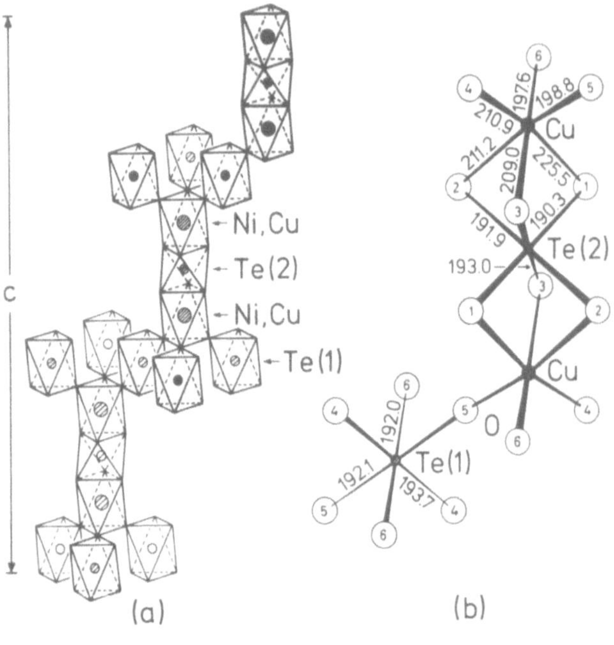

Fig. 24. Connection pattern of octahedra in the $Ba_2Ni(Cu)TeO_6$ structure [$Cs_2NaNi^{III}F_6$: Na^I and Ni^{III} in Ni^{II} and Te^{VI} sites, respectively]

type with Na and Cr in the Ni and Te positions, respectively[111]. Ba_2CuTeO_6 is a triclinic Jahn-Teller variant of this structure type in which Cu^{2+} has a strongly deformed coordination[63] (Fig. 24b). Averaging over the trigonal symmetry component of this distortion, we can estimate the tetragonal Jahn-Teller ligand-field component ($\rho \approx 16$ pm; Table 2). The extent of the Jahn-Teller distortion is small compared to that one in Ba_2CuWO_6 ($\rho = 51$ pm), though E_{JT} (1900 cm^{-1}) is nearly as great as in the elpasolite-type compound (≈ 2150 cm^{-1}). The reason is probably the face connection of the CuO_6 polyhedra with the central octahedron which is centred by the rigid Te(VI) ion and hence does not tolerate larger deformations (Fig. 24b). The outer CuO_6 polyhedra are correlated by an inversion center at the central Te position which induces the long Cu–O bond lengths to be parallel. The cooperative Jahn-Teller order may then be called ferrodistortive.

$Ba_3NiSb_2O_9$ crystallizes in the hexagonal $BaTiO_3$ type in which there are groups of two face-connected octahedra instead of three[112] (Fig. 25a). Again the NiO_6 octahedra have a trigonal ligand-field component which is quite small in this case, however (Fig. 25b). The substitution of Ni^{2+} by Cu^{2+} induces a static Jahn-Teller distortion only at low temperatures. The structure at 298 K is characterized by dynamically deformed CuO_6 octahedra[64]. This is reflected by the thermal ellipsoids of the O-atoms in $Ba_3CuSb_2O_9$ (Fig. 25c) in comparison with those in the Ni^{2+} compound (Fig. 25b). While there is a large increase of the rms displacements along the Cu–O–Sb(1) directions, it is somewhat smaller in the Cu–O–Sb(2) bridges. The radial distortion parameter which is estimated from the rms displacements ($\rho \approx 40$ pm, Table 2) lies between the ρ values of Ba_2CuWO_6 and Ba_2CuTeO_6. This finding correlates with the connection pattern of the octahedra in $Ba_3CuSb_2O_9$ (Fig. 25a) which is "intermediate" between those in the elpasolite and Ba_2NiTeO_6 structure (Figs. 1,24a).

$CsNiCl_3$ crystallizes in a hexagonal structure [$P6_3mmc$] which contains chains of $NiCl_6$ octahedra with common faces[113]. While Ni^{2+} occupies the centers of these

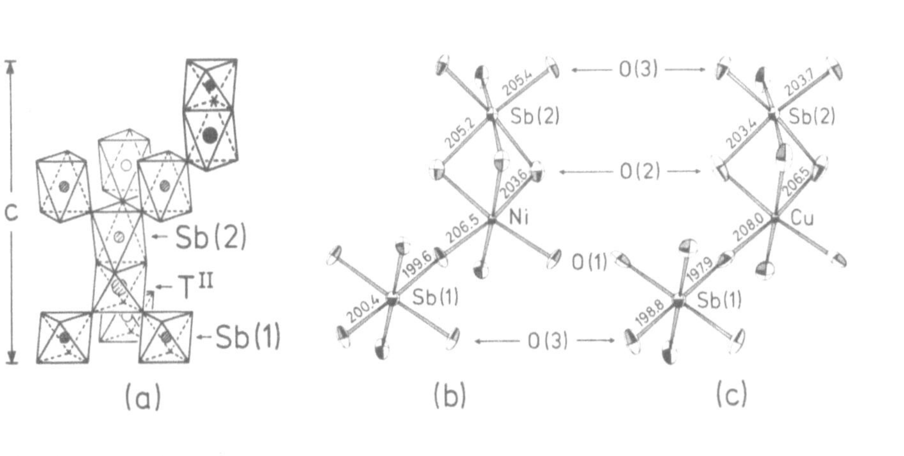

Fig. 25. Connection pattern and rms displacements of octahedra in the $Ba_3Ni(Cu)Sb_2O_9$ structure (ordered hexagonal $BaTiO_3$ type)

octahedra, the Cu^{2+} ions in the high temperature α-modification have distinctly acentric positions (Fig. 26a)[44]. The $CuCl_6$ octahedra exhibit a dynamic Jahn-Teller distortion. Lowering the temperature to 423 K a phase transition into a static β-modification is observed. The local distortion symmetry corresponds to a tetragonal elongation if only the essential symmetry component is considered[45]. The sequence of long axes along the chains is z, x′, y, z′, x, y′, corresponding to a three-sublattice geometry [$\varphi = 0°, 120°, 240°$]. The order is "disturbed", however, because the angles between the z-(y-, x-) and x′-(z′-, y′-)axes deviate from 90° by roughly 20° (compare the orientations in Fig. 26b). The cooperative Jahn-Teller order may be called antiferrodistortive. The resultant vector of this order pattern vanishes, and hence the hexagonal symmetry of the dynamic phase [$P6_3mc$] is retained [$P6_122$]. The distortion of the $CuCl_6$ octahedra in the static phase corresponds to a radial parameter ρ of 53 pm (Table 2) which is surprisingly large.

The corresponding linear-chain compounds with Cr^{2+} ions in the octahedral sites undergo two phase transitions in dependence on temperature, as was found by Maaskant et al.[65,66]. The high-temperature α-modifications of $Cs(Rb)CrCl_3$ are isostructural with the α-phase of $CsCuCl_3$ (Fig. 26a)[44]. The static γ-phase of $RbCrCl_3$ contains $CrCl_6$ octahedra which are again tetragonally elongated in first approximation (Table 2)[65]. Four sublattices Z, X′, Z, Y′ are elastically coupled in antiferrodistortive

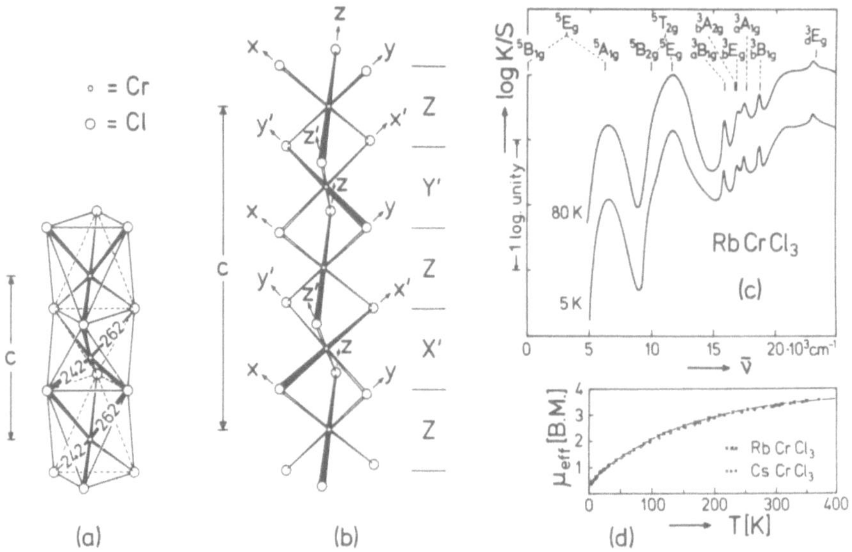

Fig. 26. Structural, spectroscopic and magnetic data of $Cs(Rb)CrCl_3$ [modified $CsNiCl_3$ type]: (a, b) $[CrCl_{6/2}]^-$ chains in α-$CsCrCl_3$ (295 K, "dynamic"; isostructural with α-$CsCuCl_3$) (a) and in γ-$RbCrCl_3$ (100 K, "static") (b) [fat lines correspond to long Cl–Cr–Cl spacings]; (c) ligand-field spectra of $RbCrCl_3$ (band energies [10^3 cm^{-1}] fitted with $\Delta = 7.80$, $B = 0.66$, $E_{JT} = 1.55$, $\delta_2 \approx 0.6$; parent quintet terms are also given); (d) $\mu_{eff}(T)$ diagram

order (Fig. 26b). Neglecting the deviations of the $CrCl_6$ polyhedra from tetragonal symmetry the four sublattices have their long axes orientated in the sequence z, x′, z, y′, with φ values 0°, 120°, 0°, 240°. It should be remarked again, however, that the order is disturbed because the angles between the long axes z and x′ (y′) deviate from 90°. The symmetry relations within this Jahn-Teller pattern resemble those proposed by Schröder and Thomas[29] in their antiferrodistortive model calculation (Fig. 18) which is based on a two-sublattice description with a disorder in one of these sublattices, however. In better agreement with the Thomas model is the cooperative Jahn-Teller order in the β-phases of $Cs(Rb)CrCl_3$. These modifications consist of only two sublattices [Z; (X′, Y′)], because a dynamic equilibration between the sublattices X′ and Y′ of the γ-phase (Fig. 26b) takes place, leading to compressed octahedra in time average (Q_B in Fig. 18)[65]. This averaging process is analogous to the "planar dynamics" which was discussed above for the cooperative Jahn-Teller order in β-$A^I_2PbCu(NO_2)_6$ complexes[33,31]. The phase transitions of $Cs(Rb)CrCl_3$ take place at the following temperatures:

$$\text{Rb:}\quad \alpha \xleftrightarrow{470\,K} \beta \xleftrightarrow{201\,K} \gamma, \quad \text{Cs:}\quad \alpha \xleftrightarrow{170\,K} \beta \overset{?}{\dashleftarrow\dashrightarrow} \gamma$$

The space group of the α-modification of $CsCrCl_3$ [$P6_3mc$[46]: Fig. 26a] seems to result from averaging over the several possibilities to realize the cooperative Jahn-Teller order which characterizes the partially dynamic β-[C 2/m] and static γ-modifications [C2], respectively[65,66].

The experimental E_{JT} energies are smaller for $Cs(Rb)CrCl_3$ [6100 (6600) cm^{-1}] (Fig. 26c) than for the layer compounds $Cs_2(Rb_2)CrCl_4$ [8200 (7500) cm^{-1}][55]. These values indicate a weaker linear Jahn-Teller coupling and also lower 2 $|\beta|$ barriers in the linear-chain chlorides. In agreement with this statement the transition temperatures to the dynamic phases lie below the decomposition temperatures for $Cs(Rb)CrCl_3$. The (pseudotetragonal) distortion of the $CrCl_6$ octahedra in γ-$RbCrCl_3$ and Rb_2CrCl_4 is comparable [$\rho \cong 37$ pm] which is unexpected because of the differing E_{JT} energies cited above. It is also surprising that the ρ value of β-$CsCuCl_3$ [53 pm] is considerably larger, though similar distortions are expected for Cu^{2+} and Cr^{2+} in identical structures and chemical coordination [$KCu(Cr)F_3$ in Tables 1 and 2]. It seems to be less difficult to verify a large Jahn-Teller distortion of $Cu(Cr)Cl_6$ octahedra in the cooperative pattern of β-$CsCuCl_3$ than in the more disordered pattern of γ-$RbCrCl_3$ (Fig. 26b).

The linear-chain compounds $Cs(Rb)CrCl_3$ are antiferrogmagnets in agreement with the antiferrodistortive four-sublattice structure of the γ-phases. Antiparallel spin-spin interactions should be indeed induced, if one follows the half-occupied d_{z^2} orbitals of Cr^{2+} along z, x′, z, y′ (Fig. 26b). The linear antiferromagnetism is reflected by the $\mu_{eff}(T)$ curve (Fig. 26d) from which a decrease of μ_{eff} from 3.4 (300 K) to 0.5 (4.5 K) is deduced[55].

VI. Various Examples

In 1967 cubic compounds $M^{II}_{1/4}Cu_{3/4}TiO_3$ [M^{II}: Cd, Ca] were described, the X-ray powder diagrams of which could easily be indexed if a perovskite structure was assumed[114]. The ligand-field spectra were not in accord with a d^9 cation in the 12-coordinated A site of this lattice type, however[18] (Fig. 1). We hence decided to perform an X-ray and neutron diffraction structure analysis with a powder sample of the Ca compound and found a unit cell which is schematically shown in Fig. 27a[115,116]. The TiO_6 octahedra are rotated around 3-fold axes in such a way that 75% of the A sites (A″) obtain a square planar coordination [bond lengths: 195.5 pm] with eight further ligands in much larger distances [277 pm (4 ×); 319 pm (4 ×)] (site symmetry: mmm) (Fig. 27b). The bond lengths in the remaining 25% A′ sites [260 pm (12 ×); symmetry: m3] are in the range usually found in regular perovskite-type lattices. The distortion of the CuO_{12} coordination can be understood as induced by the Jahn-Teller effect. A d^9 cation in a cuboctahedral ligand field (O_h) induces a strongly σ-anti-

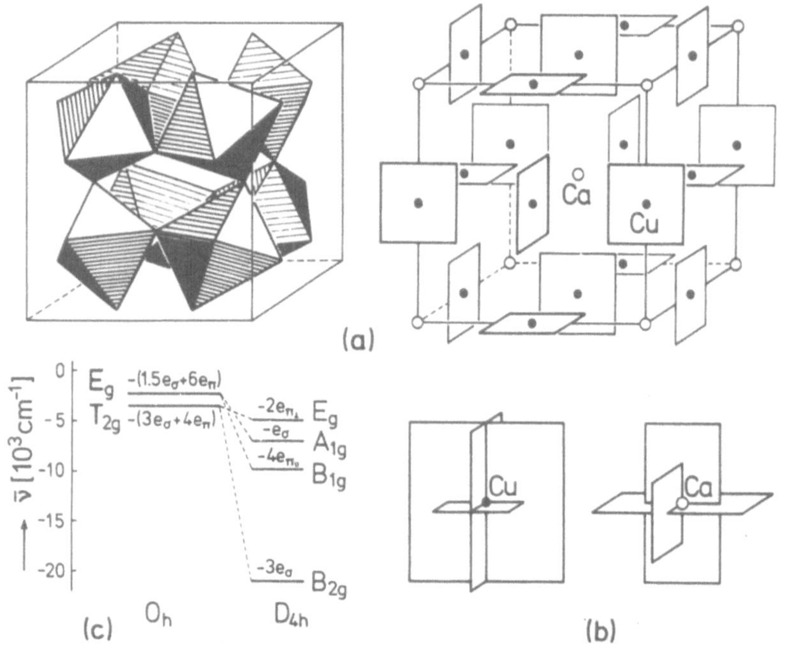

Fig. 27. The $CaCu_3Ti_4O_{12}$ structure type: (a) arrangement of TiO_6 octahedra and CuO_4 square planes in the unit cell, (b) the CuO_{4+8} and CaO_{12} polyhedra; (c) energy level diagram of CuO_{12} cuboctahedra and CuO_4 square planes (additional ligands neglected) [AOM energies, estimated for cuboctahedral and square planar Cu–O bond lengths of 260 and 196 pm, respectively; σ-diatomic overlap integrals from[123]]

bonding T_{2g} ground state which is expected to be considerably Jahn-Teller split (Fig. 27c). The angular overlap energies indicate that the σ-antibonding effect of the T_{2g} level in cuboctahedral coordination is of the same size as that one of the E_g level in octahedral coordination if the bond lengths were equal. The transition energies roughly estimated from an angular overlap calculation (Fig. 27c) closely agree with the band positions in the ligand-field spectra. The cooperative Jahn-Teller ordering pattern is of the same type as that in the low-temperature phase of $CsCuCl_3$, discussed before. There are six sublattices of CuO_4 square planes from which always three [at $\frac{1}{2}00, 0\frac{1}{2}0, 00\frac{1}{2}$ and $0\frac{1}{2}\frac{1}{2}, \frac{1}{2}0\frac{1}{2}, \frac{1}{2}\frac{1}{2}0$, respectively (Fig. 27a)] have angular parameters φ of 240°, 0° and 120°. The cooperative Jahn-Teller order orientates with respect to the 3-fold axes of the unit cell [space group: Im3] and is antiferrodistortive. The resulting distortion vector for each triple of sublattices is zero as in $CsCuCl_3$ and hence compatible with the cubic symmetry of these compounds. The observed isotropic EPR signal is obviously exchange-narrowed [18,116].

It is even possible to prepare isostructural compounds $A'_{1/4}A''_{3/4}BO_3 \equiv A'A''_3B_4O_{12}$ in which the A′ sites are empty: $\square Cu_3(Ta_2Ti_2)O_{12}$ or in which the Cu^{2+} positions A″ are partly unoccupied in addition: $\square(\square Cu_2)Ta_4O_{12} \equiv CuTa_2O_6$ [18,115,116]. The cation distribution in the A″ sites of $CuTa_2O_6$ seems still to be a problem not completely solved yet, however [115–117]. The EPR spectrum is isotropic or nearly isotropic down to 77 K, but – contrary to the other compounds discussed so far – strongly tetragonal at 4.2 K and indicates a phase transition at low temperatures [18,116].

d^4 configurated cations can also be stabilized in the $CaCu_3Ti_4O_{12}$ structure, as the high-pressure phase $NaMn_3^{III}(Mn_2^{III}Mn_2^{IV})O_{12}$ demonstrates [118]. High spin d^4 ions in a cuboctahedral coordination ($e_g^2\, t_{2g}^2$) possess a T_{2g} ground state as Cu^{2+} and may induce analogous Jahn-Teller distortions (Fig. 27c). The mentioned compound undergoes a phase transition to a monoclinic unit cell at 180 K which is obviously induced by the Mn^{3+} ions in the octahedral sites (change from a dynamic to a static Jahn-Telller distortion). In agreement with this assumption the monoclinic modification can be obtained already at 298 K if the amount of Mn^{3+} in the octahedral positions is increased: $LaMn_3^{III}Mn_4^{III}O_{12}$ [119]. A classification of the $CaCu_3Ti_4O_{12}$ lattice type from a structure-systematical point of view is given elsewhere [120].

It is puzzling that $CuSb_2O_6$ and $CuNb_2O_6$ do not adopt the $CuTa_2O_6$ structure. These compounds crystallize in Jahn-Teller distorted variants of the trirutile and columbite type [18], respectively, which are the structures of the "host lattices" $Zn(Ni)Sb_2O_6$ and $Zn(Ni)Nb_2O_6$ also. Even the substitution of the d^9 cation in $CuTa_2O_6$ by the d^4 configurated Cr^{2+} ion destabilizes the $CaCu_3Ti_4O_{12}$ type. $CrTa_2O_6$ has the same monoclinic structure as $CuSb_2O_6$ [125] which results from a deformation of the tetragonal $ZnTa_2O_6$ unit cell [trirutile type]. Hence the $CaCu_3Ti_4O_{12}$ type has to be considered as a quite singular structure which is stable under special geometric conditions only and may be considered as a perovskite variant with extremely strong Jahn-Teller distortion effects.

If the 12-coordinated A cations of the elpasolite lattice (Fig. 1) are removed, the ordered ReO_3 structure results. Compounds $T^{II}M^{IV}F_6$ [T^{II}: $Cr^{2+} \rightarrow Zn^{2+}$; M^{IV}: $Ti \rightarrow Hf, Ge \rightarrow Pb$, ...] frequently crystallize – at 298 K or above this temperature –

with cubic unit cells which are of the ordered ReO_3 structure or closely related types [126]. Second order phase transitions may occur at lower temperatures, however, which transform the cubic α- into trigonal β-modifications of the $LiSbF_6$ type [$R\bar{3}$]. In this ordered version of the VF_3 structure the 3-dimensional array of corner-connected octahedra is characterized by T–F–M bridges which stronly deviate from linearity. If T^{II} is a Jahn-Teller cation, four structure types are observed instead of two [126]:

$CuZrF_6$: $\alpha \overset{383\,K}{\longleftrightarrow} \beta \overset{353\,K}{\longleftrightarrow} \gamma$, $CuHfF_6$: $\alpha \overset{435\,K}{\longleftrightarrow} \beta \overset{298\,K}{\longleftrightarrow} \gamma$

$CrZrF_6$: $\alpha \overset{333\,K}{\longleftrightarrow} \delta \overset{\approx 150\,K}{\longleftrightarrow} \gamma$, $CrHfF_6$: $\alpha \overset{393\,K}{\longleftrightarrow} \delta \overset{\approx 200\,K}{\longleftrightarrow} \gamma$

The tetragonal structure of the δ-phases is the Jahn-Teller variant of the ordered ReO_3 type. c/a is smaller than unity, in contrast to the reverse axial ratio which is usually found in elpasolite type compounds with rigid non-Jahn-Teller cations. The (probably) triclinic γ-modification can be understood as resulting from the superposition of distortions which occur along the 3-fold axis of the $LiSbF_6$ type unit cell [host lattice effect] on the one hand and along the (pseudo-)4-fold local axes of the $Cu(Cr)F_6$ octahedra [static Jahn-Teller effect] on the other. The cooperative Jahn-Teller order pattern seems to be rather complicated, as can be deduced from the EPR spectra [127] and from preliminary neutron diffraction investigations of $CuZrF_6$ powder samples [128]. $AgSnF_6$ [129] probably belongs to the same structural family, as follows from the comparison of the spectroscopic data [122,130] with those of $CuPbF_6$ (Table 5). The E_{JT} energies and hence the extent of distortion of the $Ag(Cu)F_6$ octahedra are relatively small compared with corresponding values of compounds with the K_2NiF_4 structure, for example. While the Jahn-Teller splitting energies for Cu^{2+} and Cr^{2+} [$CrZr(Hf)F_6$: $E_{JT} = 1700\ cm^{-1}$] differ only slightly, they are significantly larger when passing to the Ag^{2+} compounds (Table 5). The ligand-field parameter Δ_0 increases from Cu^{2+} to Cr^{2+} [$CrZr(Hf)F_6$: $\Delta_0 = 7750\ cm^{-1}$] by 30% and from Cr^{2+} to Ag^{2+} by about the same percentage (Table 5).

Compounds $T^{II}Sn(OH)_6$ [T^{II}: $Mn^{2+} \rightarrow Zn^{2+}$] crystallize in a ReO_3-related structure type also [131]. Corresponding to the space group Pn3 the $T^{II}(OH)_6^{4-}$ and $Sn(OH)_6^{2}$ octahedra of the ordered ReO_3 structure [Fm3m] are slightly rotated around 3-fold axes, in a similar way as in the $CaCu_3Ti_4O_{12}$ type (Fig. 27). If T^{II} is Cu^{2+}, a static Jahn-Teller phase results which is tetragonal with $c/a > 1$ and contains elongated $Cu(OH)_6^{4-}$ octahedra ($\rho = 31$ pm) in ferrodistortive order [131]. This order is disturbed, however, because the Cu–O(H)–Sn bridges deviate from linearity and induce canting angles $2\gamma < 180°$.

High-spin ↔ low-spin transitions may be strongly influenced by the Jahn-Teller effect, as will be discussed now for d^7 configurated cations. Elpasolite type compounds $A_2^IB^ICo^{III}F_6$ [A^I, B^I: Cs, Na; K, Na; K, K] contain high-spin configurated Co^{3+} ions [$t_{2g}^4e_g^2$] and exhibit Δ values of ≅ 13100 cm^{-1}. The Racah parameter B and the nephelauxetic ratio β are calculated to be ≅ 800 cm^{-1} and ≅ 0.75 [with C/B = 4.8], respectively [133]. The Δ/B ratio is 16–17 and hence distinctly lower than the "cross-over" value of 20. Somewhat unexpected corresponding Ni^{3+} compounds [A^I, B^I: Cs, K; Rb, K] exhibit low-spin magnetic behavior, at least at lower temperatures [134]. This result is rather puzzling because the critical Δ/B ratio for the spin cross-over in the case of octahedral d^7 cations is 21 and similar Δ/B values to those of Co^{3+} should be valid for Ni^{3+}. The spectroscopic data, in particular the EPR results, confirm the conclusions from the magnetic measurements and give definite evidence in addition that the $Ni^{III}F_6$ octahedra are considerably distorted, with the symmetry of a tetragonal elongation [61]. A Jahn-Teller distortion had been suggested already before with arguments which were based on the ligand-field spectra alone [133]. We have performed energy calculations which were restricted to the split-levels of the competing $^2E_g(t_{2g}^6 e_g^1)$ and $^4T_{1g}(t_{2g}^5 e_g^2)$ levels [61]. They were based on the 4.2 K ligand-field and EPR data and yielded the simplified energy level diagram of Fig. 28a if LS coupling is neglected. In line with the discussion given above the quartet state would be lowest in O_h symmetry. Only the Jahn-Teller splitting of the ^2E_g level provides the necessary additional energy to stabilize a low-spin ground state.

Cs_2KNiF_6 and Rb_2KNiF_6 are cubic elpasolites at 298 K and show isotropic EPR signals, but undergo second order phase transitions from the dynamic phases to tetra-

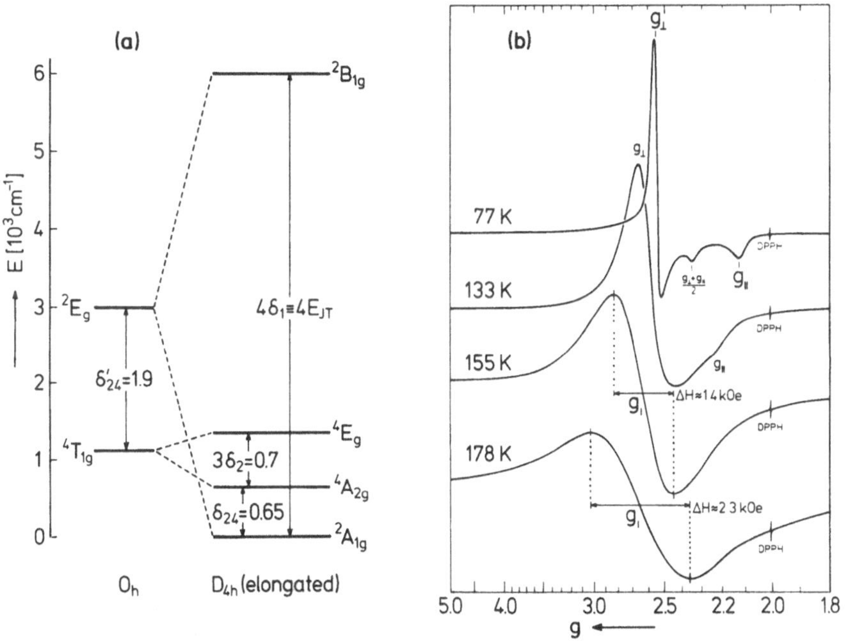

Fig. 28. Energy level diagram (a) and EPR spectra [35 Gc] (b) of the NiF_6^{3-} polyhedra in Cs_2KNiF_6: (a) energies in a hypothetical O_h and a D_{4h} (tetragonal elongation) ligand field as calculated from spectroscopic data at 4.2 K [61] [δ_{24}, δ'_{24}: doublet-quartet separation in D_{4h} and O_h ligand fields]; (b) line broadening of EPR spectra with increasing temperature [$\Delta H(203\,K) \cong 5\,kOe$]

Table 6. Calculated (first line) and experimental data (second line; Fig. 28a) for Ni–F bond lengths [pm] (D_{4h} symmetry) and energy parameters [10^3 cm^{-1}] in Cs_2KNiF_6

$\bar{d}$	ρ	E_{JT}	$2\,\lvert\beta\rvert$	$3\delta_2$	δ_{24}	δ'_{24}
190	12	2.7	2.2_5	0.4_5	0.9	1.8
–	14	1.5	–	0.7	0.6_5	1.9

gonal modifications with $c/a > 1$ at $\cong 145$ K[61,62] and $\cong 260$ K[61] (Fig. 28b). They contain elongated NiF_6 octahedra in ferrodistortive order. From the unit cell parameters of Cs_2KNiF_6 below 77 K[62] a radial distortion of $\rho \cong 14$ pm can be estimated.

As mentioned before SCFMO calculations have been performed for isolated $Ni^{III}F_6$ polyhedra[97]. The results are in remarkably good agreement with the spectroscopic findings at 4.2 K[61] (Table 6). A quartet ground state is predicted for regular octahedral ligand fields with Ni–F bond lengths of 193 pm and a doublet-quartet separation δ'_{24} which is nearly identical with the one deduced from experiment. The level with the lowest energy, however, is found for an elongated D_{4h} symmetry of the $Ni^{III}F_6$ polyhedra and corresponds to a Jahn-Teller stabilized low-spin state. The calculated splittings of the ^2E_g state are much larger than those derived from EPR data. The reason for the lower experimental E_{JT} energies is possibly the appreciable admixture of quartet contributions into the ground state wave-function via LS coupling[61] which is not taken into account in the MO calculation. This statement finds support by the comparison with stoichiometrically analogous Mn^{3+} compounds. $Cs_2KMn^{III}F_6$ for example [E_{JT} = 2200 cm^{-1}; Δ_0 = 14400 cm^{-1} [58]] exhibits a larger ground state splitting than $Cs^I_2KNi^{III}F_6$ though the Δ_0 parameters are not very different (Table 2). Also the radial distortion parameter is larger for the Mn^{3+} compound [$\rho \approx 20$ pm].

The comparatively very low transition temperatures of $Cs_2(Rb_2)KNiF_6$ made us suggest that a transition from the static to the dynamic Jahn-Teller effect may occur via the low-lying quartet states[61]. This mechanism finds support by the δ_{24} energy which is small compared with the energy barrier $2\,\lvert\beta\rvert$ (Table 6). Line broadening effects and g shifts in the EPR spectra with increasing temperature (Fig. 28b) are in agreement with this argument as well[61]. Using the experimental ligand-field and EPR results on Cs_2KNiF_6 energy calculations were performed with a *complete* d^7 basis set in D_{4h} symmetry (tetragonal elongation) also[135]. While a consistent interpretation was possible by the assumption of a $^2A_{1g}$–$^4A_{2g}$ spin equilibrium with an energy separation $\delta_{24} \approx 500$ cm^{-1} (Fig. 28a), the magnetic susceptibility data as a function of temperature could not be fitted on this basis. In particular discrepancies arise below 100 K. In this temperature range the experimental μ_{eff} value of $\cong 1.74\,\mu_B$ corresponds to a g_{eff} parameter very close to 2.0 while $\bar{g} = 2.39$ is observed by EPR spectroscopy. Spin-spin interactions via superexchange which might induce an apparent depression of the magnetic moment are not present, however. Thus the reason for the reduced magnetic susceptibility at low temperatures is unclear. Further investigations on Ni^{3+} compounds of this type are in progress.

Finally a compound Cs_2NaNiF_6 of elpasolite stoichiometry [129] which crystallizes in the Ba_2NiTeO_6 structure type with two different Ni^{3+} positions, however, was also investigated (Fig. 24). While the Ni^{3+} ions in the corner-connected octahedral sites are low-spin configurated with a similar behavior to that described above, the Ni^{3+} ions in the central positions of three face-connected octahedra have a high-spin ground state down to 4.2 K [61]. Obviously the geometric fixation of the latter Ni^{3+} sites by the neighbored NaF_6 octahedra reduces the geometrical possibilities for larger Jahn-Teller distortions, as was discussed in Chapter V. It can be suggested that the splitting of the ^2E_g state would be too small to overcome the quartet-doublet separation in O_h (Fig. 28a); thus a low-spin ground state cannot be stabilized for this Ni^{3+} site.

Acknowledgements. We have to thank many colleagues for interesting and fruitful discussions and the permission to report about results prior to publication.

VII. References

1. Jahn, H.A., Teller, E.: Proc. R. Soc. *A 161*, 220 (1937)
2. Jahn, H.A.: Proc. R. Soc. *A 164*, 117 (1938)
3. Abragam, A., Bleaney, B.: Electron Paramagnetic Resonance of Transition Ions, p. 790 ff., Clarendon Press 1970
4. Ham, F.S., in: Electron Paramagnetic Resonance, p. 16. Geschwind, S. (ed.). New York–London: Plenum Press 1972
5. Englman, R.: The Jahn-Teller Effect in Molecules and Crystals. New York: Wiley-Interscience (1971)
6. Thomas, H.: Theory of Jahn-Teller Transitions. In: Riste, T. (ed.): Electron-Phonon Interactions and Phase Transitions, p. 245. New York–London: Plenum Press 1977
7. O'Brien, M.C.M.: Proc. R. Soc. *A 281*, 323 (1964)
8. Coffman, R.E.: J. Chem. Phys. *48*, 609 (1968)
9. Bleaney, B., Bowers, K.D., Pryce, M.H.L.: Proc. R. Soc. *A 228*, 166 (1955)
10. Friebel, C., Propach, V., Reinen, D.: Z. Naturforsch. *31b*, 1574 (1976)
11. Friebel, C., Reinen, D.: Z. Naturforsch. *24a*, 1518 (1969)
12. Okazaki, A., Suemune, Y.: J. Phys. Soc. Japan *16*, 176 (1961); Okazaki, A.: J. Phys. Soc. Japan *26*, 870 (1969)
13. Friebel, C., Reinen, D.: Z. Anorg. Allg. Chem. *407*, 193 (1974)
14. Reinen, D., Weitzel, H.: Z. Anorg. Allg. Chem. *424*, 31 (1976)
15. Reinen, D.: Z. Naturforsch. *23a*, 521 (1968)
16. Grefer, J., Reinen, D.: Z. Naturforsch. *28a*, 464 (1973)
17. Schmitz-DuMont, O., Fendel, H.: Mh. Chem. *96*, 495 (1965);

Schmitz-DuMont, O., Fendel, H., Hassanein, M., Weissenfeld, H.: Mh. Chem. *97*, 1660 (1966); Schmitz-DuMont, O.: Mh. Chem. *99*, 1285 (1968)
18. Propach, V., Reinen, D.: Z. Anorg. Allg. Chem. *369*, 278 (1969)
19. Wellern, H.O., Reinen, D.: to be published; Wellern, H.O.: Thesis, Marburg 1973
20. Jørgensen, C.K.: Modern Aspects of Ligand Field Theory. Chap. 13. Amsterdam: North-Holland-Press 1970
21. Schmitz-DuMont, O., Grimm, D.: Z. Anorg. Allg. Chem. *355*, 280 (1967)
22. Wojtowicz, P.J.: Phys. Rev. *116*, 31 (1969)
23. Kanamori, J.: J. Appl. Phys. *31*, 14 S (1960)
24. Iserentant, Chr.: Verhandeling van de Koninklijke Vlamse Academie van Wetenschappen, Brüssel (1968)
25. Novák, P.: Phys. Chem. Solids *30*, 2357 (1969) and *31*, 125 (1970)
26. Englman, R., Halperin, B.: Phys. Rev. *B 2*, 75 (1970)
27. Halperin, B., Englman, R.: Phys. Rev. *B 3*, 1698 (1971)
28. Thomas, H., Müller, K.A.: Phys. Rev. Lett. *28*, 820 (1972)
29. Schröder, G., Thomas, H.: Z. Phys. *B 25*, 369 (1976); Höck, K.-H., Schröder, G., Thomas, H.: Z. Phys. *B 30*, 403 (1978)
30. Reinen, D., Friebel, C., Reetz, K.P.: J. Solid State Chem. *4*, 103 (1972)
31. Reinen, D.: Solid State Commun. *21*, 137 (1977)
32. Friebel, C.: Z. Naturforsch. *29b*, 634 (1974)
33. Friebel, C.: Z. Anorg. Allg. Chem. *417*, 197 (1975)
34. Carlin, R.L., van Duyneveldt, A.J.: Magnetic Properties of Transition Metal Compounds, p. 234, Berlin, Heidelberg, New York: Springer 1977
35. O'Connor, C.J., Sinn, E., Carlin, R.L.: Inorg. Chem. *16*, 3314 (1977)
36. Algra, H.A., de Jongh, L.J., Carlin, R.L.: Commun. Kamerlingh Onnes Lab. No. 433a (1977)
37. Reinen, D., Krause, S.: Solid State Commun. *29*, 691 (1979)
37a. Reinen, D., Krause, S.: unpublished results
38. de Jongh, L.J., Miedema, A.R.: Experiments on Simple Magnetic Model Systems. London: Taylor A. Francis LTD 1974
39. Dubler, E., Korber, P., Oswald, H.R.: Acta Crystallogr. *B 29*, 1929 (1973); Friebel, C.: Z. Naturforsch. *29b*, 295 (1974)
40. Hitchman, M.A., Waite, T.D.: Inorg. Chem. *15*, 2150 (1976); Waite, T.D., Hitchman, M.A.: Inorg. Chem. *15*, 2155 (1976)
41. Massa, W.: Z. Anorg. Allg. Chem. *415*, 254 (1975)
42. Henke, W., Reinen, D.: Z. Anorg. Allg. Chem. *436*, 187 (1977); Allmann, R., Henke, W., Reinen, D.: Inorg. Chem. *17*, 378 (1978)
43. Babel, D.: Z. Anorg. Allg. Chem. *336*, 200 (1965)
44. Kroese, C.J., Maaskant, W.J.A., Verschoor, G.C.: Acta Crystallogr. *B 30*, 1053 (1974)
45. Schlueter, A.W., Jacobson, R.A., Rundle, R.E.: Inorg. Chem. *5*, 277 (1966)
46. McPherson, G.L., Kistenmacher, T.J., Folkers, J.B., Stucky, G.D.: J. Chem. Phys. *57*, 3771 (1972)
47. von Schnering, H.G.: Z. Anorg. Allg. Chem. *400*, 201 (1973)
48. Friebel, C.: Z. Naturforsch. *29b*, 634 (1974); Friebel, C.: Z. Naturforsch. *30b*, 970 (1975)
49. Reinen, D., Weitzel, H.: Z. Naturforsch. *32b*, 476 (1977)
50. Haegele, R., Babel, D.: Z. Anorg. Allg. Chem. *409*, 11 (1974)
51. Krause, S., Reinen, D.: to be published
52. Willett, R.D.: J. Chem. Phys. *41*, 2243 (1964); Friebel, C.: unpublished results
53. Fair, M.J., Gregson, A.K., Day, P., Hutchings, M.T.: Physica (Utrecht), Sect. B, *86*, 657 (1977);

Hutchings, M.T., Gregson, A.K., Day, P., Leech, D.H.: Solid State Commun. *15*, 313 (1974)
54. Larkworthy, L.F., Trigg, J.K., Yavari, A.: J.C.S. Dalton *1975*, 1879
55. Reinen, D., Köhler, P., Massa, W.: Proceed. I, XIX.ICCC, Prague 1978, p. 153; Köhler, P., Reinen, D.: to be published
56. Müller-Buschbaum, Hk.: Angew. Chem. *89*, 704 (1977)
57. Massa, W., Steiner, M.: J. Solid State Chem., in press
58. Köhler, P., Massa, W., Reinen, D., Hofmann, B., Hoppe, R.: Z. Anorg. Allg. Chem. *446*, 131 (1978)
59. Oelkrug, D.: Structure and Bonding *9*, 1 (1971)
60. Knox, K.: Acta Crystallogr. *A 26*, 45 (1963)
61. Reinen, D., Friebel, C., Propach, V.: Z. Anorg. Allg. Chem. *408*, 187 (1974)
62. Grannec, J., Sorbe, Ph., Chevalier, B., Etourneau, J., Portier, J.: C.R. Acad. Sci. Paris, *282C*, 915 (1976)
63. Köhl, P., Reinen, D.: Z. Anorg. Allg. Chem. *409*, 257 (1974)
64. Köhl, P.: Z. Anorg. Allg. Chem. *442*, 280 (1978)
65. Crama, W.J., Bakker, M., Verschoor, G.C., Maaskant, W.J.A.: Acta Crystallogr. *B 35*, in press (1979)
66. Crama, W.J., Maaskant, W.J.A., Verschoor, G.C.: Acta Crystallogr. *B 34*, 1973 (1978)
67. Elliot, H., Hathaway, B.J., Slade, R.C.: Inorg. Chem. *5*, 669 (1966)
68. Grefer, J., Reinen, D.: Z. Anorg. Allg. Chem. *404*, 167 (1974)
69. Takagi, S., Joesten, M.D., Lenhert, P.G.: Acta Crystallogr. *B 31*, 1968 and 1970 (1975) [Errata: *B 32*, 668 (1976)]
70. Pebler, J., Backes, G., Schmidt, K., Reinen, D.: Z. Naturforsch. *31b*, 1289 (1976)
71. Takagi, S., Joesten, M.D., Lenhert, P.G.: JACS *96*, 6606 (1974); Acta Crystallogr. *B 31*, 596 (1975) and *B 32*, 2524 (1976)
72. Backes, G., Reinen, D.: Z. Anorg. Allg. Chem. *418*, 217 (1975)
73. Bertrand, J.A., Carpenter, D.A., Kalyanaraman, A.R.: Inorg. Chim. Acta *1971*, 113
74. Shannon, R.P., Prewitt, C.T.: Acta Crystallogr. *825*, 925 (1969)
75. Hathaway, B.J., Slade, R.C.: J. Chem. Soc. (A) *1968*, 85
76. Paoletti, P., Kennard, C.H.L., Martini, G.: Chem. Commun. *1971*, 768
77. Harrowfield, B.V., Weber, R.: Phys. Lett. *38A*, 27 (1972)
78. Harrowfield, B.V., Pilbrow, J.R.: J. Phys. C: Solid State Phys. *6*, 755 (1973); Harrowfield, B.V., Dempster, A.J., Freeman, T.E., Pilbrow, J.R.: Solid State Phys. *6*, 2058 (1973)
79. Harrowfield, B.V.: Solid State Commun. *19*, 983 (1976)
80. Dubler, E., Matthieu, J.P., Oswald, H.R.: Report 4th Int. Conf. on Thermal Analysis, Budapest 1974
81. Klein, S., Reinen, D.: J. Solid State Chem. *25*, 295 (1978)
82. Isaacs, N.W., Kennard, C.H.L.: JACS (A) *1969*, 386
83. Cullen, D.L., Lingafelter, E.C.: Inorg. Chem. *10*, 1264 (1971)
84. Takagi, S., Joesten, M.D., Lenhert, P.G.: Acta Crystallogr. *B 32*, 326 (1976)
85. Harrowfield, B.V.: Solid State Commun. *19*, 983 (1976)
86. Helmbold, R., Mullen, D., Ahsbahs, H., Klopsch, A., Hellner, E.: Z. Kristallogr. *143*, 220 (1976)
87. Klein, S., Reinen, D.: J. Solid State Chem., in press
88. Noda, Y., Mori, M., Yamada, Y.: Solid State Commun. *19*, 1071 (1976) and *23*, 247 (1977); Noda, Y., Mori, M., Yamada, Y.: J. Phys. Soc. Japan *45*, 954 (1978)
89. Mullen, D., Heger, G., Reinen, D.: Solid State Commun. *17*, 1249 (1975) (in this paper "β" and "β'" have to be interchanged)
90. Takagi, S., Joesten, M.D., Lenhert, P.G.: Acta Crystallogr. *B32*, 1278 (1976); Joesten, M.D., Takagi, S., Lenhert, P.G.: Inorg. Chem. *16*, 2680 (1977)
91. Mori, M., Noda, Y., Yamada, Y.: Solid State Commun. *27*, 735 (1978)
92. Kashida, S.: J. Phys. Soc. Japan *45*, 414 (1978)

93. Huiskamp, W.J.: Proceed. Low Temperature Calorimetry Conf., Ann. Acad. Sci. Fennicae *A VI*, No. 210 (1966);
Blöte, H.W.J.: J. Appl. Physics, in press
94. Dunitz, J.P., Orgel, L.E.: J. Phys. Chem. Solids *3*, 20 (1957)
95. Ammeter, J.H., Bürgi, H.B., Gamp, E., Meyer-Sandrin, V., Jensen, W.P.: Inorg. Chem. *18*, 733 (1979)
96. Clack, D.W.: Mol. Phys. *27*, 1513 (1974)
97. Clack, D.W., Smith, W.: Mol. Phys. *29*, 1615 (1975)
98. Kremer, S., Reinen, D.: in preparation
98a. Abe, H., Ono, K.: J. Phys. Soc. Japan *11*, 947 (1956);
Hathaway, B.J., Billing, D.E.: Coordin. Chem. Rev. *5*, 143 (1970)
99. Clack, D.W.: private communication; Clack, D.W., Reinen, D.: to be published
100. von Schnering, H.G.: Z. Anorg. Allg. Chem. *353*, 1 and 13 (1967)
101. Krause, S., Reinen, D.: to be published
102. Knox, K.: J. Chem. Phys. *30*, 991 (1959)
103. Khomskii, D.J., Kugel, K.J.: Solid State Commun. *13*, 763 (1973)
104. Dance, J.M., Grannec, J., Tressaud, A.: C.R. Acad. Sci. Paris, *283C*, 115 (1976)
105. Day, P.: Accounts Chem. Research, in press
106. Babel, D., Wall, F., Heger, G.: Z. Naturforsch. *29b*, 139 (1974);
Massa, W.: Inorg. Nucl. Chem. Lett. *13*, 253 (1977)
107. Grande, B., Müller-Buschbaum, Hk.: Z. Anorg. Allg. Chem. *433*, 152 (1977);
Müller-Buschbaum, Hk., Lehmann, U.: Z. Anorg. Allg. Chem. *447*, 47 (1978)
108. Demazeau, G., Pouchard, M., Hagenmuller, P.: J. Solid State Chem. *18*, 159 (1976)
109. Pausch, H., Müller-Buschbaum, Hk.: Z. Anorg. Allg. Chem., in press (1979)
110. Köhl, P., Müller, U., Reinen, D.: Z. Anorg. Allg. Chem. *392*, 124 (1972)
111. Babel, D., Haegele, R.: J. Solid State Chem. *18*, 39 (1976)
112. Köhl, P., Reinen, D.: Z. Anorg. Allg. Chem. *433*, 81 (1977)
113. Tishchenko, G.M.: Tr. Inst. Kristallogr. Akad. Nauk SSSR *11*, 93 (1955)
114. Deschanvres, A., Raveau, B., Tollemer, Fr.: Bull. Soc. Chim. Fr. *1967*, 4077
115. Reinen, D., Propach, V.: Inorg. Nucl. Chem. Lett. *7*, 569 (1971)
116. Propach, V.: Z. Anorg. Allg. Chem. *435*, 161 (1977)
117. Vincent, H., Bochu, B., Aubert, J.J., Joubert, J.C., Marezio, M.: J. Solid State Chem. *24*, 245 (1978)
118. Marezio, M., Dernier, P.D., Chenavas, J., Joubert, J.C.: J. Solid State Chem. *6*, 16 (1973)
119. Bochu, B., Chenavas, J., Joubert, J.C., Marezio, M.: J. Solid State Chem. *11*, 88 (1974)
120. Hellner, E.: Structure and Bonding, *37*, Berlin–Heidelberg–New York: Springer (1979)
121. Odenthal, R.H., Paus, D., Hoppe, R.: Z. Anorg. Allg. Chem. *407*, 144 (1974)
122. Friebel, C., Reinen, D.: Z. Anorg. Allg. Chem. *413*, 51 (1975)
123. Smith, D.W.: Structure and Bonding, *12*, 49 Berlin–Heidelberg–New York: Springer (1972)
124. Bertini, I., Gatteschi, D., Scozzafava, A.: Inorg. Chem. *16*, 1973 (1977);
Bertini, I., Dapporto, P., Gatteschi, D., Scozzafava, A.: J.C.S. Dalton, in press
125. Massard, P., Bernier, J.C., Michel, A.: J. Solid State Chem. *4*, 269 (1972)
126. Reinen, D., Steffens, F.: Z. Anorg. Allg. Chem. *441*, 63 (1978)
127. Reinen, D., Steffens, F.: to be published
128. Propach, V., Steffens, F.: Z. Naturforsch. *33b*, 268 (1978)
129. Müller, B., Hoppe, R.: Z. Anorg. Allg. Chem. *392*, 37 (1972)
130. Friebel, C.: Solid State Commun. *15*, 639 (1974)
131. Morgenstern-Badarau, I.: J. Solid State Chem. *17*, 399 (1976)
132. Balz, D., Plieth, K.: Z. Elektrochem., Ber. Bunsenges. Physik. Chem. *59*, 545 (1955)
133. Allen, G.C., Warren, K.D.: Structure and Bonding *9*, 49 (1971)
134. Alter, E., Hoppe, R.: Z. Anorg. Allg. Chem. *405*, 167 (1974)
135. Kremer, S., Reinen, D.: to be published
136. Münninghoff, G., Heger, G., Hellner, E., Reinen, D., Treutmann, W.: to be published

The Frameworks (Bauverbände) of the Cubic Structure Types*

Erwin E. Hellner**

Institut für Mineralogie der Philipps Universität,
3550 Marburg/Lahn, Germany

Table of Contents

Introduction . . . 62
Historical Survey . . . 63
"Bauverband", Point Configuration, Lattice Complex . . . 65
Coordination Polyhedra . . . 69
Representation of Invariant Lattice Complexes . . . 71
Sphere Packings in the Cubic Crystal System . . . 73
i Homogeneous Sphere Packings as "Bauverbände" . . . 73
ii Heterogenous Sphere Packings as "Bauverbände" . . . 77
Relations between Structure Types . . . 82
i Iso-, Homeo- and Heterotypism; Typism by Laves . . . 82
ii Main Classes, Subclasses and Families . . . 84
Structure Types of the I-Family . . . 97
Structure Types of the P- and F-Family . . . 100
Further Homogenous and Some Heterogeneous "Bauverbände" . . . 103
Deformed Structure Types . . . 126
Symbolism for Net Structures . . . 129
Summary and Outlook . . . 131
References . . . 132
Formula Index . . . 134

The cubic structure types are described by their frameworks (Bauverbände) in accordance with homogeneous and heterogeneous sphere packings. The Bauverbände are symbolized by large letters for invariant lattice complexes and a notation for coordination polyhedra. The homogeneous sphere packings are described by a symbolism of Fischer. A grouping of the structure types in families, in main- and subclasses, is proposed.

* Part V of the series "Verwandtschaftskriterien von Kristallstrukturtypen".
** With a contribution of F. Laves about the definitions and examples for iso-, homeo- and heterotypism in accordance with a paper given at the "Göttinger Isomorphie-Besprechung 7./8. Oct. 1943 and a report in "Die Chemie" 57, 32 (1944).

Introduction

65 years after the discovery of the X-ray diffraction of crystals there is no unique solution for the description of structure types by symbols which permit the reconstructions of the structure types on one side and which show the relations between different structure types on the other side.

This paper will give a survey of such an attempt for nearly all cubic structure types and their grouping in families main- and subclasses and homeotypic types; it will further demonstrate the extension to crystal structures with non-cubic symmetry.

Each structure type is considered to have its typical connection or construction pattern, which will be named after Laves (1930) "Bauverband"[1]. Such a "Bauverband" may be similar to a homogeneous or heterogeneous sphere packing (p.e., Frank and Kasper 1958, 1959) and will be characterised by its eigencoordination, density and – if possible[2] – by all existing coordination polyhedra which may be considered as a partitioning problem of 3-dimensional space. The points of a "Bauverband" must not be occupied by a single atom, but they may form the centres of groups of atoms (polyhedra). Moreover different points of a "Bauverband" may be occupied by different atoms or polyhedra. In this case the "Bauverband" splits up in at least two splitting parts.

Structure types with the same "Bauverband" are distributed on main classes in the following way: Two structure types belong to the same main class if the "Bauverband" is splitted up in the same way, even if the points of the Bauverband are occupied by different polyhedra. Structure types of the same "Bauverband" but with different splitting parts belong to different main classes.

If in a structure type the points of some splitting parts of the "Bauverband" are not occupied and therefore the eigencoordination reduces, but the main distances and directions between the remaining points are still preserved, the type will belong to a subclass of the corresponding main class and not to a main class of another "Bauverband".

A family is built up by all main classes and all subclasses of one "Bauverband". Main classes of the same family are different in respect to the splitting of the "Bauverband". Structure types with the same "Bauverband" but belonging to different main classes may differ in the relative size of their unit cells. For each family there exists one main class with a minimal number of points per unit cell. The index[3] of this main class is 1, whereas the index of any other main class refers to the factor of cell enlargement.

1 Plural "Bauverbände" or "Bauverbaende" or in english notation "Bauverbands".

2 In some case one may find interpenetrating coordination polyhedra and in other cases (especially in space groups with screw axis) the description by row packings (O'Keeffe and Anderson, 1977) may be more adequate.

3 In earlier papers by E. Hellner the term "order" was used instead of "index".

The "Bauverbände" are described by the symbols of invariant lattice complexes (Hermann, 1960; Donnay et al., 1966) and, if necessary, by symbols of coordination polyhedra (Donnay et al., 1964). Heterogeneous "Bauverbände" consist of two or more independent sets of equivalent points and will be characterised by two or more lattice complexes and coordination polyhedra in square brackets. Invariant lattice complexes are useful to describe "Bauverbände" because most of these complexes are homogeneous sphere packings themselves. For uni-, di- and trivariant cubic lattice complexes Fischer (1973, 1974) has given the parameter conditions for the existence of sphere packings which makes it easy to recognise the homogeneous "Bauverbände" All structure types, which do not belong to the families of the cubic invariant lattice complexes I, P and F, may be described by about 25 homogeneous and about 15 heterogeneous "Bauverbände" related to cubic sphere packings.

Historical Survey

Niggli (1919) recognised the importance of lattice complexes for the description of structure types; 1935 Hermann listed the lattice complexes in the "Internationale Tabellen zur Bestimmung von Kristallstrukturen" giving for each point position the space group and Wyckoff letter for the standard setting of the complex.

In the "Strukturbericht", Vols. *1–7* (1928–1944), the structure types were divided in respect to their chemical composition (A, B, C, . . . L, S); the number behind the letter is of historical interest only.

Laves (1930, 1956) recognising the difficulty in using a lattice-complex notation for the description of structure types made the distinction between four different homogeneous "Bauverbände": I (island), C (chain), N (net), L (lattice). Corresponding small letters are used for heterogeneous ones. Coordination numbers and distances are added.

A topological approach to find 2- and 3-dimensional frameworks is made by Wells (1954 ff.) starting with 3-connected 2- and 3-dimensional nets; he expanded his derivation up to 12-connected nets. The latest extensive application of this concept is also given by Wells (1975). The complete derivation of all homogeneous frameworks, however, has not been done so far. The properties of these networks with respect to degrees of freedom and the parameter conditions have not been given. Under the additional assumption of equal shortest distances (sphere-packing condition), this was done for the tetragonal and cubic space groups by Fischer (1971, 1973, 1974) and for the cubic interpenetrating sphere packings by Fischer and Koch (1976).

The members of an ASTM nomenclature subcommitee of Committee E-4 on crystallography published a proposal for structure-type symbols in the ASTM Bulletin (1947). Such a symbol consists of the number of atoms within the unit cell followed by a capital letter indicating the Bravais lattice of the corresponding space group. If different structure types would receive the same symbol, distinguishing small letters a, b, c ... are put in between the number and the letter. Example: Diamond 8aF, ZnS (Sphalerite) 8bF, NaCl 8cF. This dictionary listing has been differently modified by Pearson (1967, 1972) and Schubert (1964, 1977).

In two papers under the title "Complex alloy structures regarded as sphere packings", Frank and Kasper (1958, 1959) described "triangulated shells" for the coordination numbers 12, 14, 15 and 16, which were obtained in tetrahedrally close-packed structures. Pearson and Shoemaker (1969) and Shoemaker and Shoemaker (1971) found further representatives of this type in intermetallic compounds. Hermann (1960) introduced capital letters for invariant lattice complexes and an additional splitting term for lattice complexes with degrees of freedom. The usefullness of lattice complexes has been demonstrated by Bokii and Smirnowa (1963), Hellner (1965, 1976a, b, c, 1977), Loeb (1970), Sakamoto and Takahasi (1971), Niggli (1972), and Fischer and Koch (1974).

A systematic classification for structures of intermetallic compounds with respect to coordination number and coordination polyhedra has been tried by Kripyakevich (1963). Samson (1967, 1968, 1969, 1972a, b) has built up giant structures of intermetallic compounds up to 1166 atoms in the unit cell with $a_0 \sim 30$ Å with the aid of truncated tetrahedra, icosahedra, interpenetrating icosahedra, hexagonal prisms and antiprisms, pentagonal prisms, and tricapped trigonal prisms. Tetrahedral structures in all crystal systems are described in a survey by Parthé (1964).

Based on a theorem of counting by Pólya (1937), Moore (1976) proposed an approach deriving pattern inventories (closed or open structures) by discrete elements. Closed structures may be derived by planar networks, enumerating an ordering scheme of octahedral or tetrahedral population etc. It has been applied also to "bracelet and pinwheel" structures of the alkali-alkaline earth orthosilicates, phosphates and sulfates. Lima-de-Faria (1965) started a systematic derivation of simple inorganic close-packed structures by sequences of equal layers, and together with Figueiredo (1976) described finally about 800 structure types classified in five categories: atomic, group, chain, sheet and framework. Subdivisions into homogeneous and heterogeneous arrangements and packing sequences are introduced further. In the symbol Na^oCl^c the part Cl^c stands for large atoms in cubic close packing and superscript o means interstitial atoms in octahedral voids.

Crystallograms of structure-type line diagrams have been developped by Smirnova (1971, 1975a, b, 1976, 1977) to recognise relations between structure types and homologous series based on structural elements. Tolerance classes of elements of α-Fe-, Cu-, Mg- and α-La-type are discussed by Smirnova, Kurashkovskaya and Below (1977a, b, c) on the basis of miscibility properties.

On the basis of group-subgroup relations, structure types and their relations (e.g., in connection with polymorphic transitions) have been derived by Bärnighausen (1975) and by Zverdinskaya et al. (1977).

"Bauverband", Point Configuration, Lattice Complex

As mentioned above the term "Bauverband" has been introduced by Laves (1930) and can be translated as connection pattern, construction pattern or framework. There is a direct correlation between the "Bauverbände" and the n-connected nets by Wells (1954 ff.). We will stick to the German term.

A crystallographic "Bauverband" may be defined as an arrangement of points in the 3-dimensional periodic space occupied by atoms. The arrangement of points in a space group with specified size of the unit cell may be described by the parameters of one point position ("homogeneous Bauverband") or by the parameters of two or more independent point positions ("heterogeneous Bauverband").

The "Bauverband" may characterise the whole structure. In this case it may be as well a homogeneous one (α-Po, Cu, diamond, As) as a heterogeneous one (β-W-(W_3O), β-Mn, $CoAs_3$, Tl_7Sb_2). In structures of ionic compounds the "Bauverband" describes in most cases the anion framework either as homogeneous (CsCl, NaCl, CaF_2, $KSbF_6$, perovskite, garnet) or as heterogeneous ($CsBe_2F_5$, $KSbO_3$, β-$SnWO_4$, 46 $H_2O \cdot 8$ M). A pseudo-homogeneous "Bauverband" may occur in a subgroup of the space group of the homogeneous "Bauverband", if the corresponding point position splits up into several point positions of the subgroup. If a structure type belongs to a main class of a "Bauverband" which requires an enlargement of the unit cell with respect to the simplest main class, the "Bauverband" often has to be built up by several independent point positions.

Examples:

ordered $AuCu_3$ with respect to the Cu-type,
$MgCd_3$ with respect to the Mg-type ($a' = 2\,a_0$, $c' = c_0$, index 4),
$TiSi_2$ with respect to the γ-Pu-type ($a' = 3\,a_0$, $c' = c_0$, index 3),
Langbeinite with respect to the eulytite type,
$Ba_4Sb_3LiO_{12}$ with respect to the perovskite type ($a' = 2\,a_0$, index 8)

Sometimes large cations together with the anions form a pseudohomogeneous "Bauverband" as in several cubic and hexagonal perovskites.

A "Bauverband", the points of which can be described by an invariant point position, will be characterised by the symbol of the corresponding invariant lattice complex (capital letter) (Hermann 1960; Donnay et al., 1966). The coordinates of the points of the invariant cubic lattice complexes, referred to their characteristic point position, are listed in Table 1. The projections of the corresponding point pattern are presented in Fig. 1 with the z coordinates given in units of c/8.

Lattice complexes that correspond to the cubic Bravais lattices are designated by the letters P, I, F. D and W are derived from the names of the structure types "diamond" and "β-W" (W_3O-type). S and V indicate the symmetry $S_4 - \bar{4}$ and V-222, respectively, of the equipoints of these complexes. T and Y are based on properties of these complexes namely tetrahedra sharing vertices in T and trigonal planar sur-

Table 1. The 16 invariant cubic lattice complexes in their standard settings

Lattice complex	Standard Setting	
	Space group	Coordinates
P	Pm3m 1(a)m3m	000
I	Im3m 2(a)m3m	$000; \frac{1}{2}\frac{1}{2}\frac{1}{2}$
J	Pm3m 3(c)4/mmm	$0\frac{1}{2}\frac{1}{2}; \frac{1}{2}0\frac{1}{2}; \frac{1}{2}\frac{1}{2}0$
J*	Im3m 6(b)4/mmm	$J + \frac{1}{2}\frac{1}{2}\frac{1}{2}J (\equiv J')$
F	Fm3m 4(a)m3m	$000; 0\frac{1}{2}\frac{1}{2}; \frac{1}{2}0\frac{1}{2}; \frac{1}{2}\frac{1}{2}0$
D	Fd3m 8(a)$\bar{4}$3m	$F + \frac{1}{4}\frac{1}{4}\frac{1}{4}F (\equiv F'')$
^{+}Y	$P4_332$ 4(a)32	$\frac{1}{8}\frac{1}{8}\frac{1}{8}; \frac{5}{8}\frac{3}{8}\frac{7}{8}; \frac{7}{8}\frac{5}{8}\frac{3}{8}; \frac{3}{8}\frac{7}{8}\frac{5}{8}$
$^{+}Y*$	$I4_132$ 8(a)32	$^{+}Y + \frac{1}{2}\frac{1}{2}\frac{1}{2}\,^{+}Y (\equiv {}^{+}Y')$
Y**	Ia3d 16(b)32	$^{+}Y* + \bar{1}\cdot{}^{+}Y* (\equiv {}^{-}Y*)$
W	Pm3n 6(c)$\bar{4}$m2	$\frac{1}{4}0\frac{1}{2}; \frac{1}{2}\frac{1}{4}0; 0\frac{1}{2}\frac{1}{4}; \frac{3}{4}0\frac{1}{2}; \frac{1}{2}\frac{3}{4}0; 0\frac{1}{2}\frac{3}{4}$
W*	Im3m 12(d)$\bar{4}$m2	$W + \frac{1}{2}\frac{1}{2}\frac{1}{2}W (\equiv W')$
T	Fd3m 16(c)$\bar{3}$m	$\frac{1}{8}\frac{1}{8}\frac{1}{8}; \frac{1}{8}\frac{5}{8}\frac{5}{8}; \frac{5}{8}\frac{1}{8}\frac{5}{8}; \frac{5}{8}\frac{5}{8}\frac{1}{8}; \frac{1}{8}\frac{3}{8}\frac{3}{8}; \frac{1}{8}\frac{7}{8}\frac{7}{8}; \frac{5}{8}\frac{3}{8}\frac{7}{8}; \frac{5}{8}\frac{7}{8}\frac{3}{8};$ $\frac{7}{8}\frac{5}{8}\frac{7}{8}; \frac{7}{8}\frac{1}{8}\frac{3}{8}; \frac{3}{8}\frac{5}{8}\frac{3}{8}; \frac{3}{8}\frac{1}{8}\frac{7}{8}; \frac{3}{8}\frac{3}{8}\frac{1}{8}; \frac{7}{8}\frac{3}{8}\frac{5}{8}; \frac{7}{8}\frac{7}{8}\frac{1}{8}; \frac{7}{8}\frac{7}{8}\frac{5}{8}$
S	I$\bar{4}$3d 12(a)$\bar{4}$	$\frac{3}{8}0\frac{1}{4}; \frac{1}{4}\frac{3}{8}0; 0\frac{1}{4}\frac{3}{8}; \frac{1}{8}0\frac{3}{4}; \frac{3}{4}\frac{1}{8}0; 0\frac{3}{4}\frac{1}{8}; \frac{7}{8}\frac{1}{2}\frac{3}{4}; \frac{3}{4}\frac{7}{8}\frac{1}{2};$ $\frac{1}{2}\frac{3}{4}\frac{7}{8}; \frac{5}{8}\frac{1}{2}\frac{1}{4}; \frac{1}{4}\frac{5}{8}\frac{1}{2}; \frac{1}{2}\frac{1}{4}\frac{5}{8}$
S*	Ia3d 24(d)$\bar{4}$	$S + \frac{1}{2}\frac{1}{2}\frac{1}{2}S (\equiv S')$
^{+}V	$I4_132$ 12(c)222	$\frac{1}{8}0\frac{1}{4}; \frac{1}{4}\frac{1}{8}0; 0\frac{1}{4}\frac{1}{8}; \frac{3}{8}0\frac{3}{4}; \frac{3}{4}\frac{3}{8}0; 0\frac{3}{4}\frac{3}{8}; \frac{5}{8}\frac{1}{2}\frac{3}{4}; \frac{3}{4}\frac{5}{8}\frac{1}{2};$ $\frac{1}{2}\frac{3}{4}\frac{5}{8}; \frac{7}{8}\frac{1}{2}\frac{1}{4}; \frac{1}{4}\frac{7}{8}\frac{1}{2}; \frac{1}{2}\frac{1}{4}\frac{7}{8}$
V*	Ia3d 24(c)222	$^{+}V + \bar{1}\cdot{}^{+}V (\equiv {}^{-}V)$

rounding of the equipoints in Y. The complex J has a shape that reminds on the american toy "Jackstone". $^{+}$ and $^{-}$ signs as superscripts in front of the letters indicate enantiomorphic forms. The superscript* (pronounced: star) indicates a complex, that may be built up as a combination of two complexes of the same kind shifted against each other by $(\frac{1}{2}\,\frac{1}{2}\,\frac{1}{2})$ or as a combination of two enantiomorphic forms (generated by $\bar{1}$). One, two or three primes used as an abbreviation for a shifting of the

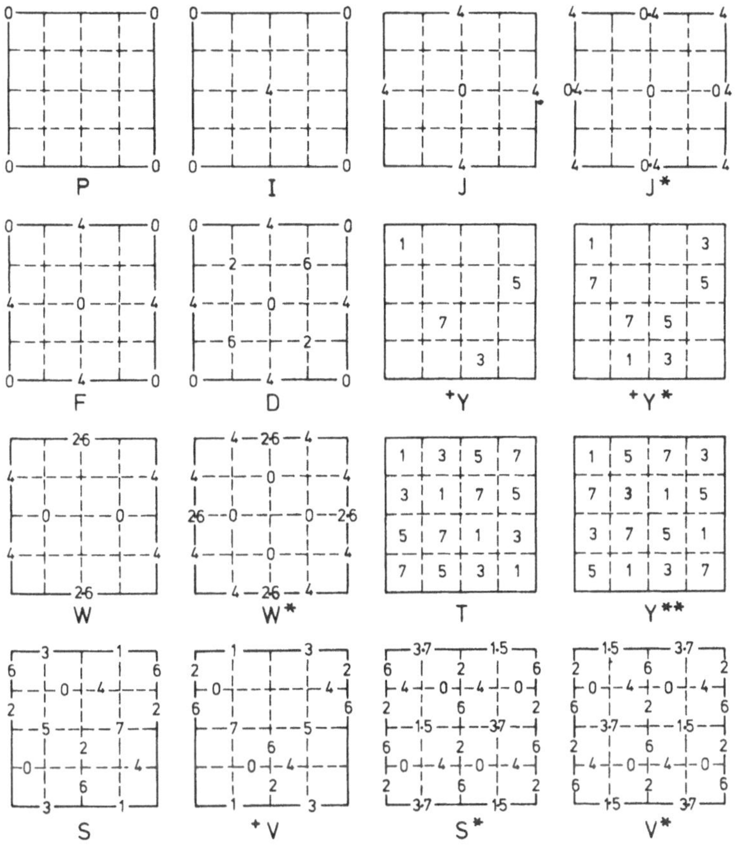

Fig. 1. The 16 invariant cubic lattice complexes in their standard settings; numbers are given in n/8 (for coordinates see Table 1)

complex by the vector $(\frac{1}{2}\frac{1}{2}\frac{1}{2})$, $(\frac{1}{4}\frac{1}{4}\frac{1}{4})$ or $(\frac{3}{4}\frac{3}{4}\frac{3}{4})$, respectively, against the standard setting.

Example: $^{+}Y + \frac{1}{2}\frac{1}{2}\frac{1}{2}\,^{+}Y = {}^{+}Y + {}^{+}Y' = {}^{+}Y^*$

$$^{-}Y + \frac{1}{2}\frac{1}{2}\frac{1}{2}\,^{-}Y = {}^{-}Y + {}^{-}Y' = {}^{-}Y^*$$

$$^{+}Y^* + {}^{-}Y^* = Y^{**}$$

The term "lattice complex" was differently interpreted since 1919, when Niggli created it in analogy to the term "face form" in crystallographic point groups (Burzlaff et al., 1974). Finally two concepts have been used to define the term lattice complex, one by Zimmermann and Burzlaff (1974) using an algebraic approach by means of triplets of linear polynoms, the second by Fischer and Koch (1974) with the aid of crystallographic point configurations[4]. Here we will follow the definition of Fischer and Koch in a simple way (omitting the formation of classes of space groups, point positions etc.): We regard an individual space group specified by its Hermann-Mauguin symbol in the usual coordinate description given in the International Tables. If one point position has degrees of freedom, e.g., Pm3m ±(x00, 0x0, 00x), each choice of the coordinate parameters within an asymmetric unit ($0.0 < x < 0.5$ in this example) results in a distinct point configuration. All possible configurations of this point position and the corresponding ones from the other space groups of the same equivalence class form the lattice complex. For the description of structure types it is necessary to know the geometrical properties of the different point configurations of the lattice complexes in dependence of the coordinate parameters and the size and the shape of the unit cell. It was the idea of Hermann (1960) to symbolise invariant lattice complexes by capital letter S (Table 1).

For all lattice complexes with degrees of freedom, it is possible to find distinct parameters to generate point configurations which belong to an invariant lattice complex. If the multiplicity of such a special point configuration is the same as that of the complex with degrees of freedom, the cooresponding invariant complex is called a limiting form of the complex under consideration (Koch, 1974). If all specialised point positions of a lattice complex with degrees of freedom belong to limiting forms the symbol of the complex under consideration consists of the symbols of all invariant limiting forms in connection with a notation for the degrees of freedom.

Example: $P2_13$ 4(a), xxx, FYxxx (= FY$\underline{\underline{x}}$). F is realised with x = 0, $^+$Y with $x = \frac{1}{8}$ and $^-$Y with $x = -\frac{1}{8}$. xxx shows the direction of the degree of freedom.

If for special coordinates a reduction of the multiplicity of the point configurations can be reached, the symbol of the lattice complex with degrees of freedom consists of the symbol of the corresponding invariant complex and a split symbol, which shows the splitting number and the degrees of freedom.[5]

Example: Pm3m 6(e) x00 P6x.

The product of the multiplicity of the invariant complex and the splitting number gives the multiplicity of the complex with degrees of freedom.

Summary. "Bauverbände" (connection pattern or frameworks) will be used to characterise structure types; they may describe the complete structure (as for most of the elements and some of the intermetallic compounds) or the anion framework

4 Wondratschek (1976) uses the term "orbit" instead of point configuration. Orbit (more exactly: crystallographic orbit of points in space groups) and crystallographic point configuration are synonyma.

5 A description of all point positions in all space groups by lattice-complex symbols is given by Fischer et al. (1973).

(as for most of the ionic coordination compounds). Sometimes large cations and the anion arrangement together form a well known "Bauverband". The "Bauverbände" may be homogeneous (one point position occupied only) or heterogeneous (two or more point positions occupied) or pseudo-homogeneous (two or more point positions can be recognised as splitting of *one* point position of a supergroup).

For the description of the "Bauverbände", symbols of invariant complexes may be used. Besides W, Y** and V* all other invariant cubic complexes may be regarded as sphere packings and therefore may be "Bauverbände" for their own. The symbols for lattice complexes with degrees of freedom do not show the geometrical properties. However the study of the conditions for homogeneous sphere packings (Hellner, 1965; Fischer, 1973; 1974) helps to classify the point configurations of a lattice complex in dependence of the parameter field and to describe the geometrical properties of these assemblages of points by symbols of invariant complexes plus symbols of coordination polyhedra.

Coordination Polyhedra

The knowledge of the coordination polyhedra of all existing voids in a "Bauverband", their size, site position and point symmetry is necessary for crystal chemical consideration. During the discussion of the "Bauverbände" and their application to structure types we will describe the existing and the occupied polyhedra to distinguish between different structure types belonging to the same main class.

Example: Garnet $Ca_3Al_2(SiO_4)_3$, $Ca_3Al_2(OH)_{12}$, Hg_3TeO_6 and $RhBi_4$ belong to the same "Bauverband" and to the same mainclass I_{222}[6 o] (cubic abbreviation) I_2[6 o].

Donnay et al. (1964) proposed symbols for the homogeneous coordination polyhedra consisting of the number of equipoints and small letters; these polyhedra are derived as duals of the 47 crystal forms and of 9 noncrystallographic forms which can be grouped into 30 topologically distinct forms.

Small letters are introduced for the following surroundings:

Prismatic	r	cube	8 c
pyramidal	y	octahedron	6 o
planar	l	tetrahedron	4 t
antiprismatic	a		
scalenohedron	s	icosahedron	12 i
		cuboctahedron	12 co

The same symbols are used for distorted coordination polyhedra and for those built up by several independent positions (heterogeneous coordination polyhedra).

Example: 4t stands for an ideal tetrahedron but is used also for a tetragonally or orthorhombically distorted tetrahedron; in addition 4t is used if the tetrahedron is built up by a trigon 3y and a single point 11 which lies above or below the plane of the trigon. The oxygen tetrahedra of the SO_4 group in alum $KAl(SO)_2 \cdot 12\,H_2O$ are built up in such a way: In general a coordination polyhedron consists of nonequivalent points obeying only the point symmetry of the site. Heterogeneous coordination polyhedra may be found which are known as face forms, but not as their duals, i.e. as homogeneous coordination polyhedra, for instance the pentagondodecahedron 20 pd = 12 i + 8 c.

In Table 2 symbols of coordination polyhedra are listed which are mainly used in this paper. To describe clusters a set of polyhedra each inclosing the former may be used.

Table 2. Symbols of coordination polyhedra

Symbols for coordination polyhedra	
1 l	One point
2 l	Dumbbell, colinear with origin
2 y	Dumbbell
3 l	Trigon, colinear with origin
3 y	Trigon
4 l	Square, also rectangle
4 t	Tetrahedron, tetragonal or rhombic, also distorted, also combined by 3 y + 1 l
6 r	Trigonal prism (+ pinakoid)
6 y	Hexagon
6 l	Hexagon, colinear with origin
6 a	Trigonal antiprism
6 o	Octahedron
8 l	Octogon, colinear with origin, also truncated tetragon
8 y	Octogon
8 r	Tetragonal prism, also rhombic prism
8 c	Cube (= hexahedron)
8 a	Tetragonal antiprism, also rhombic antiprism
12 r	Hexagonal prism, also ditrigonal prism
12 tt	Truncated tetrahedron
12 co	Cuboctahedron (= cube + octahedron), also tetragonal distorted
12 i	Icosahedron
12 a	Hexagonal antiprism, also ditrigonal antiprism
24 c < o	Cuboctahedron (= truncated octahedron)
24 c > o	Cuboctahedron (= truncated cube)
24 cod	Rhombicuboctahedron (= cube + octahedron + rhomb-dodecahedron)
5 by	Trigonal bipyramide
5 py	Tetragonal pyramide, also rhombic pyramide
9 tco	Truncated cuboctahedron (a trigon is cut off)

Example: $Mg(Mo_6Cl_8)Cl_6$, Pn3, a = 12.706 Å, z = 4

$$\begin{array}{ccccc} & Mo_{24}, & Cl_8, Cl_{24}, & Cl_{24}, & Hg_4 \\ \frac{1}{4}\frac{1}{4}\frac{1}{4}F & (6\,o & ,2\,e,6\,a & ,6\,o\ ) & +\frac{3}{4}\frac{3}{4}\frac{3}{4}F \\ & & 8\,c & & \end{array}$$

$$F''\ (6\,o, 8\,c, 6\,o) + F'''$$

The 24 Mo-atoms form octahedra (6 o) around the points of $\frac{1}{4}\,\frac{1}{4}\,\frac{1}{4}$ F. The first 32 Cl-atoms together build up cubes around $\frac{1}{4}\,\frac{1}{4}\,\frac{1}{4}$ F. The faces of the cubes are centered by the Mo-atoms. The remaining 24 Cl-atoms form octahedra, which enclose the smaller Mo-octahedra and Cl-cubes; the octahedra are parallel. They also may be described as octahedra around the Hg-atoms on $\frac{3}{4}\,\frac{3}{4}\,\frac{3}{4}$ F; this is not expressed by the symbolism.

Representation of Invariant Lattice Complexes

In Table 1 and Fig. 1 the symbol, the coordinates of equipoints, the characteristic space group and a drawing of the projection is given for each invariant cubic lattice complex in its standard representation. Other representations may be congruent or enantiomorphous to the standard representation.

In the first case a vector in front of the symbol gives the shift against the standard setting. Enantiomorphous representations are distinguished by a plus or minus sign in front of the letter (p.e. $^{+}$Y, $^{-}$Y).

Other representations may be obtained by enlarging the unit cell, i.e., in the cubic system all axes have to be changed by the same factor. This factor is used as subscript in the lattice-complex symbol.

Example: 8-pointer $P_{222} = P_2$ in Fm3c 8(a)

16-pointer $I_{222} = I_2$ in Fd3c 16(a) or in Ia3d 16(a)

64-pointer $\frac{1}{8}\,\frac{1}{8}\,\frac{1}{8}\,P_{444} = P'_4$ in Ia3d as a combination of Y** and [P'_4 – Y**] (Fig. 2). Y** is the symbol of the invariant complex Ia3d 16(b) $\frac{1}{8}\frac{1}{8}\frac{1}{8}$. [$P'_4$ – Y**] is the symbol of the remaining part of P'_4 and of the configuration of the point position Ia3d 48(g) $\frac{1}{8}, x, \frac{1}{4} - x$ with $x = \frac{3}{8}$ lying in the region $0.2835 < x < 0.4005$ of the univariant sphere packing 4/4/c5.

48-pointer $W_{222} = W_2$ in Ia3d as a combination of S* and V*; Fig. 2.

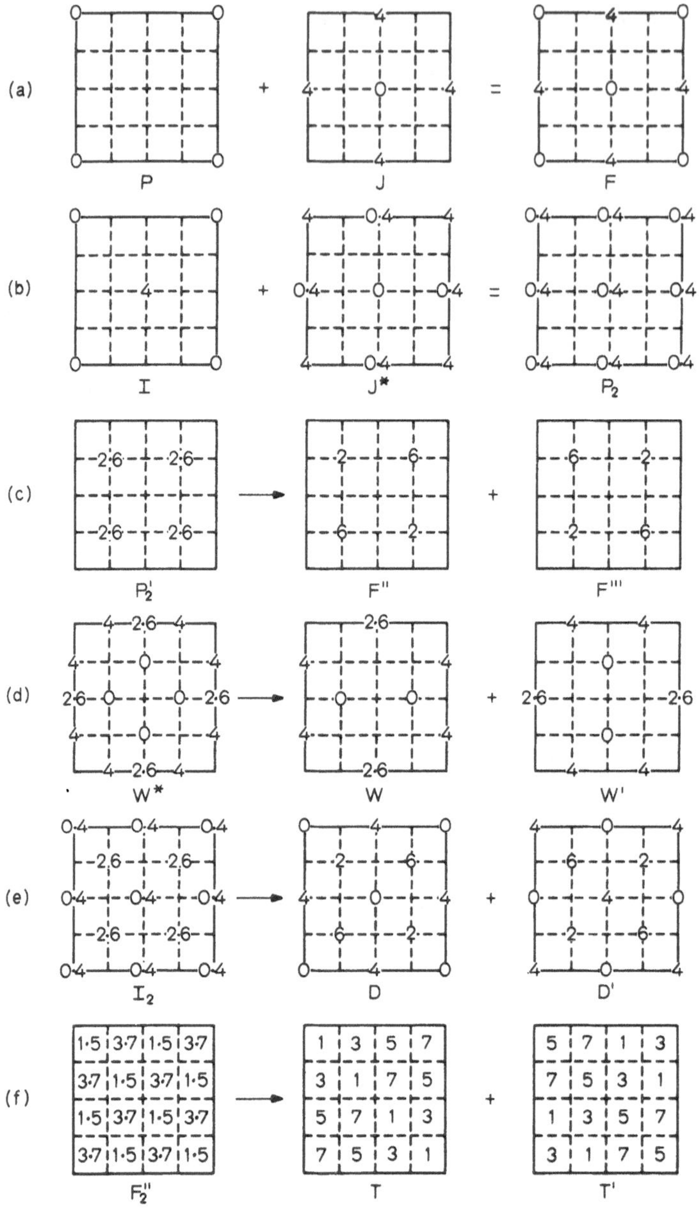

Fig. 2. Some relations between invariant lattice complexes

In Fig. 2 combinations of complexes are shown; they are important for a classification scheme; in addition we will find structure types which can be described by

$I_{666} = I_6$ and $P_{888} = P_8$. Most of the cubic invariant lattice complexes are known as limiting forms too (Koch, 1974).

Examples:	$P2_13$ 4(a), xxx	$x = 0$	F
		$x = \frac{1}{8}$	^+Y
		$x = \frac{1}{4}$	$\frac{1}{4}\frac{1}{4}\frac{1}{4}F = F''$
		$x = \frac{3}{8}$	$\frac{1}{2}\frac{1}{2}\frac{1}{2}{}^-Y = {}^-Y'$
		$x = \frac{1}{2}$	$\frac{1}{2}\frac{1}{2}\frac{1}{2}F = F'$
		$x = \frac{5}{8}$	$\frac{1}{2}\frac{1}{2}\frac{1}{2}{}^+Y = {}^+Y'$
		$x = \frac{3}{4}$	$\frac{3}{4}\frac{3}{4}\frac{3}{4}F = F'''$
		$x = \frac{7}{8}$	^-Y
	Pm3m 8(g), xxx	$x = \frac{1}{4}$	$\frac{1}{4}\frac{1}{4}\frac{1}{4}P_{222} = P_2'$
	Ia3d 48(g), $\frac{1}{8}, x, \frac{1}{4} - x$	$x = \frac{1}{4}$	$\frac{1}{4}\frac{1}{4}\frac{1}{4}W_2 = W_2'$

The knowledge of the limiting forms is one important step forward in the ability of structure-type descriptions.

Sphere Packings in the Cubic Crystal System

Fischer (1971) gave the following definition:
"A 3-dimensional arrangement of spheres is called a *sphere packing* if

1) the arrangement is periodic in three independent directions,
2) every two spheres of the arrangement may be connected by a sequence of spheres in contact ("Zerfallsverbot"). A sphere packing is called homogeneous if all its spheres are equivalent by the symmetry of one of the 230 space groups.

A *sphere-packing point* is the center of an arbitrary sphere out of a homogeneous sphere packing."

i Homogeneous Sphere Packings as "Bauverbände"

In the cubic system the conditions for the occurrence of a homogeneous sphere packing may be given by conditions for the coordinates of one sphere-packing point. The number of free parameters – within a certain parameter range – gives the number of degrees of freedom for the sphere-packing type. Fischer (1973, 1974) derived the parameter conditions for all homogeneous sphere packings with cubic symmetry. Moreover he formed topological types of sphere packings and characterized them by the number of nearest neighbors and by the length of the smallest mash[6]. Different sphere-packing types which correspond to each other in both these features are distinguished by a numeration (c1, c2, c3, ...).

Examples of cubic invariant complexes regarded as sphere packings are:

P	Pm3m	1(a)	000	6/4/c1
I	Im3m	2(a)	000	8/4/c1
F	Fm3m	4(a)	000	12/3/c1
J	Pm3m	3(c)	$0\frac{1}{2}\frac{1}{2}$	8/3/c2
T	Fd3m	16(c)	$\frac{1}{8}\frac{1}{8}\frac{1}{8}$	6/3/c2

The examples show that 6 is the number of nearest neighbors for both the P- and the T-complex; but P and T belong to different sphere-packing types because they differ in the length of the smallest mash, namely 4 and 3, respectively. An analogous difference arises for I and J. On the other hand J and T may be regarded as subsets of F and $\frac{1}{8}\frac{1}{8}\frac{1}{8}F_{222}$, respectively, with different numbers of nearest neighbors but still with the smallest mash of F in common.

If the point configurations of a sphere packing with one degree of freedom lies between the point configurations of two limiting complexes one may give both symbols to characterize these arrangements of points.

Examples:	$P2_13$	4(a)xxx	x = 0	F	12/3/c1
			$0 < x < \frac{1}{8}$	$F^{+}Yxxx$	6/3/c1
			$x = \frac{1}{8}$	^{+}Y	6/3/c1
	$P4_332$	8(c)xxx	x = 0	D	4/6/c1
			$-\frac{1}{8} < x < 0$	$D^{-}Y^{*}xxx$	3/10/c1
			$x = -\frac{1}{8}$	$^{-}Y^{*}$	3/10/c1

The point configurations of point positions with degrees of freedom in general may be described by symbols of invariant lattice complexes plus those for coordination polyhedra following the lattice-complex symbol in parentheses. These polyhedra

6 This is the minimal number of contacts one has to pass to reach the starting sphere again without using a contact twice.

describe the arrangement of equipoints around the points of an invariant lattice complex. Brackets instead of parentheses are used if special parameters or a special parameter range for which sphere packings occur is meant.

Examples:	Pm3m 6(e)	x00	$0 < x < 0.293$	P(6 o)	
			$x = 0.293$	P[6 o] J′[2 ℓ] = P[6 o]	5/3/c3
			$0.293 < x < \frac{1}{2}$	$\frac{1}{2}\frac{1}{2}\frac{1}{2}$J(2 ℓ) = J′(2 ℓ)	
	$I\bar{4}3m$ 8(c)	xxx	$0 < x < \frac{3}{16}$	I(4 t)	
			$x = \frac{3}{16}$	I[4 t]	9/3/c2
			$\frac{3}{16} < x < \frac{1}{4}$	I(4 t) P_2'xxx	6/4/c1
			$x = \frac{1}{4}$	$\frac{1}{4}\frac{1}{4}\frac{1}{4}P_{222} = P_2'$	6/4/c1

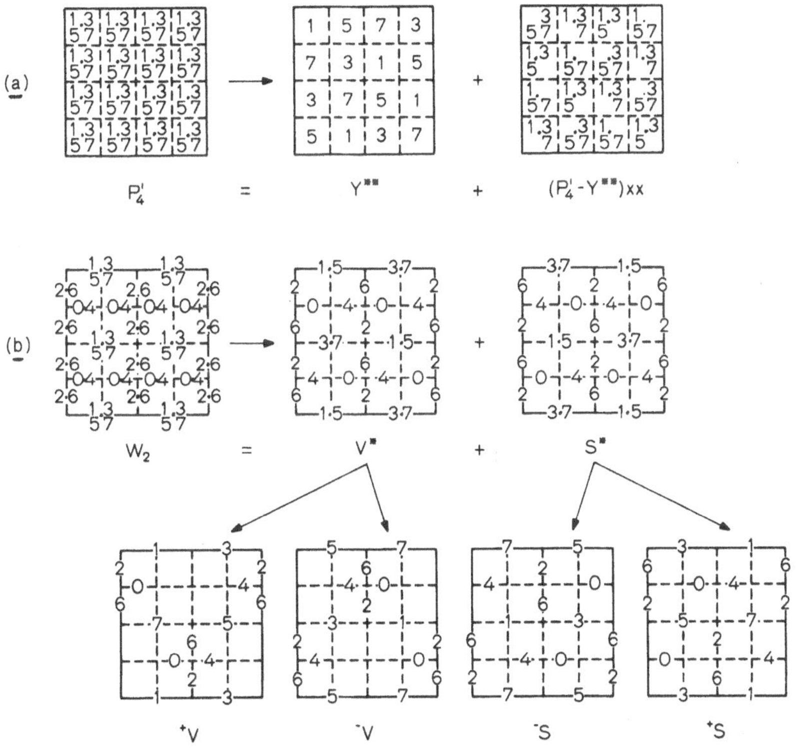

Fig. 3.
(a) The splitting of P_4' in Ia3d and
(b) the splitting of W_2 in Ia3d and subgroups.

In Pm3m 6(e) is only one sphere packing without degrees of freedom with x = 0.293; for all other values of x one finds a dumbbell around J′ or an octahedron around P, respectively. The symbol P[6 o] in brackets is chosen for abbreviation instead of P[6 o] J′[21]. In I$\bar{4}$3m 8(c)xxx there is a parameter range $\frac{3}{16} < x \leqslant \frac{1}{4}$ un the asymmetric unit for the existence of the sphere-packing type 6/4/c1 with an increase of the number of nearest neighbors up to 9 and change of the smallest mash at $x = \frac{3}{16}$. Of course the total parameter range of 6/4/c1 is $\frac{3}{16} < x < (\frac{1}{2} - \frac{3}{16} =) \frac{5}{16}$. Some point configurations may be regarded as generated by removing the points of one lattice complex from those of another one. In this case they may be characterized by symbols constructed as differences of the two corresponding lattice complexes. If a sphere packing is formed, the difference formula is put into brackets.

Examples:	Ia3d	48(g)	$\frac{1}{8}, x, \frac{1}{4} - x$	$x = \frac{3}{8}$ [7]	$[\frac{1}{8}\frac{1}{8}\frac{1}{8}P_{444} - Y^{**}]$ $= [P_4' - Y^{**}]$	4/4/c5
	$P4_332$	24(e)	xyz	$x = y = \frac{1}{8}$ $z = \frac{3}{8}$	$[\frac{3}{8}\frac{3}{8}\frac{3}{8}F_{222} - {}^{-}Y^*]$ $= [F_2''' - {}^{-}Y^*]$	9/3/c1
	I$\bar{4}$3m	24(g)	xxz	$x = \frac{3}{8}$ $z = \frac{1}{8}$	$[\frac{1}{8}\frac{1}{8}\frac{1}{8}F_{222} - I(4\,t)]$ $= [F_2'' - I(4\,t)]$	9/3/c4
	$P4_332$	12(d)	$\frac{1}{8}, x, \frac{1}{4} - x$	$x = \frac{3}{8}$ [8]	$[\frac{5}{8}\frac{5}{8}\frac{5}{8}T - \frac{1}{2}\frac{1}{2}\frac{1}{2}{}^{+}Y]$ $= [T' - {}^{+}Y']$	4/3/c1

In Table 3 those "Bauverbände" are listed which may be derived from invariant complexes, in Table 4 those which have coordinate parameters other than n/8 and are described by a symbol of an invariant complex plus a symbol of a coordination polyhedron. In the first column the symbol for the "Bauverband" as a sphere packing, the Fischer symbol and the corresponding radius of the sphere are listed; the second column gives the point positions and the third one the parameter if necessary. In the large fourth column the coordination polyhedra which exist in the sphere packing are listed. The lattice-complex symbol describes the point position for the central points of the voids. The corresponding coordination polyhedra , the vertices of which are formed by the original sphere-packing points, are described by the symbols listed in Table 2. Provided the sphere of the original sphere packing touch each other, there is a greatest sphere that may be put into a void. The radius of this sphere is given underneath the description of the void. In addition the product of the volume of one co-

7 $x = \frac{3}{8}$ lies in the 1-dimensional parameter range $0.2965 < x < \frac{1}{2}$ for 4/3/c1.

8 $x = \frac{3}{8}$ lies in the 1-dimensional parameter range $0.2835 < x < 0.4005$ for 4/4/c5.

ordination polyhedron with the multiplicity of the corresponding point position is tabulated, which gives the percentage of space filling for this kind of voids.

The sum of the normalised volumes of all different coordination polyhedra, that occur in a point configuration, should be 1 if all coordination polyhedra are listed and no overlapping takes place.

It is not always an easy task to obtain this result because the voids of sphere packings which occur in space groups with screw axis are not always clear to delimitate. In this case the voids may be described more adequately by rod-shaped units than by polyhedra. Another case of overlapping of coordination polyhedra is shown e.g. by I[4 t] if one accepts the coordination number 8 for the voids around J*. In this case the volume of the tetrahedra around W* have to be subtracted to obtain 1 for the sum of the volumes. The latter fact shows clearly that an interpretation of a structure by coordination polyhedra alone is difficult. It becomes easier if one uses only small polyhedra like tetrahedra and octahedra, because the packing of large coordination polyhedra is more complicated.

In the fifth column is listed the density of the sphere packing itself (a) and the total density of the configuration of spheres (b) under the condition that all voids of the sphere packing are filled with other spheres. The listed value under (c) gives the sum of the volumes of all coordination polyhedra which partition the sphere packing; it should be 1 under normalized conditions ($a_0 = 1$). The sixth and the last column give examples of compounds where the sphere packing exists as a "Bauverband".

ii Heterogeneous Sphere Packings as "Bauverbände"

Heterogeneous "Bauverbände" which may be described as two or more crystallographic point configurations will be discussed with the aid of drawings in a later section. Each of these point configurations by itself or all these point configurations together may follow a sphere packing condition. Here we may give a few *examples*:
[I + W] describes the total A15-β-W-W_3O(Cr_3Si)-type.
[$^-$V + $D\underline{\underline{x}}$] are occupied by the F in $CsBe_2F_5$ to form tetrahedra around Be.

W*[$\bar{4}$ t_c][9] forms the homogeneous "Bauverband" of oxygen in sodalite and hauyne. With a different parameter it also appears in α-Mn, Tl_7Sb_2 and γ-brass. Together with J*(2 ℓ) and an expanded, tetragonal deformed tetrahedron (4 t) it builds up the O-framework in $NaSbO_3$, $KSbO_3$, $La_4Re_6O_{19}$ etc. forming octahedra (6 o) around Sb and Re.

Each of these point configurations by itself or all these point configurations together may follow a sphere packing condition.

---

9 c as a subscript at (4 t) means "corner or vertex-sharing".

Table 3. Space partition of some homogeneous "Bauverbände" as invariant lattice complexes

Sphere pack., Fischer symb., r_{sp}	Space group, Position	Site position and coordination polyhedron r: Radius of the sphere in the coordination polyhedron V: Volume of the coordination polyhedron · multiplicity 4	6	8	12	24	Space filling a) Sphere pack. b) $V_a + 4/3\,\pi\,\Sigma r_i^3$ c) ΣV_i	Example	Structure formula Description Remarks
I 8/4/c1 r = 0.4330	Im3m 2(a)000	W*(4 t) r = 0.1260 $V = 0.08\overline{3} \times 12$ = 1.0000					a) 0.6802 b) 0.7807 c) 1.000	W β-W (W_3O)	W_2 I O_2, W_6 I, W
P 6/4/c1 r = 0.5000	Pm3m 1(a)000			P′(8 c) r = 0.3660 V = 1.0000 × 1 = 1.0000			a) 0.5236 b) 0.7290 c) 1.0000	CsCl	Cl_1, Cs_1 P, P′
F 12/3/c1 r = 0.3536	Fm3m 4(a)000	P_2'(4 t) r = 0.0794 $V = 0.041\overline{6} \times 8$ $= 0.333\overline{3}$	F′(6 o) r = 0.1464 $V = 0.166\overline{6} \times 4$ $= 0.666\overline{6}$				a) 0.7405 b) 0.8099 c) 1.0000	NaCl Li_2O	Cl_4, Na_4 F, F′ O_4, Li_8 F, P_2'
J 8/3/c2 r = 0.3536	Pm3m 3(c)0$\frac{1}{2}\frac{1}{2}$		P′(6 o) r = 0.1464 $V = 0.166\overline{6} \times 1$ $= 0.166\overline{6}$		P(12 co) r = 0.3536 $V = 0.833\overline{3} \times 1$ $= 0.833\overline{3}$		a) 0.5554 b) 0.7536 c) 1.0000	$CaTiO_3$ Perovs-kite	O_3, Ca_1, Ti_1 J, P, P′
T 6/3/c2 r = 0.1768	Fd3m 16(c)$\frac{1}{8}\frac{1}{8}\frac{1}{8}$	D(4 t) r = 0.0397 V = 0.0052 × 8 $= 0.041\overline{6}$			D′(12 tt) r = 0.2165 V = 0.1198 × 8 $= 0.958\overline{3}$		a) 0.3702 b) 0.7124 c) 1.0000	SiO_2 Cristo-balite $MgCu_2$	O_{16}, Si_8 T, D Cu_{16}, Mg_8 T, D′

D 4/6/c1 r = 0.2165	Fd3m 8(a)000	D′(4 t) r = 0.0397 $V = 0.041\overline{6} \times 8$ $= 0.333\overline{3}$	T′(6 o) r = 0.1768 $V = 0.041\overline{6} \times 16$ $= 0.666\overline{6}$		a) 0.3401 b) 0.7124 c) 1.000	NaTl	Na_8, Tl_8 D, D′
W* 4/4/c1 r = 0.1768	Im3m 12(d)$\frac{1}{4}0\frac{1}{2}$			I(24 cod) r = 0.3822 V = 0.5000 × 2 = 1.0000	a) 0.2777 b) 0.7454 c) 1.0000	Ir_3Sn_7	Sn_{12}, Sn_{16}, Ir_{12} W*, I(8 c), J*(21)
J* 4/6/c2 r = 0.2500	Im3m 6(b)0$\frac{1}{2}\frac{1}{2}$	I(6 o) r = 0.2500 $V = 0.166\overline{6} \times 2$ $= 0.333\overline{3}$	P_2'(6 o) r = 0.1830 $V = 0.083\overline{3} \times 8$ $= 0.666\overline{6}$		a) 0.3927 b) 0.7290 c) 1.0000	Mg_3P_2	Mg_6, P_4 J*, F″

Table 4. Space partition of some homogenous Bauverbände with special parameter conditions

Sphere packing Fischer symbol r_{sp}	Space group Position	Parameter	Site position and coordination polyhedron r: Radius of the sphere in the coordination polyhedron V: Volume of the coordination polyhedron · multiplicity 4		6		8	9	12	24/36	Space filling a) Sph. pack. V_{a_3} b) $V_a + 4/3\,\pi\, \bar{o}\, r_i$ c) ΣV_i	Example Parameter in the structure	Structure formula Description Remarks
[± $Y_{\underline{\underline{x}}}$] 7/4/c1 r = 0.2676	Pa3 8(c)xxx	x = 0.1545	$P'_2\underline{\underline{x}}$ (4 t); r = 0.0364 V = 0.0220 × 8 = 0.1760	[F''_2 – ± Y] xyz (4 t) r = 0.0748 V = 0.0197 × 24 = 0.4721	F′(6 o) r = 0.1412 V = 0.0880 × 4 = 0.3519						a) 0.6422 b) 0.7331 c) 1.0000	Fe Si x_{Fe} = 0.1358 x_{Si} = 0.156	Fe_4; Si_4 $^+Y_{\underline{\underline{x}}}$, $^-Y_{\underline{\underline{x}}}$
[T″ – ⁻Y′] 4/3/c1 r = 0.1768	$P4_132$ 12(d) $\frac{1}{8}$, x, $\frac{1}{4}$ + x	x = $\frac{1}{8}$			⁻Y′ (6 a) r = 0.1768 V = 0.04167 × 4 = 0.166̄			D(12 tt – 3 y) r = 0.2378 V = 0.1417 = 0.833̄			a) 0.2777 b) 0.8209 c) 1.000	$(Sn_2F_3)Cl$ $P2_13$ x = 0.375 y = 0.135 z = 0.119	F_{12}, Sn_4 Sn_4, Cl [T″ – ⁻Y′]$_{\underline{\underline{x}}}$, $\underbrace{^-Y'_{\underline{\underline{x}}}, {}^+Y'_{\underline{\underline{x}}}}$, $F_{\underline{\underline{x}}}$ $^{\pm}Y_{\underline{\underline{x}}}$
[F''_2 – I (4 t)] 9/3/c4 r = 0.1768	I$\bar{4}$3m 24(g)xxz	x = 0.375 = $\frac{3}{8}$ z = 0.125 = $\frac{1}{8}$	J*(4 t) r = 0.0397 V = 0.0052 × 6 = 0.0313	W*(4 t) r = 0.0397 V = 0.0052 × 12 = 0.0625	J_2xxz(5 py) V = 0.0104 × 24 = 0.2500			$P'_2\underline{\underline{x}}$(9 tco) V = 0.0521 × 8 = 0.4167	I(12 tt) r = 0.2378 V = 0.1198 × 2 = 0.2396		a) = 0.555	Tetrahedrite $(Cu, Ag)_3(Sb, As)S_3$	S_{24}, $(As, Cu)_{12}$, $(Ag, Cu)_{12}$, $(Sb, As)_8$ [F''_2 – I(4 t)], I(6 o), W*, P_2 or [F''_2 – I(4 t)], F'_2
I[4 t] 9/3/c2 r = 0.2652	I$\bar{4}$3m 8(c)xxx	x = 0.1875	I(4 t) r = 0.0596 V = 0.0176 × 2 = 0.0352	J*(4 t) r = 0.1447 V = 0.0293 × 6 = 0.1758	W*(4 t) r = 0.1046 V = 0.0111 × 12 = 0.1328	$P'_2\underline{\underline{x}}$(4 t) r = 0.0596 V = 0.0176 × 8 = 0.1406	[F''_2 – I(4 t)] (4 t) r = 0.0073 V = 0.0215 × 24 = 0.5156				a) 0.625 b) 0.767 c) 1.0000	Si F_4 x = 0.165	F_8, Si_2 I[4 t], I
	other partition:		I(4 t) r = 0.0596 V = 0.0176 × 2 = 0.02516	$P'_2\underline{\underline{x}}$(4 t) r = 0.0596 V = 0.0176 × 8 = 0.1406	– W*(4 t) r = 0.1046 V = 0.0111 × 12 = 0.1328		J*(4 t + 4 t) r = 0.1447 V = 0.1595 × 6 = 0.9570				a) 0.625 c) 1.0000	Tl_3VS_4 x = 0.171	$\mathring{S}_8$, V_2, Tl_6 I[4 t], I, J*
D[4 t] 4/3/c6 r = 0.0973	Fd3m 32(e)xxx	x = 0.0688	D(4 t) r = 0.0219 V = 0.0009 × 8 = 0.0069					D′(12 tt) r = 0.1841 V = 0.0829 × 8 = 0.6634	T′(12 a) r = 0.0324 V = 0.0206 × 16 = 0.3296		a) 0.1235 b) 0.3351 c) 0.9999	Fe_3W_3C Fd3m x = – 0.175 = $\frac{3}{4}$ + 0.075	Fe_{32}, $(W_{32}Fe_{16})$, C_{16}, Fe_{16} D′[4 t], Ð[6 o], T, T′
W*[4 t] 6/3/c6 r = 0.1464	Im3m 24(h)oxx	x = 0.3535	W*(4 t) r = 0.0329 V = 0.0030 × 12 = 0.0355		P'_2(6 a) r = 0.1433 V = 0.0072 × 8 = 0.0573		J*(2 I (8 a)) r = 0.0836 V = 0.0265 × 12 = 0.31812	I(12 co) r = 0.3535 V = 0.2945 × 2 = 0.5890			a) 0.316 b) 0.815 c) 1.0000	Sodalite $Na_4Cl[AlSiO_4]_3$ P$\bar{4}$3n	O_{24}, Si_6, Al_6, $(ClNa_4)_2$ W′[4 t], W, W′, I(· 4 t) x = 0.3513; y = 0.359̄9; z = 0.0615
[P'_4 – Y**] (4 t) 3/6/c6 r = 0.0941	Ia3d 96(h)xyz	x = 0.1400 y = 0.0300 z = 0.1000	[P'_4 – Y**] [4 t] r = 0.0289 V = 0.0009 × 48 = 0.0438	I_2(6 o (4 t)) r = 0.0221 V = 0.0010 × 96 = 0.0188	I_2(6 a) r = 0.0806 V = 0.0040 × 16 = 0.0647	J_2^*(6 a) r = 0.0576 V = 0.0049 × 48 = 0.2334		Y**(12 i) r = 0.1506 V = 0.0353 × 16 = 0.5641			a) 0.3348 b) 0.6431 c) 1.0048	HT-Leucite $K[AlSi_2O_6]$ Ia3d	O_{96}, $(Al, Si)_{48}$, K_{16} [P'_4 – Y**] xyz [4 t], [P'_4 – Y**] $\underline{x}$, Y** x = 0.1467; y = 0.030; z = 0.1178
P[6 o] 5/3/c 3 r = 0.2071	Pm3m 6(e)xoo	x = 0.2929			P(6 o) r = 0.0858 V = 0.0335 × 1 = 0.0335					P′(24 co) r = 0.5000 V = 0.9665 × 1 = 0.9665	a) 0.2232 b) 0.7671 c) 1.0000	CaB_6 x = 0.29	B_6, Ca_1 P[6 o], P
I[6 o] [5/3/c 3]^2 r = 0.2071	Im3m 12(e)xoo	x = 0.2929	W*(4 t) r = 0.1176 V = 0.0143 × 12 = 0.1714	$J_2\underline{x}$ (4 t) r = 0.1174 V = 0.0143 × 24 = 0.3430	I(6 o) r = 0.0860 V = 0.0335 × 2 = 0.0671	P'_2(6 a) r = 0.1492 V = 0.0523 × 8 = 0.4185					a) 0.4465 b) 0.8074 c) 1.0000	Tl_7Sb_2 Im3m x = 0.29	Sb_{12}, Tl_{24}, Tl_{16}, Tl_2 I[6 o], W*[4 t], I[8 c], I
D[6 o] 8/3/c 3 r = 0.1326	Fd3m 48(f)xoo	x = 0.1875		D(6 o) r = 0.0549 V = 0.0088 × 8 = 0.0703	D′(6 o) r = 0.1799 V = 0.0407 × 8 = 0.3255	T (6 o) r = 0.0549 V = 0.0088 × 16 = 0.1406	T′(6 a) r = 0.0361 V = 0.0081 × 16 = 0.1302	F'''_2(6 o) r = 0.0361 V = 0.010 × 32 = 0.3333			a) 0.4686 b) 0.6898 c) 0.9999	$RbNiCrF_6$ Fd3m x = 0.187	F_{48}, Rb_8, $(Ni, Cr)_{16}$ D[6 o], D′, T

I[8 c] 4/4/c 2 r = 0.1585	Im3m 16(f)xxx	x = 0.1585			I(8 c) r = 0.1160 V = 0.0319 x 2 = 0.0637	W*(8 c) r = 0.2290 V = 0.0437 x 12 = 0.5245	J*(8 r) r = 0.2500 V = 0.0686 x 6 = 0.4118			a) 0.267 c) 1.0000	Ir_3Sn_7 x = 0.156	Sn_{16}, Ir_{12}, Sn_{12} I[8 c], J*(21), W*
F[8 c] 6/3/c 4 r = 0.1464	Fm3m 32(f)xxx	x = 0.1465	P_2'(4 t) r = 0.0329 V = 0.0030 x 8 = 0.0237			F(8 c) r = 0.1072 V = 0.0251 x 4 = 0.1005			F'(24 co) r = 0.2633 V = 0.2190 x 4 = 0.8758	a) 0.4210 b) 0.7487 c) 1.0000	$TlPd_3O_4$ x_0 = 0.1413	O_{32}, Pd_{24}, Tl_4, Tl_4 F[8 c], J_2, F, F' Pd has a planar surrounding J_2(4 l)
F[12 co] 5/3/c 4 r = 0.1179	Fm3m 48(h)oxx	x = $0.166\overline{6}$					F(12 co) r = 0.1179 V = 0.0309 x 4 = 0.1235	P_2'(12 tt) r = 0.1585 V = 0.0355 x 8 = 0.2840	F'(24 cod) r = 0.2548 V = 0.1681 x 4 = 0.5926	a) 0.3291 b) 0.7673 c) 1.000	UB_{12} x_B = 0.166	B_{48}, U_4 F[12 co], F'
D[12 co] 6/3/c 14 r = 0.0884	Fd3m 96(g)xxz	x = 0.1250 z = 0.0		T(6 o) r = 0.366 V = 0.0026 x 16 = 0.0417	J_2^*(6 r) r = 0.0749 V = 0.0026 x 48 = 0.1250		D(12 co) r = 0.0884 V = 0.0130 x 8 = 0.1942	D'(12 tt) r = 0.2178 V = 0.0391 x 8 = 0.3125	T'(12 a) r = 0.1911 V = 0.0260 x 16 = 0.4167	a) 0.2777 c) 1.000	Northurpite $Na_3MgCl(CO_3)_2$ $x = \frac{1}{2} - 0.125$	O_{96}, Na_{48}, C_{32}, Mg_{16}, Cl_{16} D'[12 co], J_2^*x, $F_2'''\underline{x}$, T', T $z = \frac{1}{2} - 0.025$
D'[12 tt] 5/3/c 11 r = 0.0973	Fd3m 96(g)xxz	x = 0.1812 z = 0.0436		T(6 a) r = 0.0891 V = 0.0038 x 16 = 0.0606	J_2^*(6 r) r = 0.0093 V = 0.0035 x 48 = 0.1669		D(12 co) r = 0.1626 V = 0.0383 x 8 = 0.3063	D'(12 tt) r = 0.1309 V = 0.0200 = 0.1598	T'(12 r) r = 0.0093 V = 0.0192 x 16 = 0.3065	a) 0.3704 b) 0.7540 c) 1.0000	$Mg_3Cr_2Al_{18}$ x = 0.1824 z = 0.0436	Al_{96}, Mg_{16}, Mg_8, Al_{48}, Cr_{16} D'[12 tt], T', D', D[6 o], T
I[12 i] 9/3/c 3 r = 0.1854	Im3 24(g)0yz	y = 0.3010 z = 0.1873	I($\overline{12}$ i (5 py)) r = 0.0128 V = 0.0123 x 24 = 0.2950	P_2'(6 o) r = 0.0773 V = 0.0242 x 8 = 0.1935	W*x (6 r) r = 0.1923 V = 0.0238 x 12 = 0.2855		I(12 i) r = 0.1691 V = 0.1130 x 2 = 0.2260			a) 0.641 b) 0.751 c) 1.000	WAl_{12} y = 0.184 z = 0.309	Al_{24}, W_2 I[12 i], I
..nI[12 i] 7/3/c 1 r = 0.1563	Pm3n 24(k)0yz	y = 0.2803 z = 0.1733	W(4 t) r = V = 0.0049 x 6 = 0.0296	P_2'(6 o) r = 0.0900 V = 0.0203 x 8 = 0.1626			W'(12 co) r = 0.1549 V = 0.1044 x 6 = 0.6262	I(12 i) r = 0.1563 V = 0.0908 x 2 = 0.1815		a) 0.523 b) 0.678 c) 1.000	UH_3 y = 0.155 z = 0.310	H_{24}, U_6, U_2 ..nI[12 i], W', I
..cP_2[12 i] 8/3/c 4 r = 0.1094	Fm3c 96(i)0yz	y = 0.1761 z = 0.1141	J_2'(4 t) r = 0.0266 V = 0.0013 x 24 = 0.0308	W_2^*(4 t) r = 0.0374 V = 0.0013 x 96 = 0.1209			P_2(12 i) r = 0.1004 V = 0.0234 x 8 = 0.1873		P_2'(24) r = 0.1846 V = 0.0826 x 8 = 0.6610	a) 0.5263 b) 0.7939 c) 1.0000	$NaZn_{13}$ y = 0.18063 z = 0.11924	Zn_{96}, Zn_8, Na_8 ..cP_2[12 i], P_2, P_2'

Relations between Structure Types

With the knowledge of invariant lattice complexes and coordination polyhedra it is possible to describe homogeneous and heterogeneous Bauverbände and the positions of occupied voids with their coordination polyhedra. The symbols are easy to understand and the structures can be reconstructed from them. Examples are given in Tables 3 and 4 by Hellner (1965), Niggli (1972) and Koch and Fischer (1974). The question of iso-, homeo- and heterotypic structures and the classification principle have still to be discussed, however.

i Iso-, Homeo- and Heterotypism

The terms mentioned above have been discussed at a special meeting in 1943 and published under the title "Über die Göttinger Isomorphie-Besprechung" (1944). A paper on this subject was given by Laves (1944); in the following his translation of special topics of this paper is given (private communication, 1978):

Typism by Laves

After these introducing general remarks it shall be tried to find conditions for calling crystal structures "equal", "similar" or "different". Obviously, as in many classifications, there may occur special cases of linking character that cannot be handled without arbitrariness but these do not encroach upon the principles to be proposed:

Isotypism (*Examples*: NaCl–PbS–TiC–LaBi–NaSeH (high temp. modification); CaF_2–$PbMg_2$–$AuIn_2$; Au–Ne). Two crystals A and B have the "same atomic arrangement" and are called *isotypic* if they satisfy the following conditions:

1) A and B should have the same symmetry (space group).
2) To each lattice complex of A occupied by atoms, ions, rotating groups (or groups of preferred directions that are distributed randomly according to symmetry conditions) exists a congruent lattice complex of B which is occupied by atoms, ions or groups with equal (or "nearly equal") occupation factor.
3) The axial ratios and special parameter values of the lattice complex of A and B must not deviate from each other in such a way that the "Bauverbände" are changed "essentially".
   The condition 3) is necessary because without it e.g. CO_2 and FeS_2 were considered to be isotypes, a conclusion I would like to exclude.

Homeotypism. There are several kinds of similarity between different crystal structures:

1) Differences of axial ratios or parameters that exceed the ones mentioned under isotypism 3) "essentially".

Examples: Mg–Zn, Mg–Cd.

2) Differences in symmetry, as far as the characteristics of their "Bauverbände" are not changed "essentially". –

Examples: Cu–γ-Mn; α-quartz–β-quartz.

3) Differences with respect to order/disorder. Two crystals A and B may be considered as homeotypic if the condition 2) for isotypism is not satisfied in the following way; At least *one* n-fold lattice complex in A corresponds to the sum of several (p-, q-, r- ...fold) lattice complexes in B with the condition $n = p + q + r \ldots$.

Examples: Quartz (Si_2O_4)–$AlAsO_4$; Ag–$AuCu_3$; $FeCO_3$–$MgCa(CO_3)_2$; diamond-sphalerite (ZnS); α-Fe–NiAl.

4) Differences with respect to cell content (the number of atoms per unit cell). Crystals A and B are considered as homeotypic even then, if only parts of the "Bauverbände" in A and B are equal from a topological point of view. These parts should be composed of those bondings, however, which are believed to be the most important for the structure types under consideration.

Examples: Li_3PO_4–Mg_2SiO_4–Fe_2PO_4; NiTe–$NiTe_2$; α-carnegieite, Na($AlSiO_4$)-cristobalite ($SiSiO_4$) = SiO_2.

5) Differences in the topology of the "Bauverbände" but not in the coordination numbers.

Examples: (Cubic close packed) Cu-(hexagonal close packed) Mg; sphalerite (ZnS) – wurtzite (ZnS); cristobalite (SiO_2) – tridymite (SiO_2); $MgCu_2$–$MgNi_2$–$MgZn_2$.

6) Combinations of the five subdivisions 1) to 5).

Examples: spinel ($MgAl_2O_4$) – corundum (Al_2O_3) = combination of 4) and 5); BPO_4-tridymite (SiO_2) = combination of 3) and 5).

Heterotypism. If the relations between crystals A and B cannot be classified as *isotypic* or *homeotypic* they are called *heterotypic.*

Example: Cu–diamond.

Footnote (also by Laves): Some differences of opinion still exist with respect to the question how to make a distinction between homeotypism and isotypism. Originally it was my personal feeling that the cases 1)–4) should be considered as sub-divisions of isotypism and, analogously, if there is miscibility, belonging to isomorphism, but as "isotypism (or isomorphism) in a wider sense". Accordingly I compiled a memo distributed during the "Diskussionstagung". In this memo only examples 5) were listed under homeotypism. During the discussion O'Daniel advocated the opinion to consider the cases 1)–4) as cases of homeotypism, referring to an earlier paper [O'Daniel u. Tscheischwill: Z. Krist, *104*, 124 (1942)]. Following O'Daniels proposal I transferred 1)–4) from isotypism to homeotypism. However, Neuhaus informs me by letter right now (1944) that he dislikes the idea of grouping together 1)–4) and 5) for considering them as cases of homeotypism. He would be inclined to agree with my original sub-division (as proposed in Göttingen).

This translation of the definition for "isotypism", "homeotypism" and "heterotypism" by Laves is given in such detail because the original paper is not generally known and a discussion on this subject should be encouraged. The term "Bauzusammenhang" in the original paper (Laves, 1944) is replaced by "Bauverband" in this translation.

The following expansion of the term homeotypism has been introduced by Hellner (1965):

7) Differences with respect to a replacement of points by polyhedra: If a point of a "Bauverband" is replaced by a coordination polyhedron, this structure will also be called homeotypic with the original structure.

Examples:	NaCl	FeS_2
	Cu	CO_2
	BiF_3	alum $KAl(SO_4)_2 \cdot 12\,H_2O$

The description of these structure types by symbols will be given below.

ii Main Classes, Subclasses and Families

All structure types with analogous "Bauverbände" belong to the same family and are subdivided into main classes. In addition those structure types are assigned to this family, the "Bauverbände" of which may be derived from one of the main classes of this family by removing the points of one or more splitting parts but preserving part of the connexions in such a way that the shortest bondings come from the original "Bauverband". These structure types are subdivided into subclasses of the main class.

Examples for "Bauverbände":

I	W-type
P	α-Po-type
F	Cu-type
$^{\pm}Yxxx = {}^{\pm}Y\underline{\underline{x}}$	S_2-dumbbells in pyrite FeS_2
I[6 o]	Sb in Tl_7Sb_2
F[8 c]	O in $TlPd_3O_4$
F[12 co]	B in UB_{12}
[I + W]	β-W-(W_3O-)Cr_3Si-type
W*[4 t_c]	O in hauÿne, sodalite

Main classes of the same family are corresponding to each other with respect to the number and position of the points of the "Bauverband", but they differ in their splitting parts and (or) in their indices.

Examples for different main classes of the same family:

F-family:	F^P = P, J	$AuCu_3$-type (ordered)
	F_2^T = T, T′	PtCu

I-family:	I^P = P, P′	CsCl-type
	I_2^D = D, D′	NaTl
	I_2^P = P, J, P′, J′, F″xxx, F‴xxx	α-BiF_3
	I_3 = I, I(6 o), I(8 c), W*(4t)	Tl_7Sb_2
	I_6^F = F, ...	$Li_{22}Pb_5$

The superscripts P, F, D and T indicates the characteristic splittings of the complexes. Subscript 2 stands as abbreviation for 222 and indicates complexes with the index 8, subscript 3 for 333 and the index 27, 6 for 666 and the index 216.

Examples for subclasses:

J as a subclass of F	O in ReO_3
P[6 o] as a subclass of I[6 o]	B in CaB_6
D as a subclass of I_2	Diamond
T as a subclass of F_2''	O in cristobalite
[F_2''-I(4 t)] as a subclass of F_2''	O in $Zn_4O(BO_2)_6$ and S in tetrahedrite

For each main and subclass several homeotypic structure types may exist (Tables 5b, 6, 7b). A structure type may belong to different families if different parts of the structure are considered to be the important "Bauverband".

Examples:	CsCl		I^P = P, P′ or P + P′
	NaCl		P_2^F = F, F′ or F + F′
	Tl_7Sb_2		I_3^I = I, I[6 o], I[8 c], W*[4 t_c]
		or	I[6 o] + I, I(8 c), W*(4 t_c)
		or	I[8 c] + I, I(6 o), W*(4 t_c)
		or	W*[4 t_c] + I, I(6o, 8c)
		or	[I(·6o, 8c) + W*(4 t_c)]

Table 5a. Main classes and "addition structure types" of the I family of index 1 (*left part*). Main and subclasses of the I-family of Index 8 (*right part*)

I-FAMILY

Im3m	Im3m	Pn3m	Pm3n	Pm3m	$I2_13$	Fd3m	Fm3m	$F\bar{4}3m$	Im3m	Pn3m	$P\bar{4}3m$	Pm3m
1. Index						8. Index						
I I W	$I+P'_2$ I,P'_2 Hg_4Pt	$I+F''$ I,F'' Cu_2O	I+W I,W W_3O-Cr_3Si	I^P P,P' CsCl	$I_2^{P_2\underline{\underline{x}}}$ $P_2\underline{\underline{x}},P'_2\underline{\underline{x}}$ [CoU]	I_2^D D,D' NaTl	I_2^F F,F',P'_2 BiF_3	$I_2^{F''}$ F,F',F'',F''' —	I_2^I I,J^*,P'_2 —	$I_2^{I,F''}$ I,J^*,F'',F''' —	$I_2^{P,F''}$ $P,J,P',J',F''_{\underline{\underline{x}}},F'''_{\underline{\underline{x}}}$ α-BiF_3	I_2^P $P,J,P',J',P'_2\underline{\underline{x}}$ $Fe_{13}Ge_3$
						I_2-D D C Diamond	I_2-F' F,P'_2 CaF_2	$I_2^{F''}-F'''$ F,F',F'' CuSbMg	I_2-I J^*,P'_2 Pt_3O_4	I_2-I,F''' J^*,F'' Mg_3P_2	I_2-P',J,F''' P,J',F'' Sulvanite Cu_3VS_4	
								I_2-F',F''' F,F'' ZnS Sphalerite	I_2-J^* I,P'_2 Hg_4Pt	I_2-J^*,F''' I,F'' Cu_2O		

Table 5b. Representatives of the main classes and homeotypes of the I-family of index 1

Class	Symbol	Chemical composition	Space group Unit cell	Structure formula Nomenclature of the structure type	References Remarks
Main	I	W	Im3m a = 3.15	W_2 I	1.15
Homeo t.	t I	Pa	I4/mmm a = 3.925 c = 3.238	Pa_2 t I	16.133 with c/a = 0.82 hexagonal closed packed nets (f) H are formed with c.n. 10 + 4
Homeo t.	I(· 4 t)	SiF_4	I$\bar{4}$3m a = 5.41	$\{SiF_4\}_2$ I(· 4 t)	2.37 The F-atoms form a sphere-packing with X_{Sp} = 0.1875 $X_{Str.}$ = 0.17
Homeo t.	o I(· 4 l)	$CuCl_2 \cdot 2H_2O$	Pbmn a = 7.38 b = 8.04 c = 3.72	$\{CuCl_2(H_2O)_2\}_2$ o I(· 4 l)	4.13
Homeo t.	I(· 12 i)	WAl_{12}	Im3 a = 7.580	$\{WAl_{12}\}_2$ I(· 12 i)	18.30 The Al-atoms form a sphere packing X_{Sp} = 0.302; Y_{Sp} = 0.185; X_{Al} = 0.309; Y_{Al} = 0.184
Main	I^P	CsCl	Pm3m a = 4.110	Cs_1, Cl_1 P; P′	1.74
Homeo t.	I^P(2 l)	CsCN	Pm3m a = 4.29	$Cs_1, \{CN\}_1$ P; P′(2 l$\circlearrowleft$)	9.138
Homeo t.	r I^P (2 l)	CsCN	R$\bar{3}$m a = 4.23 α = 86° 21′	$Cs_1, \{CN\}_1$ r P; P′(2 l)	9,138
Homeo t.	t I^P(2 l)	NH_4CN	P4/mcm a = 4.17 c = 7.62	$\{NH_4\}_2, \{CN\}_2$ t P_c; P'_c(2 l)	9.137 actually of index 2
Homeo t.	I^P(· 3 l)	NH_4NO_3 (1)	Pm3m a = 4.40	$\{NH_4\}_1, \{NO_3\}_1$ P; P′(3 l$\circlearrowleft$)	2.69
Homeo t.	t I^P(· 3 l)	NH_4NO_3 (2)	P$\bar{4}2_1$m a = 5.74 b = 5.00	$\{NH_4\}_2, \{NO_3\}_2$ t P_v; P'_vz(· 3 l)	2.69 actually of index 2

Table 5b. Continuation

Class	Symbol	Chemical composition	Space group Unit cell	Structure formula Nomenclature of the structure type	References Remarks
Homeo t.	$rI^P(\cdot 3y)$	$KBrO_3$	$R\bar{3}m$ a = 4.62 $\alpha = 86° \pm 6'$	K, BrO_3 rP; P′(· 3 y)	2.658
Homeo t.	$I^P(6o)$	CaB_6	Pm3m a = 4.15	$Ca_1, \{B_6\}_1$ P; P′(6 o)	2.37 The B-atoms form a sphere-packing $x_B = 0.29$ $x_{Sp} = 0.293$
Homeo t.	$rI^P(\cdot 6c)$	$BaSiF_6$	$R\bar{3}m$ a = 4.76 $\alpha = 97°38'$	Ba, SiF_6 P; P′(· 6 o)	9.196
Homeo t.	$tI^P(\cdot 4t, \cdot 4t)$	$BeSO_4 \cdot 4H_2O$	$I\bar{4}c2$ a = 8.02 c/a = 1.34 c = 10.75	$\{Be(H_2O)_4\}_4, \{SO_4\}_4$ $tP_{vc}(\cdot 4t)$; $P'_{vc}(\cdot 4t)$	2.91 actually of index 4
Homeo t.	$tI^P(\cdot 6o, \cdot 6o)$	$NiSnCl_6 \cdot 6H_2O$	$R\bar{3}$ a = 7.09 $\alpha = 96°45'$	$\{Ni(H_2O)_6\}_1, \{SnCl_6\}_1$ rP(· 6 o); P′(· 6 o)	2.102

If in CsCl all points are considered to be equal as is the case for some intermetallic compounds, for example, it is assigned to the I-family as a main class; if one considers the Cl-ions as building up the "Bauverband" by itself and the Cs-ions as occupying the voids (8 c), then the CsCl-type belongs to the P-family as a main class.

If in NaCl all occupied positions are regarded as equal the structure type belongs to a main class of index 8 of the P-family; if the Cl-ions are regarded as building up the "Bauverband" by itself and the Na-ions as occupying the octahedral voids (6 o), then NaCl has to be classified into the F-family. Tl_7Sb_2 with 54 atoms in the unit cell had been interpreted in the past as an I-type of index 27; but the deviations of the coordinates from the ideal values n/6 suggest to regard the "Bauverband" of the structure as an interpenetrating sphere packing I[6 o], or as I[8 c] or as W*[4 t_c] (the comparison of the parameter in the idealized structure I_3 with those which fulfill the sphere-packing conditions for these "Bauverbände" will be given below). To get all possible crystal-chemical relation of Tl_7Sb_2 it has to be ranged in main classes of all four different families.

The "Bauverband" is homogeneous for three of them and heterogeneous for one. Of course, three other heterogeneous "Bauverbände" may be constructed as combination of two of the homogeneous ones.

Table 6. Representatives of main and subclasses as well as homeo types of the I family of index 8

Class	Symbol	Chemical composition	Space group Unit cell	Structure formula Nomenclature of the structure type	References Remarks
Main	$I_2^{P_2\underline{\underline{x}}}$	CoU	$I2_13$ a = 6.357	U_8, Co_8 $P_2\underline{\underline{x}}; P_2'\underline{\underline{x}}$	13.94 $x_U = 0.035, x_{Co} = 0.294$; in the supergroup $I\bar{4}3d$ exists the sphere packing 5/5/c1 with x = 0.0366 symbolized as $[I_2\underline{\underline{x}}]$
Homeo t.	I_2^D	NaTl	Fd3m $a = 7.47_2$	Na_8, Tl_8 D; D′	3.19
Homeo t.	$o\,I_2^D(\cdot 6\,o, \cdot 4\,l)$	$Pt(NH_3)_2Br_2 \cdot Pt(NH_3)_2Br_4$	Fddd a = 16.39 b = 15.45 c = 11.08	$\{Pt(NH_3)_2Br_4\}_8, \{Pt(NH_3)_2Br_2\}_8$ o D(· 6 o); D′(· 4 l)	11.483
Homeo t.	$I_2^D(6\,o) + T$	$16\,Al(F, OH)_3 \cdot 6\,H_2O$	Fd3m a = 9.849	$(F, OH)_{48}, ((H_2O)_{3/4})_8, Al_{16}$ D(6 o), D′, T	11.281 with r = 0.1875 sphere-packing D[6 o] 8/3/c3
Homeo t.		$RbNiCrF_6$	Fd3m a = 10.21	$F_{48}, Rb_8, (Ni, Cr)_{16}$ D(6 o), D′, T	38A,205; $x_F = 0.187$
Homeo t.	$I_2^D(6\,o) + T, T'$ $= I_2^D(6\,o) + F_2^T$	Koppite-pyrochlore $R_2Q_2X_7$ R = Ca, Ce, Na, K, Th, Y Q = Nb, F_2, Ti, Ta X = O, OH, F	Fd3m a = 10.37	$X_{48}, X_8, R_{16}, Q_{16}$ D(6 o), D′, T, T′	2,58 $x_X = 0.19$
Homeo t.	$I_2^D(6\,o, 4\,t) + T$	$NiTi_2$	Fd3m a = 11.278	$Ti_{48}, Ni_{32}, Ti_{16}$ D(6 o), D′(4 t), T	22,889; 27,269; 28,20 $x_{Ti} = 0.1825; x_{Ni} = 1/2 + 0.090$; J[6 o],4/3/c6; $x_{Sp} = 0.0688$
Homeo t.	$I_2^D(6\,o, 4\,t) + F_2^T$	Fe_3W_3C	Fd3m a = 11.06	$(W_{32}Fe_{16}), Fe_{32}, C_{16}, Fe_{16}$ D(6 o), D′(4 t), T, T′	3,71; 22,889 $x_W = 0.195; x_{Fe} = 0.075$

Table 6. Continuation

Class	Symbol	Chemical composition	Space group Unit cell	Structure formula Nomenclature of the structure type	References Remarks
Homeo t.	$I_2^D(\cdot 4t, 12co) +$ $F_2'''\underline{\underline{x}}, T', J_2^*x$	Tychite $Na_6Mg_2SO_4(CO_3)_4$	Fd3m a = 13.90	$S_8, O_{32}^I, O_{96}^{II}, C_{32}, Mg_{16}, Na_{48}$ D, D(4t), D′(12co), $F_2'''\underline{\underline{x}}$, T′, J_2^*x or D(·4t), $F_2'''(\cdot 3l)$, T′, J_2^*x or homeo I_4-D', T	2,98 $x_0^I = 0.065$ $x_0^{II} = 1/2 - 0.140$ $z_0^{II} = 1/2 - 0.040$ $x_C = 3/8 + 0.015$
Homeo t.	$I_2^D(6o, \cdot 12tt)$ $+ F_2^T$	$Mg_3Cr_2Al_{18}$	Fd3m a = 14.53	$Al_{48}^I, Mg_8, Al_{96}^{II}, Cr_{16}, Mg_{16}$ D(6o), D′, D′(12tt), T, T′ with x = 0.125 [D(6o) + T] form a heterogeneous sphere packing	18,14; 22,8 $x_{Al}I = 0.1407$ $x_{Al}II = 0.1834$; $z_{Al}II = -0.0488$ 5/3/c11: x = 0.1812 z = − 0.0436
Homeo t.	$I_2^D((4t, 6o)4t)$ $+ T'$	$Ba_2Ge_4S_{10}$	Fd3m a = 14.899	$Ge_{32}, S_{48}^I, S_{32}^{II}, Ba_{16}$ D(4t), D(6o), D′(4t), T′ D(4t, 6o), D′(4t), T′ Cluster S_6, Ge_4, S_4 D(6o, 4t, 4t), T′	39A,22 $x_{Ge} = 0.0830$ $x_{SI} = 0.1731$ $x_{SII} = 1/2 - 0.0821$
Sub-	I_2-D	Diamond	Fddd a = 3.567	C_8 D	1.19
Homeo t.	$D'(6o) + F_2^T$	$AgSbO_3$	Fd3m a = 10.25	O_{48}, Ag_{16}, Sb_{16} D′(6o); T; T′	6,120 $x_0 = 1/2 - 0.21$ 8/3/c3: x = 1/2 − 0.1875
Homeo t.	$D'(12co) +$ $F_2^T, F_2'''\underline{\underline{x}}, J_2^*x$	Northupite $Na_3MgCl(CO_3)_2$	Fd3m a = 14.08	$O_{96}, C_{32}, Cl_{16}, Mg_{16}, Na_{48}$ D′(12co); $F_2'''\underline{\underline{x}}$; T; T′; J_2^*x $F_2'''(\cdot 3l)$, F_2^T; J_2^*x homeo $I_4-D, D' \equiv I_4-I_2^D$	2,80 $x_0 = 1/2 - 0.125$ $z_0 = 1/2 - 0.025$ D[12co]6/3/c14: x = 0.125, z = 0.0

Main	I_2^F	BiF_3	Fm3m a = 5.85	Bi_4, F_4, F_8 F; F′ P_2	2.22 Cu_2AlMn is isotype
Homeo t.	I_2^F(· 3 l)	$RbNO_3$	Pa3 a = 8.74	$Rb_4, Rb_4, \{NO_3\}_8$ F; F′; $P'_2\underline{\underline{x}}$(3 l)	15,277
Homeo t.	I_2^F(· 6 o, · 6 o, · 4 s)	$KAl(SO_4)_2 \cdot 12\,H_2O$ Alum	Pa3 a = 12.13_3	$\{K(H_2O)_6\}_4, \{Al(H_2O)_6\}_4, \{SO_4\}_8$ F(· 6 o); F′(· 6 o); $P'_2\underline{\underline{x}}$(· 4 t)	3,108
Sub-	$I_2 - F'$	CaF_2	Fm3m a = 5.45	Ca_4, F_8 F; P'_2	1,148 listed also as P'_2 + F and for Li_2O as F + P'_2
Homeo t.	$I_2 - F'$(· 3 l)	$Pb(NO_3)_2$	Pa3 a = 7.84	$Pb_4, \{NO_3\}_8$ F; P'_2(· 3 l)	2,73
Homeo t.	$I_2 - F'$(· 6 o, · 3 l)	$Zn(BrO_3)_2 \cdot 6\,H_2O$	Pa3 a = 10.31_6	$\{Zn(H_2O)_6\}, \{BrO_3\}_8$ F(· 6 o); $P'_2\underline{\underline{x}}$(· 3 y)	4,63
Homeo t.	$I_2 - F'$(2 l)	$N_2H_6Cl_2$	Pa3 a = 7.87	$\{N_2\}_4, Cl_8$ F(2 l); $P'_2\underline{\underline{x}}$	11,269
Main	$I_2^{F''}$	no representative		F; F′; F″; F‴	
Sub-	$I_2^{F''} - F'$	CuSbMg	F$\bar{4}$3m a = 6.16_4	Cu_4, Sb_4, Mg_4 F; F″; F‴	8,28
Homeo t.	m $I_2^{F''} - F'$	ZrO_2 Baddelyite	$P2_1/a$ a = 5.21 b = 5.26 c = 5.37 $\beta = 80°32'$	Zr_2, Zr_2, O_4, O_4 m A;A′; F″xyz; F‴xyz	4,9
Homeo t.	r $I_2^{F''} - F'$	LaOF	R$\bar{3}$m a = 7.11_8 $\alpha = 33.01°$	La, O, F r $F_{\underline{\underline{x}}}, F''_{\underline{\underline{x}}}, F'''_{\underline{\underline{x}}}$	15,167
Main	I_2^I	no representative	Im3m	I; J*; P'_2	
Sub-	$I_2^I - I$	Pt_3O_4	Im3m a = 6.238	Pt_6, O_8 J*; P_2	8,147
Sub-	$I_2 - J^*$	Hg_4Pt	Im3m a = 6.18	Pt_2, Hg_8 I; P'_2	17,227

Table 6. Continuation

Class	Symbol	Chemical composition	Space group Unit cell	Structure formula Nomenclature of the structure type	References Remarks
Sub-	I_2–I; F‴	Mg_3P_2	Pn3m a = 5.92	Mg_6, P_4 J*; F″	2,40
Sub-	I_2–J*; F‴	Cu_2O	Pn3m a = 4.26	O_2, Cu_4 I; F″	1,153
Main	I_2^P	$Fe_{13}Ge_3$	Pm3m a = 5.763	$Fe_1, Fe_1, Fe_3, Ge_3, Fe_8$ P; P′; J; J′; $P'_{2\underline{\underline{x}}}$	27,209
Main	I_2^P	α-BiF_3	P$\bar{4}$3m a = 5.86_5	$F_1, F_3, F_1, F_3, F_4, Bi_4$ $\underbrace{P; J;}_{F}$ $\underbrace{P'; J';}_{F'}$ $F''_{\underline{\underline{x}}}$; $F'''_{\underline{\underline{x}}}$	12,163
Sub-	I_2^P–P′; J; F‴	Cu_3VS_4	P$\bar{4}$3m a = 5.37	V_1, Cu_3, S_4 P; J′; $F''_{\underline{\underline{x}}}$	3,94

Table 7a. Main and subclasses of the I-family of index 64

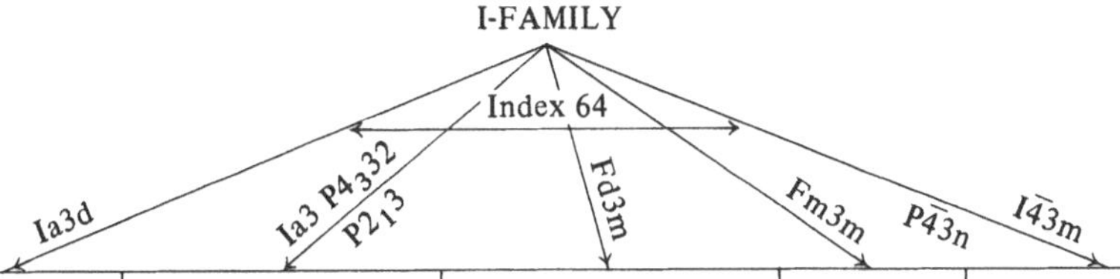

I_4^{Y**} I_2, J_2^*x, $Y**$, $[P_4'-Y**]x$ – –	$I^{Y**}_{4\,\underline{\underline{x}}}$ P_2, P_2', J_2x, $J_2'x$, $Y**x$, $[P_4'-Y**]xyz$ – –	I_4^D D, D', J_2^*x, T, T', $F_2'''\underline{\underline{x}}$ – –	I_4^F F, F', P_2', J_2, $J_2'x$, $F_2''\underline{\underline{x}}$, $F_{\underline{\underline{x}}}''$ – –	$I_4^{I(4t)}$ I, J^*, $J_2\underline{xz}$, $P_2'\underline{\underline{x}}$, W^*, $I(6o)$, $I(4t)$, $[F_2''-I(4t)]$, $I(4t^-)$, $[F_2'''-I(4t^-)]$ – –
I_4-Y** Eglestonite $(Hg_2)_3Cl_3O_2H$	$I_4^{Y*x}-F_2'^{P_2'}$ $AlLi_3N_2$	I_4-D' CoMnSb	I_4-F, P_2 Th_6Mn_{23}	$I_4-F_2'^{W*}$ TlOF
	$I_4-F_2^{P_2}$, $Y**$ Mn_2O_3	$I_4-T_2^D$ Fe_3W_3C	I_4-J_2, $F'(8c)$ Sulfohalite $2Na_2SO_4\cdot$ $NaCl\cdot NaF$	$I_4-F_2'''^{I(4t)}$ Li_7VN_4
	I_4^Y-D', T, J_2^* $LiFe_5O_8$	I_4-D', T Tychite $[Na_6Mg_2SO_4(CO_3)_4]$	I_4-F', J_2 Co_9S_8	$I_4-I(4t), F_2'$ Na_6PbO_4
	$I_4-F', T', J_2^*, F_2'''$ or D^F, T'^Y, F''' α-Carnegieite $NaAlSiO_4$	$I_4-I_2^D, T'$ $NiTi_2$		$I_4-F_2, F_2'''+I$ or $D_2^{W*}+I$ Chalcopyrite β-$CuFeS_2$
		I_4-D, F_2''' Koppit-Pyrochlore $(Na, Ca)_2Nb_2O_6F$		$I_4-I(4t), F_2, F_2''$ or $D_2^{W*}-I(4t)$ Tetrahedrite Cu_3SbS_3
		I_4-D, T', F_2''' or D', T, J_2^*x $16\,Al(OH, F)_3\cdot$		
		I_4-D', J^*, T or D, T', F_2''' Mg_2AlO_4 Spinel		
		$I_4-I_2^D, F_2''^T$ or J_2^*, F_2''' Sb_2O_3		
		$I_4-D', J_2^*, T', F_2'''$ or D, T Cristobalite SiO_2		

Table 7b. Main and subclasses of the I-family of index 64

Class	Symbol	Chemical Composition	Space group Unit cell	Structure formula Nomenclature of the structure type	References Remarks
Main	I_4^{Y**}		Ia3d	16(a)000, 48(f)$x0\frac{1}{4}$, 16(b)$\frac{1}{8}\frac{1}{8}\frac{1}{8}$, 48(g)$\frac{1}{8}$, x, $\frac{1}{4}-x$ I_2, J_2^*x, Y^{**}, $[P_4'-Y^{**}]\,\underline{x}$ x = 0.000 or, x = 0.375 x = 0.250 P_4 (over I_2, J_2^*x); P_4' (over Y^{**}, $[P_4'-Y^{**}]\,\underline{x}$)	
Sub	I_4-Y^{**}	Eglestonite $(Hg_2)_3Cl_3O_2H$	Ia3d a = 16.036	H_{16}, O_{32}, Cl_{48}, Hg_{96} I_2, $I_2(21)$, $[P_4'-Y^{**}]\,\underline{x}$, $J_2^*(21)$ $I_2(\cdot 21)$ x = 0.3629	TMPM 23 (1976), 105 also in P-family (64. index)
Main	$I_4^{Y\underline{\underline{x}}**}$	–	Ia3	P_2, P_2', $J_2\,x$, J_2', $Y\underline{\underline{x}}^{**}$, $[P_4'-Y^{**}]xyz$	
Sub	$I_4^{Y**}-F_2^{P_2'}$	$AlLi_3N_2$	Ia3 a = 9.480	N_8, N_{24}, Al_{16}, Li_{48} P_2, J_2x, $Y\underline{\underline{x}}^{**}$, $[P_4'-Y^{**}]xyz$ x = 0.205, x = 0.115, x = 0.160 y = 0.382 z = 0.110	11.20 also in F-family (8. index); with fluoride type; Li_5GeN_3 isotyp;
Sub	$I_4-F_2^{P_2}\,Y^{**}$	Mn_2O_3	Ia3	Mn_8, Mn_{24}, O_{48}, P_2, $J_2'\,x$, $[P_4'-Y^{**}]xyz$ x = 0.035, x = 0.378 y = 0.167 z = 0.379	2,38; 20,273; 27,497 also in P-family (64. index and subclass $P_4'-Y^{**}$),
Sub	I_4^Y-D', T, J_2^*	$LiFe_5O_8$	$P4_332$ a = 8.331	Fe_8, Li_4, Fe_{12}, O_8, O_{24}, $D\underline{\underline{x}}$, $^+Y'$, $[T'-{}^+Y']\underline{x}$, $^-Y^*\underline{x}$, $[F_2''-Y^*]xyz$	13,245; 16,244 also in F-family (8. index and as $F_2''^{Y^*}$ homeotypic to F_2).
Sub	$I_4-F', T'J_2^*, F_2'''$ or D^F, T^{+Y}, F'''	α-Carnegieite $NaAlSiO_4$	$P2_13$ a = 7.38_2	Si_4, Al_4, Na_4, O_4, O_{12} $F\underline{\underline{x}}$, $F\underline{\underline{x}}''$, $F\underline{\underline{x}}'''$, $^+Y\underline{\underline{x}}$, $[T-{}^+Y]xyz$	2,158; 11,475; also in F-family (8. index and subclass $T = F_2''-T'$)
Main	I_4^D		Fd3m	8(a)000, 8(b)$\frac{1}{2}\frac{1}{2}\frac{1}{2}$, 48(f)xoo, 16(c)$\frac{1}{8}\frac{1}{8}\frac{1}{8}$, 16(d)$\frac{5}{8}\frac{5}{8}\frac{5}{8}$, 32(e)xxx, D, D', J_2^*x, T, T', $F_2'''x$, x = 0.250, x = 0.375, I_2^D (over D, D'); F_2'' (over T, T') P_4 (over D, D', J_2^*x); P_4' (over T, T', $F_2'''x$) I_4	
Sub-	I_4-D'	CoMnSb	Fd3m a = 11.746	Co_8, $(Co_{1/2})_{48}$, Mn_{16}, Mn_{16}, Sb_{32}, D, J_2^*x, T, T', $F_2'''\,\underline{\underline{x}}$, x = 0.2363, x = 0.3813,	38A,14
Sub-	$I_4-I_2^D$	Fe_3W_3C	Fd3m a = 11.06	$(W_{32}Fe_{16})$, C_{16}, Fe_{16}, Fe_{32}, D(6 o) J_2^*x, T, T', D'(4 t) $F_2''\,\underline{\underline{x}}$, x = 0.195, x = 0.425,	3,71; 22,889

Homeotype		Northurpite $Na_3MgCl(CO_3)_2$	Fd3m a = 14.08	Na_{48}, Cl_{16}, Mg_{16}, C_{32}, O_{96}, J_2^* x, T, T′, F_2''' $\underline{x}$, D′(12 co) F_2''' $\underline{\underline{x}}$(3 1), x = 0.225, x = 0.41, x = 0.408	2,80
Sub- & Homeotype	I_4 – D′, T	Tychite $Na_6Mg_2SO_4(CO_3)_4$	Fd3m a = 13.90	S_8, O_{32}, Na_{48}, Mg_{16}, C_{32}, O_{96}, D, D(4 t), J_2^* x, T′, F_2''' $\underline{x}$, D′(12 co) F_2''' $\underline{\underline{x}}$(3 1) x = 0.065, x = 0.0225, x = 0.39, x = 0.393,	2,98
Sub-	$I_4 - I_2^D$, T	$NiTi_2$	Fd3m a = 11.278	Ti_{48}, Ti_{16}, Ni_{32}, D(6 o) J_2^* x, T, D′(4 t) F_2''' $\underline{\underline{x}}$, x = 0.190, x = 0.410	22,889; 27,296; 28,20
Sub-	$I_4 - F_2'''$	Sb_2O_4	Fd3m a = 10.24	O_8, O_8, O_{48}, Sb_{16}, Sb_{16}, D, D′, J_2^* x, T, T′ x = 0.23,	3,54; also in the P-family (64. index)
Sub-	I_4 – D, F_2'''	Koppite, Pyrochlore $(Na, Ca)_2Nb_2O_6F$	Fd3m a = 10.393–10.40	F_8, O_{48}, $(Na, Ca)_{16}$, Nb_{16}, D, J_2^* x, T, T′ $x = \frac{1}{2} - 0.187$ till $\frac{1}{2} - 0.170$,	2,58; 33A,190 also in the P-family (64. index Subclass P_4 – D′) $Tl_2Ta_2O_6$ homeotypic
Sub-	$I_4 - P_2^D, F_2'^T$	Senarmontite Sb_2O_3	Fd3m a = 11.15	O_{48}, Sb_{32}, D(6 o) J_2^* x, F_2''' $\underline{\underline{x}}$, x = 0.190, x = 0.365,	1,245; 9,164 also in P-family (index 64. and Subclass $P_4 - I_2^D$).
Homeotype	$I_4 - I_2^D$, F_2''',	$AgSbO_3$	Fd3m, a = 10.32,	O_{48}, Ag_{16}, Sb_{16}, D′(6 o) J_2^* x, T, T′ x = 0.290,	6,120
Sub-	I_4 – D, T, F_2'''	16 $Al(F, OH)_3 \cdot 6\,H_2O$	Fd3m a = 9.829	$(\frac{3}{4} \cdot H_2O)_8$, $(F, OH)_{48}$, Al_{16}, D′, D(6 o) J_2^* x, T x = 0.1875,	11,281; $RbNiCrF_6$, (38A, 205) isotypic also in P-family (index 64 and Subclass $P_4 - I_2^D$).
Sub-	I_4 – D′, T, J_2^* or D, T′, F_2''' $\underline{\underline{x}}$	$MgAl_2O_4$	Fd3m, a = 8.509,	Mg_8, Al_{16}, O_{32}, D, T′, F_2''' $\underline{\underline{x}}$, x = 0.387	1,350; 2,474; 16,244; 33A,272; also in F-family (8. index)
Homeotype	I_4 – D′, T, J_2^*	$K_2Zn(CN)_4$	Fd3m, a = 12.529	Zn_8, $K_{16}(CN)_{32}$ D, T′, F_2'''(2 1)	29,293
Sub-	I_4 – D′, T′, J_2^*, F_2''' or D, T	Cristobalite SiO_2	Fd3m a = 7.13	Si_8, O_{16} D, T	1,169; also in F-family (8. index and Subclass $F_2'' - T' \equiv T$).
Main-	I_4^F	–	Fm3m	4(a)000, 4(b)$\frac{1}{2}\frac{1}{2}\frac{1}{2}$, 24(d)0$\frac{1}{4}\frac{1}{4}$, 8(c)$\frac{1}{4}\frac{1}{4}\frac{1}{4}$, 24(e)xoo, F, F′, J_2, P_2', J_2'x [F, F′, J_2] = F_2; [P_2', J_2'x] = F_2' 32(f)xxx, 32(f)xxx, F′(8c), F(8c), x = 0.375, x = 0.125 = P_4'	F(8c) in Fm3m 32(f)xxx forms sphere packings with x = 0.1465 and $x = \frac{1}{2} - 0.1465 = 0.3535$.

Table 7b. Continuation.

Class	Symbol	Chemical Composition	Space group Unit cell	Structure formula Nomenclature of the structure type	References Remarks
Sub-	$I_4 - F, P_2'$	$Mn_{23}Th_6$	Fm3m a = 12.523	Mn_4, Mn_{24}, Th_{24}, Mn_{32}, Mn_{32}, F′, J_2, $J_2'x$, F′(8c), F(8c), x = 0.203, x = 0.378, x = 0.178	16,113; the unoccupied F changes $J_2'x$ in F(6 o) and the unoccupied P_2 changes F(8c) in $P_2'(4\,t)$ by parameter changes.
Sub-	$I_4 - J_2$, F′(8c)	Sulfohalite $2\,Na_2SO_4 \cdot NaCl \cdot NaF$	Fm3m a = 10.068	F_4, Cl_4, S_8, Na_{24}, O_{32}, F, F′, P_2', $J_2'x$ F(8c) $P_2'(4\,t)$, x = 0.2232, x = 0.1665	3,118; 33A,377; also addition composed of $I_2(+J_2'x)$; $I_2 = F + F' + P_2'(\cdot 4\,t)$
Sub-	$I_4 - F, J_2$, F′(8c)	Co_9S_8	Fm3m a = 9.928	Co_4, S_8, S_{24}, Co_{32} F′, P_2', $J_2'x$ F(8c) x = 0.2591; x = 0.1260	4,26; 27, 169; also in P-family (8. index $F_2'^{P_2'} + F'$; F(8c))
Main-	$I_4^{I(4\,t)}$	–	I$\bar{4}$3m	2(a)000, 6(b)0$\frac{1}{2}\frac{1}{2}$, 24(g)xxz, 8(c)xxx, 12(d)$\frac{1}{4}\frac{1}{2}$0, 12(e)xoo, 8(c)xxx, 24(g)xxx, 8(c)xxx, 24(g)xxz I, J*, $J_2\,\underline{xz}$, $P_2'\,\underline{\underline{x}}$, W*, I(6 o), I(4 t⁺), $[F_2''-I(4\,t^+)]\underline{xz}$, I(4 t⁻), $[F_2'''-I(4\,t^-)]\underline{xz}$ x = 0.250, x = 0.250, x = 0.250, x = 0.125, x = 0.375 x = 0.125, x = 0.125 z = 0.000 z = 0.125 z = 0.375 (I, J* = P_2; I … P_2' = F_2; W*, I(6 o) = F_2'; I(4 t⁺), [F_2''–I(4 t⁺)] = F_2''; I(4 t⁻), [F_2'''–I(4 t⁻)] = F_2'''; all = I_4)	
Sub-	$I_4 - F_2'^{W*}$	TlOF	I$\bar{4}$3m a = 10.78	Tl_2, Tl_6, Tl_{24}, $(O,F)_8$, $(O,F)_{24}$, $(O,F)_8$, $(O,F)_{24}$, I, J*x, $J_2\,\underline{xz}$, I(4 t⁺), $[F_2''-I(4\,t^+)]$, I(4 t⁻), $[F_2'''-I(4\,t^-)]$, x = 0.2633, x = 0.145, x = 0.391, x = 0.363, x = 0.100 z = 0.0570, z = 0.145, z = 0.345	38A.221; also in P-family (index 64 and $P_4^{I(4t)}$ class)
Sub-	$I_4 - F_2'''^{I(4\,t)}$	Li_7VN_4	P$\bar{4}$3n a = 9.604	V_2, Li_6, Li_{24}, Li_8, V_6, Li_6, Li_{12}, N_8, N_{24}, I, J*, J_2 xyz, $P_2'\,\underline{\underline{x}}$, W, W′, I(6 o), I(4 t⁺), $[F_2''-I(4\,t)]$xyz x = 0.250, x = 0.250, x = 0.250, x = 0.115, x = 0.375, y = 0.250, y = 0.375, z = 0.000, z = 0.125,	23,174; also in F-family (8. index and $F_2''^{I(4\,t)} + P_4^W$ antifluoride type)
Sub-	$I_4 - I(4\,t), F_2'$	Na_6PbO_4	I$\bar{4}$3m a = 11.91	O_2, O_6, O_{24}, Pb_8, Na_{24}, Na_{24}, I, J*, $J_2\,\underline{xy}$ I(4 t⁺), $[F_2'-I(4\,t^+)$, $F_2''-I(4\,t^-)]$, x = 0.256, x = 0.1252, x = 0.107 x = 0.134, z = 0.039, z = 0.670 z = 0.409,	39A,214; also in F-family (8. index)and $F_2^I + [P_4'-I(4\,t^-)]$ class)
Sub-	$I_4 - F_2', F_2''' + I$ or $D_2^{W*} + I$	Chalcopyrite β-$CuFeS_2$	I$\bar{4}$3m a = 10.605	M_2, M_8, M_{12}, M_{12}, S_8, S_{24}, I, $P_2'\,\underline{\underline{x}}$, W*, I(6 o), I(4 t⁺), $[F_2''-I(4\,t^+)]$, x ~ 0.250, x ~ 0.250, x ~ 0.125, x ~ 0.375, z ~ 0.125,	20,93; also in F-family (of 8. index) Talnakhite (38 A, 79) isotype.
Sub-	$I_4 - I(4\,t), F_2, F_2''$ or $D_2^{W*} - I(4\,t)$	Tetrahedrite Cu_3SbS_3	I$\bar{4}$3m a = 10.3908	Sb_8, Cu_{12}, Cu_{12}, S_{24}, $P_2'\,\underline{\underline{x}}$, W*, I(6 o), $[F_2'''-I(4\,t^-)]$, x = 0.270, x = 0.220, x = 0.117, z = 0.360,	11,345; 29,15; 37A,4; also in F-family (8. index and $F_2'''-I(4\,t)$ subclass)

Structure Types of the I-Family

In a recent paper (Hellner, 1977) it has been shown that the subsequent coordination shells of the I complex are (8 c), (6 o), (12 co), (24 cod), (8 c) and (6 o). The normalized distances from the central point are $\frac{1}{2}\sqrt{3}$, 1.0, $\sqrt{2}$, $\frac{1}{2}\sqrt{11}$, $\sqrt{3}$ and 2.0 respectively. This consideration makes plausible the assignment of different structure types to different families; the I-family with the nearest neighbors (8 c), the P-family with the neighbors (6 o) and the F-family with the neighbors (12 co). Besides the main classes, subclasses and homeotypes of the I-, P- and F-families some further homogeneous "Bauverbände" contained in Table 4 as $^{\pm}Y\underline{\underline{x}}$, I[4 t]. W*[4 t], I_2[6 o], I[12 i] and some heterogeneous ones as [I + W], W*x[6 o], [$^{+}$V + D], [$^{-}$Y(31) + D] will be discussed.

I-Types (Index 1)

In Table 5a the two main classes of index 1 as I and I^P with their representatives are listed; subclasses cannot exist. However three different "Additionsstrukturen" of the I complex are added. In $PtHg_4$ the "Bauverband" is built up by the Pt-atoms, in Cu_2O by the O-atoms, while the Hg-atoms and the Cu-atoms, respectively, occupy voids with coordination number 2. In W_3O the O-atoms occupy the points of an I-lattice and the W-atoms the points of the lattice complex W. Each O-atom possesses 12 nearest W-neighbors. The structure types of $PtHg_4$ and Cu_2O may also be assigned to subclasses of index 8 in the I-family if all atoms are considered to be equivalent. The anticuprite type $Cd(CN)_2$ will be found in the F-family. [I + W] represents a heterogeneous "Bauverband" in addition. In Table 5b homeotypes are listed for the main classes I and I^P. The Pa structure is tetragonally distorted. It may be regarded also as hexagonally closed packed nets with a two-layer sequence (f) and coordination number 10 + 4. (f) stands for the sequence of a tetragonally distorted F complex with $c/a = \sqrt{3}$. In the (100) and (010) planes exist hexagonal closed packed nets with a two-layer sequence named (f). Homeotypes with isolated and centered coordination polyhedra (• 4 t) and (• 12 i) have the representatives SiF_4 and WAl_{12}, respectively. The F atoms and the Al atoms form the sphere packings I[4 t] and I[12 i], respectively, with interesting coordination polyhedra (Table 4). The orthorhombic structure of $CuCl_2 \cdot 2H_2O$ with a planar surrounding (4 ℓ) for the Cu-atoms are listed too; the enlargement of a and b is caused by the (4 ℓ)-polygon, lying perpendicular to the c-axis. Some homeotypes are listed below the main type I^P-CsCl. Cubic ones exist for structures with rotating (Ω) CN- or NO_3-groups; rhombohedrally and tetragonally distorted ones are given too; subscript c indicates a doubling of the unit cell in c-direction, v a doubling in the a, b-plane with $\vec{a}' = \vec{a} + \vec{b}$, $\vec{b}'' = \vec{a} - \vec{b}$. Furthermore some structure types are listed where tetrahedra or octahedra are arranged around the points of P and P'.

$I_2(I_{222})$-Types (Index 8)

Eight different kinds of splitting of I_2 (16 points per unit cell) are shown in Table 5a The first one occurs in space group $I2_13$ with splitting parts P_2xxx and $\frac{1}{4}\frac{1}{4}\frac{1}{4}P_2$xxx. It is identical with I^P (index 1) for special x-coordinates ($x_1 = 0$, $x_2 = \frac{1}{4}$). However, in CoU a small deviation from this ideal values has been found ($x_1 = 0.035$, $x_2 = 0.294 = \frac{1}{4} + 0.044$). In supergroup $I\bar{4}3d$ of $I2_13$ the two different splitting parts are equivalent [point position 16(c)xxx] and the sphere-packing type 5/5/c1 exists without degrees of freedom (x = 0.0366 and x = 0.2866, respectively), which may be symbolized by $I_2\underline{\underline{x}}$ (= I_{222}xxx). With $x = \frac{1}{8} = 0.125$ the limiting form Y^{**} is reached, which may be interpreted as the penetrating sphere packings $^{+}Y^{*}$ and $^{-}Y^{*}$.

In connection with splitted I_2 Pearson (1972) has listed Pu_2C_3 and Ag_3AuTl_2 (petzite) as "Additionsverbindungen" or – as he calls them – "filled-up derivatives". In Pu_2C_3 ($I\bar{4}3d$) the Pu-atoms occupy I_2xxx [16(c)xxx, x = 0.050] while the C-atoms [24(d)x0$\frac{1}{4}$] form dumbbells with the centers at ^{+}S. The symbol $I_2Y\overset{**}{\underline{\underline{x}}}$, $^{+}S(21)$ shows that it is homeotype to the anti-Th_3P_4-type, which may better be described as a heterogeneous "Bauverband" [$I_2 Y\overset{**}{\underline{\underline{x}}} + {}^{+}S$] ($x_p = 0.083$). Eulytite $Bi_4Si_3O_{12}$ [$I_2Y\overset{**}{\underline{\underline{x}}} + {}^{+}S(4\,t)$] in $I\bar{4}3d$ and langbeinite $K_2Mg_2(SO_4)_3$ [$I_2^F Y\overset{**}{\underline{\underline{x}}} + {}^{+}S(4\,t)$] in $P2_13$ may be considered as homeotypes of Th_3P_4.

The next main class is generated by the splitting of I_2 in D and D′. A representative is NaTl; the only subclass is I_2–D′ = D, in which only 4 of the 8 nearest neighbors of the I complex are preserved. In the first two homeotypes of I_2^D only the points of D are replaced by octahedra (6 o), while the points of D′ are preserved. In the following homeotypes there are polyhedra (4 t), (6 o), 12 co), (12 tt) around D and D′. In $Mg_3Cr_2Al_{18}$ the Al-atoms occupy D(6 o) with x = 0.1497. This is far off the parameter of the sphere packing 8/3/c3, D[6 o] with x = 3/16 = 0.1875. However D(6 o) + T with $x = \frac{1}{8}$ forms a heterogeneous sphere packing. In the last example $S_6Ge_4S_4$-clusters are described. Homeotypes of the subclass D are listed with octahedra or cuboctahedra around the points of D.

The third main class of Table 5a is generated by a splitting in F, F′ and $\frac{1}{4}\frac{1}{4}\frac{1}{4}P_2$. There is one subclass I_2–F′ which represents the CaF_2-type. For both cases the homeotypes are listed in Table 6.

The next main class I_2 is splitted into four different F-lattices. No representative has been found for this case. The subclasses F, F″, F″ and F, F″ are related to the CaF_2-structure and the diamond structure, respectively.

The main classes with splitting parts I, J*, $\frac{1}{4}\frac{1}{4}\frac{1}{4}P_2$ and I, J*, F″, F‴ show interesting subclasses which contain structure types like Pt_3O_4, Hg_4Pt, Mg_3P_2 and Cu_2O. The last two columns of Table 5a contain main classes and subclasses of I_2 with splitting into complexes with degrees of freedom like $\frac{1}{4}\frac{1}{4}\frac{1}{4}P_2$xxx, F″xxx and F‴xxx, therefore the number of the nearest neighbors splits up into 6 + 1 + 1 and 3 + 3 + 1 + 1, respectively. Cu_3VS_4 is listed as structure type from a subclass of $I_2^{P,F''}$ but will also be found as an addition compound (= "Additionsverbindung") in the F-family, namely as F″xxx + P, J′.

$I_3 (\equiv I_{333})$-Types (Index 27)

Two structures are to be discussed, Tl_7Sb_2 as nearly of I_3-type and γ-brass Cu_5Zn_8 as structure type from the subclass I_3–I. Besides the points of I, in I_3 all other points have degrees of freedom. The free coordinates should be n/6.

	Point position in Im3m	Point position in I$\bar{4}$3m	For ideal I_3	For homogeneous sphere packing	In Tl_7Sb_2 Im3m	In Cu_5Zn_8 I$\bar{4}$3m
I	2(a)000	2(a)000	x = 0	x = 0	x = 0	Unoccupied
I(8 c)	16(f)xxx	8(c)xxx	x = 0.166$\bar{6}$	x = 0.1875	x = 0.17	I(4 t^+) x = 0.1089 I(4 t^-) x = – 0.1720
I(6 o)	12(e)x00	12(e)x00	x = 0.333$\bar{3}$	x = 0.2929	x = 0.29	x = 0.3558
W*(4 t_c)	24(h)xx0	24(g)xxz	x = 0.333$\bar{3}$	x = 0.3535	x = 0.35	x = 0.3128, z = 0.0366

The deviations of the parameters in Tl_7Sb_2 from the ideal values, especially for I(6 o) and W*(4 t_c) show that parts of the structure may be described as sphere packings by itself in a good approximation. Morral and Westgren (1934) consider Tl_7Sb_2 as the ideal type of γ-brass. Pearson (1972) requests a confirmation of the parameters by single crystal work. The different parameters in γ-brass (Cu_5Zn_8 in this case) may be caused by the voids connected with the unoccupied I complex which may be responsible for the splitting of I[8 c] in two parts I(4 t); but besides the additional degrees of freedom for the points of W*[4 t_c], there is a drastic parameter change in I(6 o) and W*(4 t_c) which may be interpreted as formation of octahedra between these two configurations and as formation of a heterogeneous "Bauverband". This will be discussed in detail in a later paragraph. The description of γ-brass by means of 2-dimensional nets (Schubert, 1964) seems to be inadequate.

$I_4 (\equiv I_{444})$-Types (Index 64)

Five different main classes are listed in Table 7a. Their characteristic splitting parts are Y*, Y$\overset{*}{x}\overset{*}{x}$x, D, F and I(4 t), respectively. For all these main classes of the I_4-types no representative could be found. Several representatives of subclasses, however, exist and demonstrate the importance of these combinations of complexes and their voids. In Table 7b for the representative structures the values of the free coordinates are listed, to allow a survey about the deviations from the ideal structure. The structure types are listed here under the assumption that all atoms or atomic groups (polyhedra) like SO_4, CO_3, Hg_2, H_2O of a structure form the "Bauverband"; this is probably adequate for intermetallic compounds like Th_6Mn_{23}, for nitrides or sulfides. On the other side one may consider also the framework of the O or S or N for itself as "Bauverband" and the whole structure as an addition compound. In this case these structures belong to other families than the I-family.

Examples: spinel,	F-family,	Index 8,	$\frac{3}{8}\frac{3}{8}\frac{3}{8}F_2xxx + D, T'$
Sb_2O_4,	P-family,	Index 64,	$P_4^D + T, T'$

Most of these possibilities are contained in the last column of Table 7b. In some subclasses the complexes which describe the unoccupied sites form well-known structure types by themselves.

Chemical compound	Voids
Th_6Mn_{23}	F, P_2'-Type (CaF_2)
Northurpite	D, D′-Type (NaTl)
Tychite	D′, T-Type ($MgCu_2$)
16 $Al(OH,F)_3 \cdot 6\,H_2O$ $RbNiCrF_6$	D, T′, $\frac{3}{8}\frac{3}{8}\frac{3}{8}F_2$-Type (spinel)

$I_6(\equiv I_{666})$-Types (Index 216)

Three compounds should be mentioned here: $Li_{22}Pb_5$ (F23, a = 20.08 Å) with 432 atoms per unit cell forms a complete I_6-type. $Cu_{41}Sn_{11}$ (a = 17.98 Å) and $NiCuSn_3$ (a = 18.011 Å) crystallize in the (I_6–I_2)-type ($F\bar{4}3m$), i.e., the type of γ-brass with a unit cell enlarged by a factor of 2 in each direction.

$I_8(\equiv I_{888})$-Types (Index 512)

For the structure of magnussonite, $(Mn, Mg)_9As_6MnO_{18}Cl$, the F, P_2' (= F, $\frac{1}{4}\frac{1}{4}\frac{1}{4}P_2$)-type ($CaF_2$) has been found by P. B. Moore (private communication). In this case the unit cell, referred to the CaF_2-structure, is enlarged in each direction by a factor of 4, and there are vacancies within $\frac{1}{16}\frac{1}{16}\frac{1}{16}P_8(= P_8')$.

Structure Types of the P- and F-Family

As an analogous detailed treatment of the P- and F-family would be beyond the scope of this paper, only examples of structure types for these families are discussed. Table 8a contains some structure types which may be described either in the F-family (index 1) as addition compounds or as belonging to main classes or subclasses of the P-family (index 8). An hypothetical structure type with symmetry Pm3m is listed too. The first three structure types of Table 8b are closely related to the perovskite structure. The first two only differ in the degrees of freedom for some splitting parts. The second one allows a tilting of the octahedra, built up by the points of J_2' xyz $(= \frac{1}{4}\frac{1}{4}\frac{1}{4} J_2$ xyz). F and F' are the central points of the octahedra. In the third case however, J_2' splits up into W* and I(6 o) and the central points of the octahedra split up also into I and I*. The structures of $CaCu_3Ti_4O_{12}$ and $NaMn_7O_{12}$ (Im3) with the "Bauverband" I[12 i] are derived from the perovskite type too. This distortion will be discussed below.

Two representatives of the NaCl structure type are shown in Table 8(b), namely InCl and Li_2TiO_3, with very different types of splitting and different possibilities for the deformation of the octahedra. A further perovskite-like structure (index 64) is listed below:

K_3TlF_6, potassium hexafluorothallate, Fd3, a = 17.86 Å, Z = 32

8K	8(a) 000				D	I_2
8K	8(b) $\frac{1}{2}\frac{1}{2}\frac{1}{2}$				D'	
16Tl	16(c) $\frac{1}{8}\frac{1}{8}\frac{1}{8}$				T	F_2''
16Tl	16(d) $\frac{5}{8}\frac{5}{8}\frac{5}{8}$				T'	
32K	32(e) xxx	$x = \frac{3}{8}$			F_2'''x	P_4'
48K	48(f) x00	$x = \frac{1}{4}$			$J_2^*(+I_2) = P_4$	
96F	96(g) $x_1 y_1 z_1$	$x_1 = 0.015$	$y_1 = 0.105$	$z_1 = 0.140$	D[12 co]	J_4 xyz
96F	96(g) $x_2 y_2 z_2$	$x_2 = \frac{1}{2} + x_1$	$y_2 = \frac{1}{2} + y_1$	$z_2 = \frac{1}{2} + z_1$	D'[12 co]	

Table 8a. Description of some structure types belonging to the F-family as "addition types" or to the P-Family as subclasses or main classes (NaCl)

Composition	Spacegroup		Structure-formula	Description in the F-Family Index 1	P-Family Index 8
$CaTiO_3$ Perovskite	Pm3m	a = 3,84	Ca_1, O_3, Ti_1 P , J , P'	F + P'	P_2 - J'
ReO_3	Pm3m	a = 3,734	O_3, Re_1 J , P	F−P + P'	P_2 - P, J'
NbO	Pm3m	a = 4,2097	O_3, Nb_3 J , J'	F−P + J'	P_2 - P, P'
U_4S_3	Pm3m	a = 5,505	S_3, U_1, U_3 J , P', J'	F−P + P', J'	P_2 - P
$A_1B_3C_1D_3$	Pm3m	hypothetic	C_1, D_3, A_1, B_3 P , J , P', J'	F + P', J'	P_2^P
NaCl	Fm3m	a = 5,6406	Cl_4, Na_4 F , F'	F + F'	P_2^F

Table 8b. Description of structures belonging to the F-family as "addition types" or to the P-family as sub- and mainclasses.

Composition	Spacegroup	Structureformula	Description in the F-Family Index 8	Description in the P-Family Index 64
Ba_2CaWO_6	Fm3m a = 8,390	Ba_8, O_{24}, W_4, Ca_4 $P_2', J_2'x, F, F'$	$F_2'x + P_2^F$	$P_4x - J_2$
NaK_2AlF_6 Elpasolith	Pa3 a = 8,09₂	K_8, F_{24}, Al_4, Na_4 $P_2'xxx, J_2'xyz, F, F'$	$F_2'xyz + P_2^F$	$P_4xyz - J_2$
$Ba_4Sb_3LiO_{12}$	Im3m a = 8,217	$Ba_8, O_{12}, O_{12}, Li_2, Sb_6$ $P_2', W^*, I(6o), I, J^*$	$F_2' + P_2^I$	$P_4 - J_2$
K_2PtCl_6	Fm3m a = 9,73	K_8, Cl_{24}, Pt_4 $P_2', J_2'x, F$	$F_2'x + F$	$P_4x - F', J_2$
KPF_6	Pa3 a = 7,69	F_{24}, P_4, K_4 $J_2'xyz, F, F'$	$[F_2' - P_2']xyz + P_2^F$	$P_4 - P_2', J_2$
Mg_6MnO_8	Fm3m a = 8,381	$O_8, O_{24}, Mn_4, Mn_{24}$ $P_2', J_2'x, F, J_2$	$F_2'x + F, J_2$	$P_4 - F'$
Ca_3PJ_3	I4₁32 a = 12,315	P_8, J_{24}, Ca_{24} $^+Y^*, [F_2''{}^+Y^*], [F_2'''{}^-Y^*]$	$F_2''xx + [F_2'' - Y]xx$	$P_4' - Y^*$
$NbF_{2,5}$	Im3m a = 8,19₀	F_6, F_{24}, Nb_{12} $J^*, J_2, I(60)x$	$[F_2 - I] + I(60)$ or $[F_2 - I] + [F_2' - P_2', W^*]$	$P_4 - I, P_2', W^*$
JnCl	P2₁3 a = 12,368	$Cl_4, Cl_4, Cl_{12}, Cl_{12}$ $\underbrace{\underbrace{F''\underline{\underline{x}}, F''\underline{\underline{x}}}_{P_2'xxx}, \underbrace{F(3y), F(3y')}_{J_2'xyz}}_{F_2'xyz}$ $Jn_4, Jn_4, Jn_{12}, Jn_{12}$ $\underbrace{\underbrace{F\underline{\underline{x}}, F'\underline{\underline{x}}}_{P_2xxx}, \underbrace{F''(3y), F''(3y)}_{J_2xyz}}_{F_2xyz}$	$F_2'xyz + F_2xyz$	P_4xyz
Li_2TiO_3	Fd3m a=8,301	$O_{32}, Li_{16}, (LiTi)_{16}$ $F_2'''\underline{\underline{x}}, T, T'$	$F_2'''\underline{\underline{x}} + F_2''$	$P_4\underline{\underline{x}}$

The ideal parameters for the sphere packing D[12co] in Fd3m are $x = 0, y = z = \frac{1}{8}$, the properties of this "Bauverband" may be taken from Table 4. D[12co] and D′[12co] together form the configuration of J_4, the "Bauverband" of the anions; it is of special interest that the K atoms occupy as well the 12-coordinated cuboctahedral as the 6-coordinated octahedral voids. The symbol for the K_3TlF_6 structure is P_4, P_4', J_4 xyz.

The compound $Sr_3Al_2O_6$ (Pa3, a = 15.84) is considered to have a perovskite-like structure (index 64) in addition, with holes in the "Bauverband" of the anions and in the cation positions. (v. Schnering, private communication).

Further Homogeneous and Some Heterogeneous "Bauverbände"

In the following we will deal with some homogeneous "Bauverbände" listed in Table 4.

$^{\pm}Y_{\underline{\underline{x}}}$-Types

The description of this "Bauverband" is given in Table 4. Figure 4 shows the (7 + 6) = 13 neighbors and the coordination polyhedron of one point. The "Bauverband" forms octahedral voids around F (Fig. 4), the vertices of which belong to three octahedra. Tetrahedral voids exist around P_2' x and $[F_2'' - {}^{\pm}Y]$ xyz. The "Bauverband" $^{\pm}Y_{\underline{\underline{x}}}$ with x = 0.1545 builds up a sphere packing with a density of 64%, that is a relatively close packed sphere packing. The sulfur atoms in pyrite FeS_2 and the oxygen-atoms in CO_2 form this "Bauverband" ($x_S = 0.116$, $x_O = 0.11$); the deviation from the sphere-packing parameter causes dumbbells around F with the possible description F(2l) too. In pyrite the Fe atoms are situated in the octahedral voids, in CO_2, forming stretched molecules, the carbon atoms are the centers of dumbbells. A splitting of the "Bauverband" $^{\pm}Y_{\underline{\underline{x}}}$ in $^{+}Y_{\underline{\underline{x}}}$ and $^{-}Y_{\underline{\underline{x}}}$ ($P2_13$) is found in the FeSi-type ($x_{Fe} = 0.1358$; $x_{Si} = 0.156$) and in ZrOS ($x_O = \frac{1}{2} + 0.1535$, $x_S = \frac{1}{2} - 0.1665$).

The structures may be described as

FeS_2 pyrite			$^{\pm}Y_{\underline{\underline{x}}} + F'$	or	$F(2\ell) + F$
CO_2			$^{\pm}Y_{\underline{\underline{x}}} + F$	or	$F(\cdot 2\ell)$
FeSi	$^{+}Y_{\underline{\underline{x}}}, {}^{-}Y_{\underline{\underline{x}}}$	=	$^{\pm}Y_{\underline{\underline{x}}}^{Y}$	or	$F(2\ell)$
ZrOS	$^{+}Y_{\underline{\underline{x}}}, {}^{-}Y_{\underline{\underline{x}}}, F'$	=	$^{\pm}Y_{\underline{\underline{x}}}^{Y} + F'$	or	$F(2\ell) + F'$

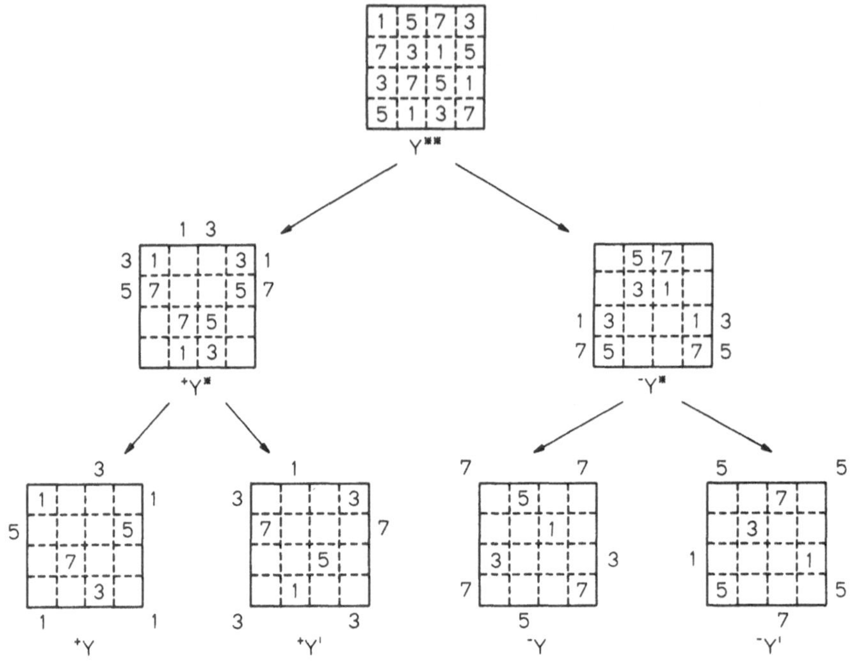

Fig. 3 (c). The splitting of Y** in subgroups of Ia3d.

I[4t]-Types

The "Bauverband" I[4t] consists of tetrahedra around I. With the parameter x = 0.1875 = 3/16 there exists a sphere packing with 9 nearest neighbors and a density of 63%.

Two representatives may be mentioned here:

$$SiF_4 \; I[4t] + I = I[\cdot 4t]$$
$$Tl_3VS_4 \; I[4t] + I, J^* = I[\cdot 4t] + J^*$$

In both cases the tetrahedra around I are centered by atoms (Fig. 5). In Tl_3VS_4 the thallium ions occupy the J* position; the first coordination sphere is a rhombically deformed tetrahedron and the second one a tetrahedron too. Both together form a distorted rhombic antiprism (8a).

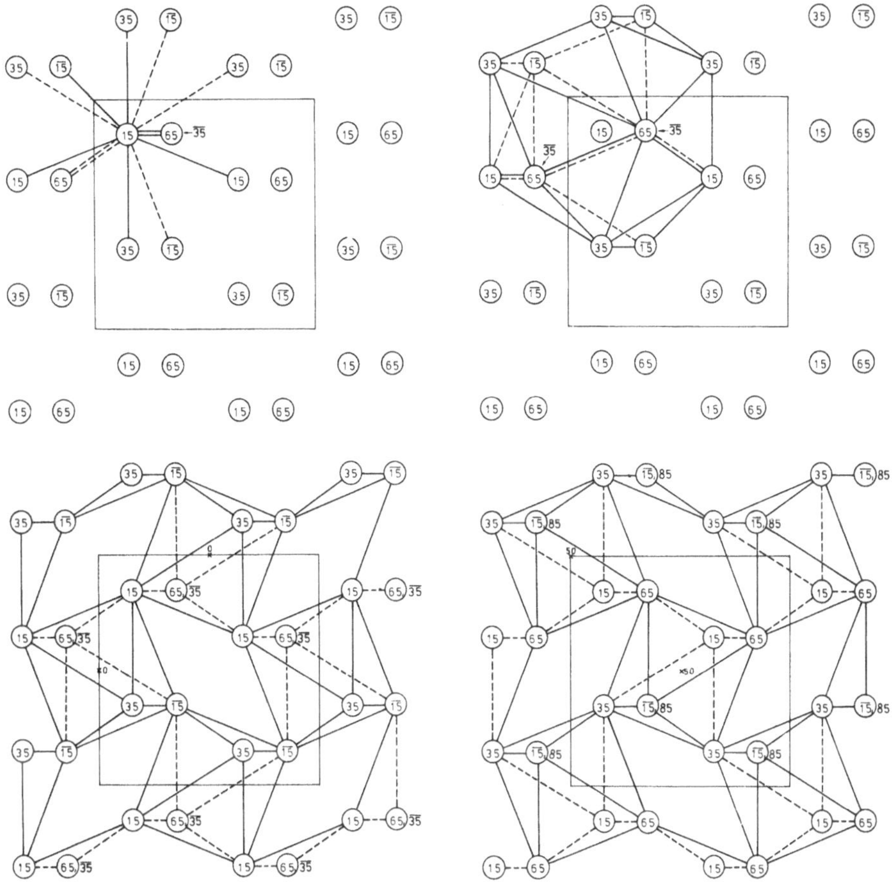

Fig. 4. Self-coordination number (7 + 6) and the polyhedron of the "Bauverband" [$^{\pm}Y_{\underline{x}}$] in Pa3(8 c), x = 0.1545 (approximately the S position in pyrite or the O position in CO_2). The *lower part* of the figure shows F′(6 o) in z = 0.0 (*left*) and 0.50 (*right*) with the octahedra in [$^{+}Y_{\underline{\underline{x}}}$] (Fe positions in pyrite)

From Table 4 it may be seen that a space partition can be constructed which consists of tetrahedra around five different positions I, J*, W*, $P_2'x$, [F_2'' – I(4 t)]; if one accepts, however, the coordination number 8 for the voids of J* (the Tl position in the second structure) then the deformed rhombic antiprism (8 a) overlap with common tetrahedra W*(4 t). This shows that a description of structure types by combinations of coordination polyhedra is not always possible. Figure 5 shows the three polyhedra discussed.

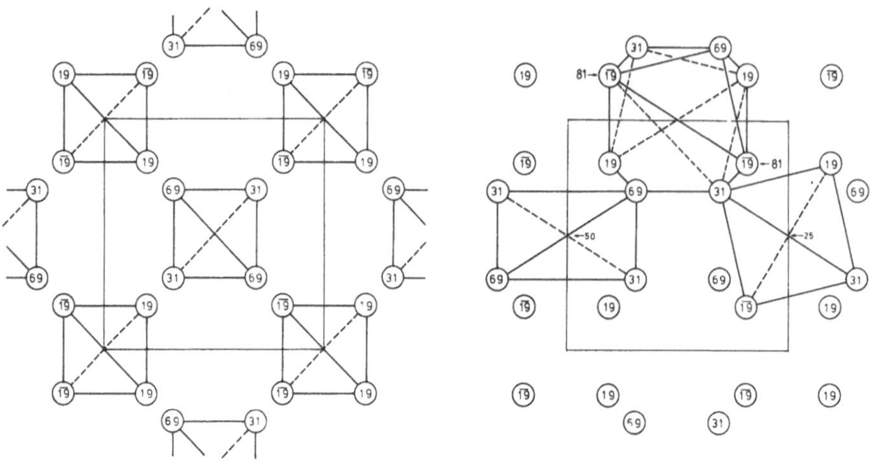

Fig. 5. „Bauverband" I[4 t]; 9/3/c2 in I$\bar{4}$3m(8 c)xxx, x = 0.1875 ~ 19/100.

I(4 t) the tetrahedra around Si in SiF_4 around V in Tl_3VS_4 (*left*)

J*(4 t) orthorhombically deformed tetrahedron in $\frac{1}{2}0\frac{1}{2}$ or

J*(4 t + 4 t) in $0\frac{1}{2}\frac{1}{2}$ (position of Tl in Tl_3VS_4),

W*(4 t) tetragonally deformed tetrahedron around $\frac{1}{2}1\frac{1}{4}$ (*right*)

I[6 o]-Types

I(6 o) J*(2 ℓ) describes the point position Im3m 12(e) x00. In the parameter range $0 < x < 0.293$ there exist octahedra around I and in $0.293 < x < \frac{1}{2}$ there exist dumbbells around J*. With the special parameter x = 0.293 there occur two interpenetrating sphere packings P[6 o] and P′[6 o] building up the I[6 o] (Fischer & Koch 1974). I(6 o) with $x = \frac{1}{4}$ may be looked at as part of $F_2'(=\frac{1}{4}\frac{1}{4}\frac{1}{4}F_2)$. In Table 4 Tl_7Sb_2 is listed as a structure type which contains this "Bauverband".

The subclass P[6 o] is realised in CaB_6 where the boron atoms form octahedra around P with the same distance within the octahedra and between nearest octahedra.

An important "Bauverband" of index 8 is I_{222}[6 o] = I_2[6 o] with the Fischer symbol 6/3/c 38 which occur in Ia3d 96(b)xyz with x = 0.0526, y = 0.1482, z = – 0.0388. There exist several other sphere packings with similar parameters. The "Bauverband" I_2[6 o] with slight parameter changes are occupied by 0 in garnet $Ca_3Al_2[SiO_4]_3$, by Bi in $RhBi_4$ and by 0 in Hg_3TeO_6. The coordination polyhedra (6 o) around I_2 for Mg and Te, (8 c) around V* for Ca and Rh and (6 o) around W_2' for Hg are drawn in Fig. 6. As the parameters for the interpenetrating sphere packings in I_2[6 o] are x = 0, y = 0.2929/2 = 0.1465, z = 0, from this point of view one may describe the idealised structure of the O-atoms in garnet as I[6 o].[1]

1 For more detailed informations see: Physik Daten/Physics Data *16*-1 (1979).

[T″ − ⁻Y′]- and the Heterogeneous [⁺V + Dx]- and [⁻Y(3 ℓ) + Dx]-Types

The point positions 12(d) $\frac{1}{8}$, x, $\frac{1}{4}$ + x of space group $P4_132$ or $\frac{1}{8}$, x, $\frac{1}{4}$ − x of space group $P4_332$, respectively, is a subset of 48(g) $\frac{1}{8}$, x, $\frac{1}{4}$ − x of Ia3d. With x = 0 and x = $\frac{1}{2}$, there exist the limiting forms ⁺V and ⁻V, respectively. With x = $\frac{1}{8}$ the configuration [T″ − Y′] is reached. It is related to the sphere-packing type 4/3/c1, the parameter range of which is $0 \leqslant x < 0.2035$ in $P4_132$. At x = 0.2035 the number of nearest neighbors increases from 4 to 6 and a sphere packing of type 6/3/c17 is built up, which may be described by ⁻Y[3 ℓ]. The configurations of ⁺V, [T″ − Y′] and ⁻Y[3 ℓ] are demonstrated by Fig. 7a. The "Bauverband" [T″ − Y′] is found in the structure of $(Sn_2F_3)Cl$ ($P2_13$, x = 0.375, y = 0.135, z = 0.119).

⁺V or ⁻Y(3 ℓ) in combination with the 8-pointer $D_{\underline{\underline{x}}}$ = Dxxx (point position $P4_132$, 8(c) xxx) give rise to two heterogeneous sphere packings described by [⁺V + $D_{\underline{\underline{x}}}$] and [⁻Y(3 ℓ) + $D_{\underline{\underline{x}}}$] respectively (Fig. 7b). In [⁺V + $D_{\underline{\underline{x}}}$] tetrahedra are formed; this is realised by the fluorine atoms in $CsBe_2F_5$, the Be atoms occupying the central points of these tetrahedra.

The heterogenous "Bauverband" [⁻Y(3 ℓ) + $D_{\underline{\underline{x}}}$] may describe the β-Mn structure and in subgroup $P2_13$ the Au_4Al structure. The parameter for ⁻Y(3 ℓ) with x = 0.206 is very near by the sphere-packing parameter. The high coordination numbers (Fig. 7c) of each of the two different kinds of Mn atoms explain the high density (72.4%) of this structure.

J_2-, I[12 i]-, [F_2'' − I(4 t)]- and W*[4 t]-Types

These four invariant "Bauverbände" exist in

Im3m	24 (h) 0xx,
Im3	24 (h) 0yz and
I$\bar{4}$3m	24 (g) xxz.

In Im3m(h), 0xx, there exists J_2 as limiting form with x = $\frac{1}{4}$ and the sphere packing W*[4 t] with x = 0.3535. In the range $\frac{1}{4} < x < 0.3535$ sphere packings of type 4/4/c3 are built up. They are found also in the two-dimensional parameter ranges of point position Im3(h), 0yz, and I$\bar{4}$3m(g), xxz (Fig. 8a). The coordinations of the two sphere packings without degrees of freedom 9/3/c3 and 6/3/c10 of Im3(h) are very similar. Therefore both of them may be described as I[12 i]. In I$\bar{4}$3m the sphere packing 9/3/c4 with x = $\frac{3}{8}$, z = $\frac{1}{8}$ can be considered as an F complex of the index 8 namely F_2''(= $\frac{1}{8}\frac{1}{8}\frac{1}{8}$ F_{222}) with a subtraction of 8 points arranged as I(4 t); the total symbol therefore is [F_2'' − I(4 t)]. In the following survey the Fischer symbols and the densities of the sphere packings are given. In addition the coordination polyhedra for the voids of each sphere packing are listed including the planar surrounding (4 l) which does not give a contribution to the volume and is not included in Table 4.

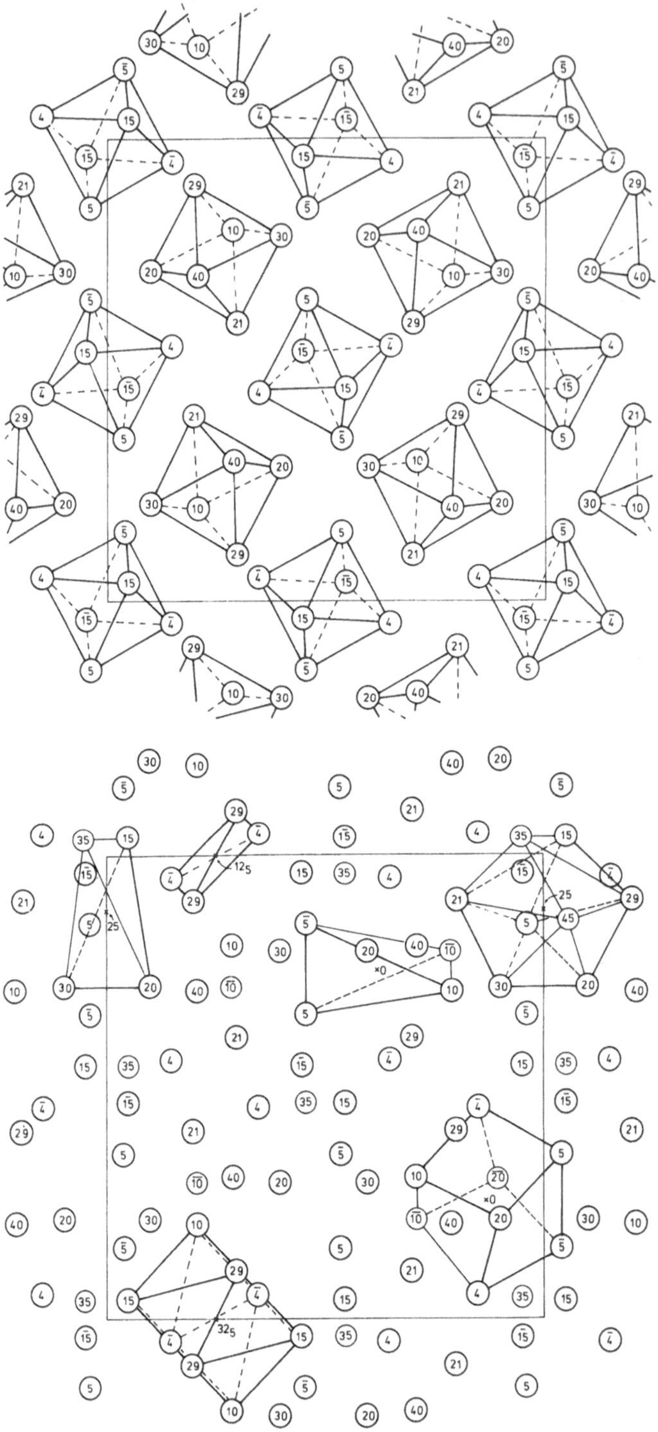

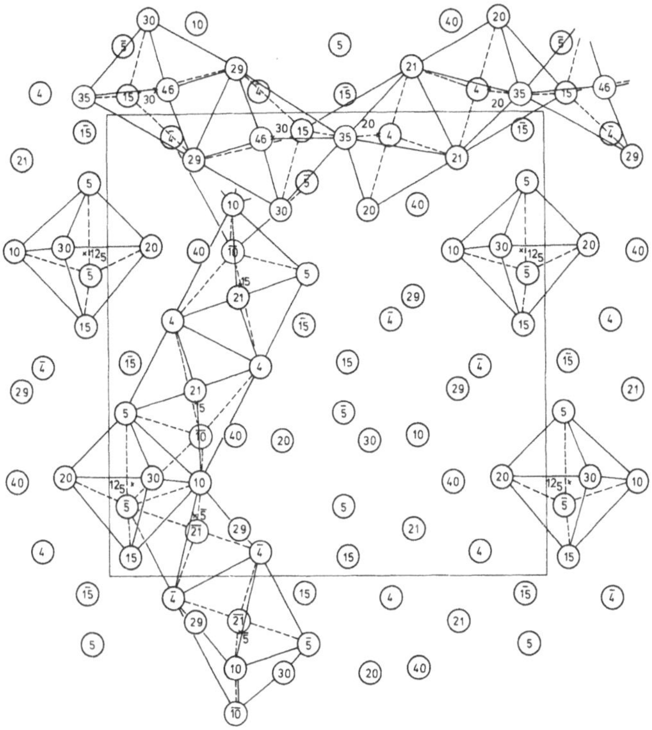

Fig. 6. "Bauverband" I_2[6 o] with the sphere-packing condition 6/3/c38 in Ia3d 96(g),
x = 0.0526 y = 0.1482 z = – 0.0388
x ~ 5/100 y ~ 14/100 z ~ – 4/100

I_2(6 o) shows the octahedra around the Al in garnet or Te in Hg_3TeO_6 *(upper left)*

V*(4 t) shows the tetrahedra of the first coordination shell in three positions and the (4 t + 4 t) surrounding in other three positions (Ca in garnet or Rh in $RhBi_4$ occupy V*) *(lower left)*

W_2'(6 o) shows the octahedra around Hg in Hg_3TeO_6. Rows of edge-connected octahedra along ~ 0 y $\frac{1}{4}$ and ~ x $\frac{1}{4}$ 0; the rows extending in ~ $\frac{1}{4}$ 0 z and ~ $\frac{3}{4}$ 0 z are represented by one octahedron only *(right)*

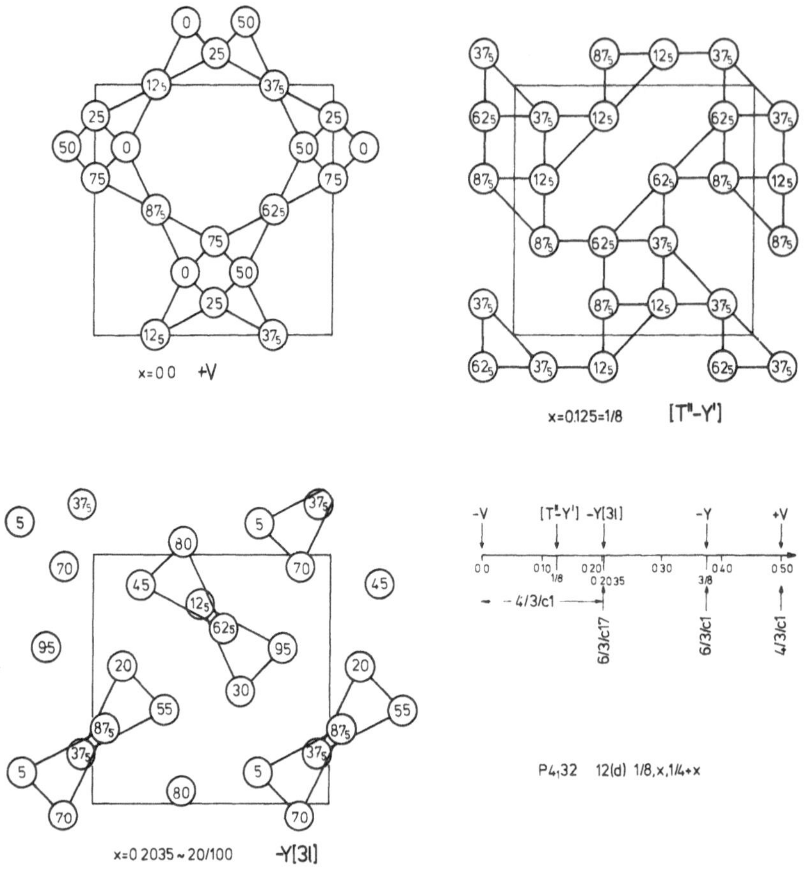

Fig. 7a. Representation of the parameter field of $P4_132$ 12(d) $\frac{1}{8}, x, \frac{1}{4} + x$ and the drawings of three configurations at x = 0.0 (^+V), $x = \frac{1}{8}$ ([T'' – $^-$Y'] and at x = 0.2035 ($^-$Y[3 1]). The latter one represents an invariant sphere-packing 6/3/c17, the other two are limiting complexes of the univariant sphere-packing 4/3/c1

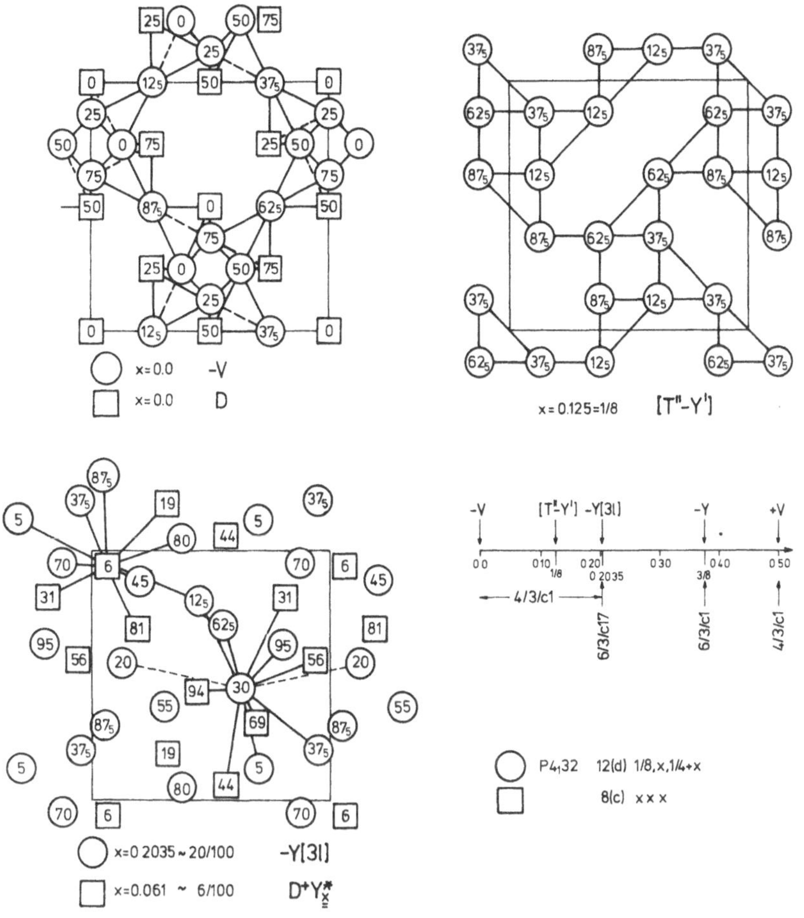

Fig. 7b. The heterogeneous "Bauverbände" [$^+V + D_{\underline{\underline{x}}}$] and [$^-Y(3\,1) + D^+Y^*_{\underline{\underline{x}}}$], constructed by adding $D_{\underline{\underline{x}}}$ twice. The first "Bauverband" describes the fluorine positions in $CsBe_2F_5$, the second the β-Mn structure. [T'' – Y'] represents the fluorine position in $(Sn_2F_3)Cl$. In [$^-Y(3\,1) + D^+Y^*_{\underline{\underline{x}}}$] the 12 neighbors of a $D^+Y^*_{\underline{\underline{x}}}$ point 6 and the 12 + 2 neighbors of a $^-Y(3\,1)$ point are shown

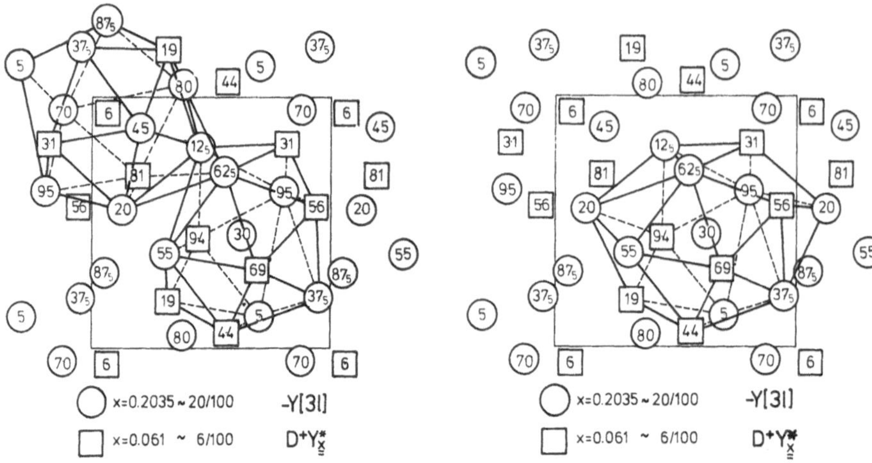

Fig. 7c. The coordination polyhedra (c.n.12) for a point of $D^{+}Y^{*}_{\underline{\underline{x}}}$ and of $^{-}Y(3\,l)$ (*left*); the coordination polyhedra (c.n.14) for $^{-}Y(3\,l)$ (*right*)

J_2	I[12 i]	[F″ – I(4 t)]	W*[4 t_c]
8/3/c2	9/3/c3	9/3/c4	6/3/c6
0.555	0.641	0.555	0.316
I(12 co)	I(12 i)	I(12 tt)	I(12 co)
J*(12 co)	J(*4 l)	J*(4 t)	J*(21(8 a))
P_2'(6 o)	P_2(6 o)	P_2'x(9 tco)	P_2'(6 o)
W*(4 t)	W*x(6 r)	W*(4 t)	W*(4 t)
	I(12i (5py))	J_2 xyz(5py)	

In Fig. 8b–8e some of the polyhedral voids of these four "Bauverbände" are illustrated. The density of J_2 and [F_2'' – I(4 t)] amounts to $\frac{3}{4} \cdot 74.1\% = 55.5\%$, with 74.1% beeing the density of F. With respect to this value, there is an increase of density for I[12 i] by about 15% and a decrease for W*[4 t_c] by 43%. While the polyhedra around I in the first three "Bauverbände" change from (12 co) via (12 i) to (12 tt) with a corresponding small increase of their volumes from 0.104 via 0.113 to 0.120 the cuboctahedra (12co) around I in W*[4 t_c] have a volume, which is enlarged by a factor of 2.8 (V = 0.295). Therefore the I position in W*[4 t_c] can not be occupied by a single ion but only by groups of atoms or ions (Fig. 8b).

A more drastic difference exists at the J* position where the coordination polyhedra change from (12 co) to (4 ℓ) (or 4 ℓ + 4 ℓ) or to (4 t) or to (2 ℓ(8 a)) respectively (Fig. 8c). As a consequence the polyhedra around P_2' and at W* (Fig. 8d,e) change also and additional polyhedra are necessary for I[12 i] and [F_2'' – I(4 t)] to fill the 3-dimensional space completely (Table 4). In addition to WAl_{12}, the "Bauverband"

I[12 i] has been found recently in the deformed perovskites $NaMn_7O_{12}$ (high pressure modification) and $CaCu_3Ti_4O_{12}$ (Reinen and Propach (1971)) for the O atoms. A complete description may be given for

$$NaMn_7O_{12}: \quad O_{24}, \quad Na_2, Mn_6^{I}, Mn_8^{II},$$
$$I[12\,i], I, \quad J^*, \quad P_2' \quad = I[12\,i] + P_2^{I}, P_2' \text{ or } I[\cdot 12\,i] + J^*, P_2'.$$

The sulfur atoms in tetrahedrite occupy the "Bauverband" $[F_2'' - I(4\,t)]$ (Table 4). As the only Bauverband for all O-atoms in

sodalite	$Na_4(AlSiO_4)_3Cl$
hauÿne	$Na(Ca, K)\,(AlSiO_4)_3 \cdot (SO_4)$
zinkmetaborate	$Zn_4O(BO_2)_6$

$W^*[4\,t_c]$ has been found.

A complete description for sodalite is given in Table 4, with the abbreviated symbol $W^*[\cdot 4\,t_c] + \underline{I}(\cdot 4\,t)$. An ordered distribution of Si and Al over the tetrahedral sites its possible in $P\bar{4}3n$; the tetrahedra around I have the composition $ClNa_4$.

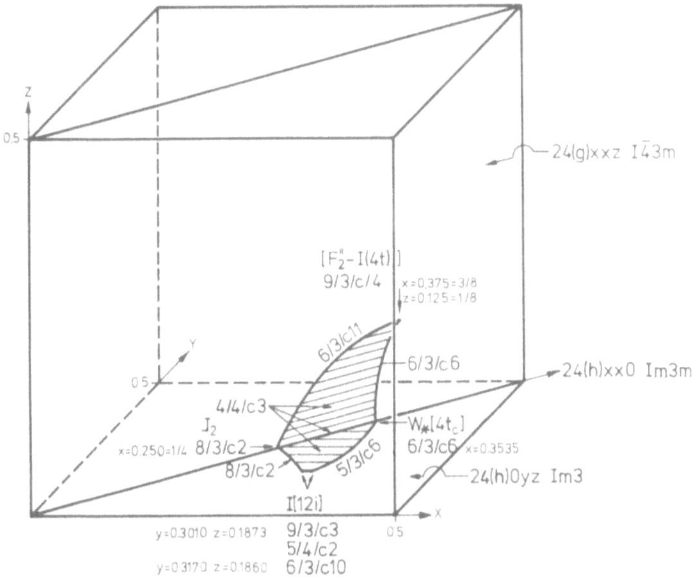

Fig. 8a. Parameter field for invariant sphere-packings in
Im3m 24(h) 0xx with J_2, $W^*[4\,t_c]$,
Im3 24(h) 0yz with I[12 i],
$I\bar{4}3m$ 24(g) xxz with $[F_2''-I(4\,t)]$
[after Fischer (1973)] and their connections by uni- and divariant sphere-packings

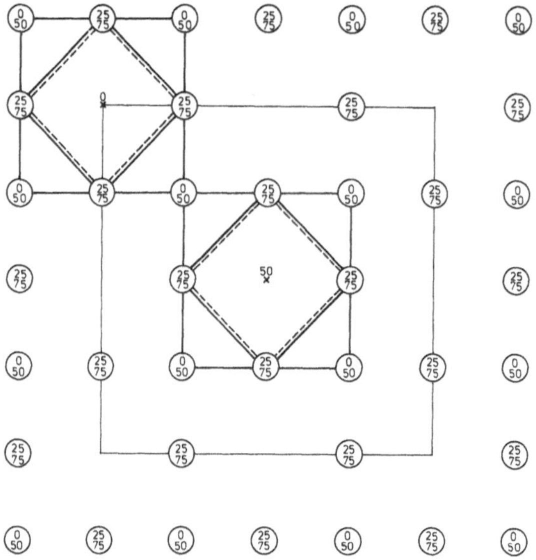

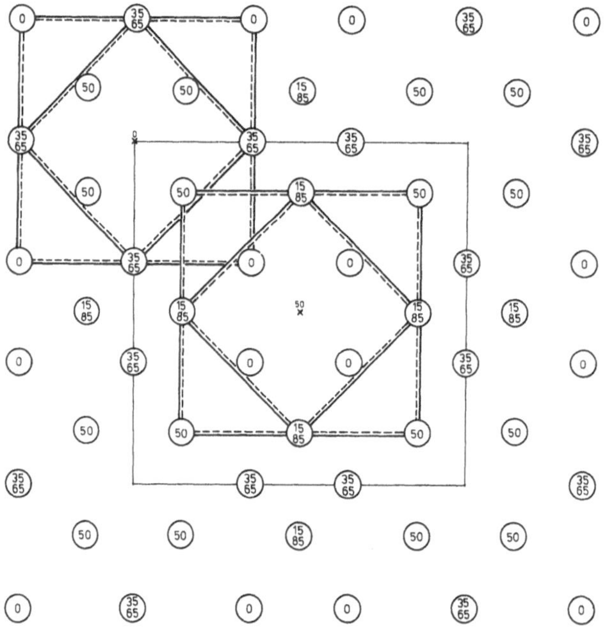

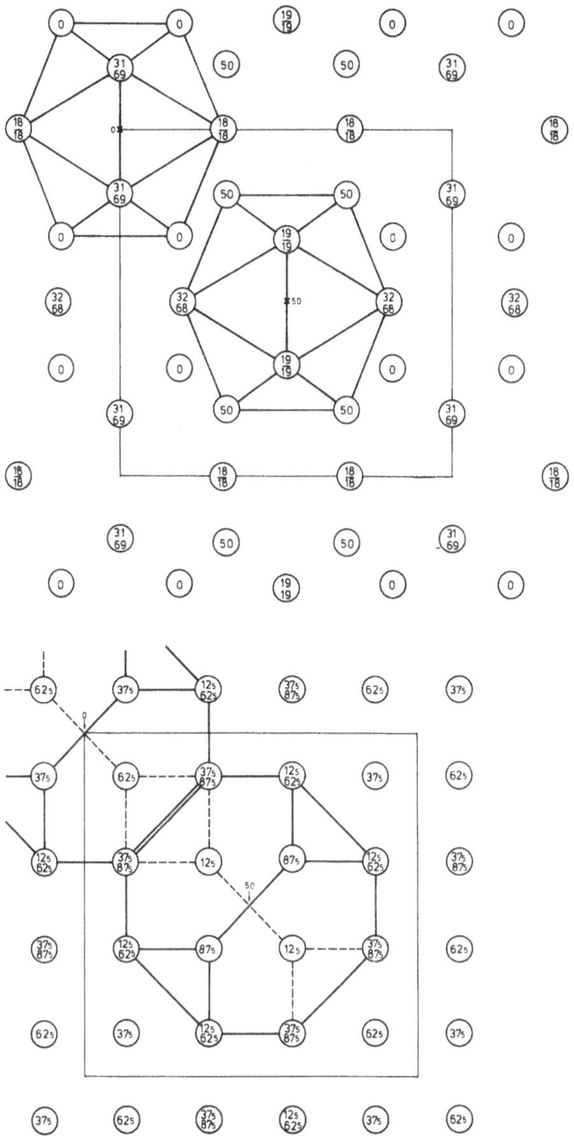

Fig. 8b. First coordination shell of the I complex:
I(12 co) in J_2 (*upper left*; Ca^{2+} position in perovskite of index 8)
I(12 i) in I[12 i] (*upper right*; Na^+ position in $NaMn_7O_{12}$)
I(12 co) in W*[4 t_c] (*lower left*; occupied by $ClNa_4$-groups as I(· 4 t) in sodalite; the second coordination shell consists of (24 co) with $d_1/d_2 = 0.92$
I(12 tt) in [F_2''–I(4 t)] (*lower right*; occupied by $(Ag, Cu)_6$ groups I(6 o) in tetrahedrite)
J_2, I[12 i] and W*[4 t_c] are the O-positions in the compounds mentioned and [F_2''–I(4 t)] is the S-position in tetrahedrite.

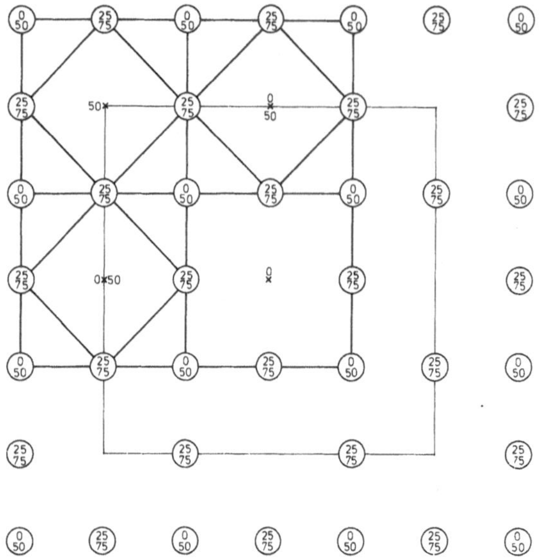

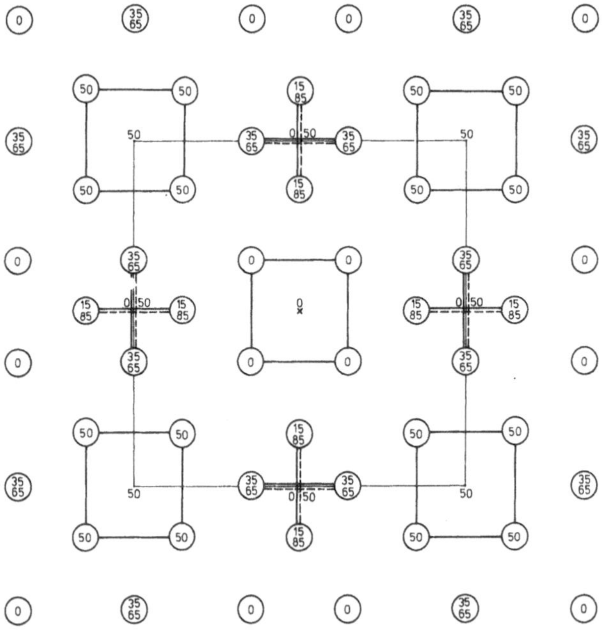

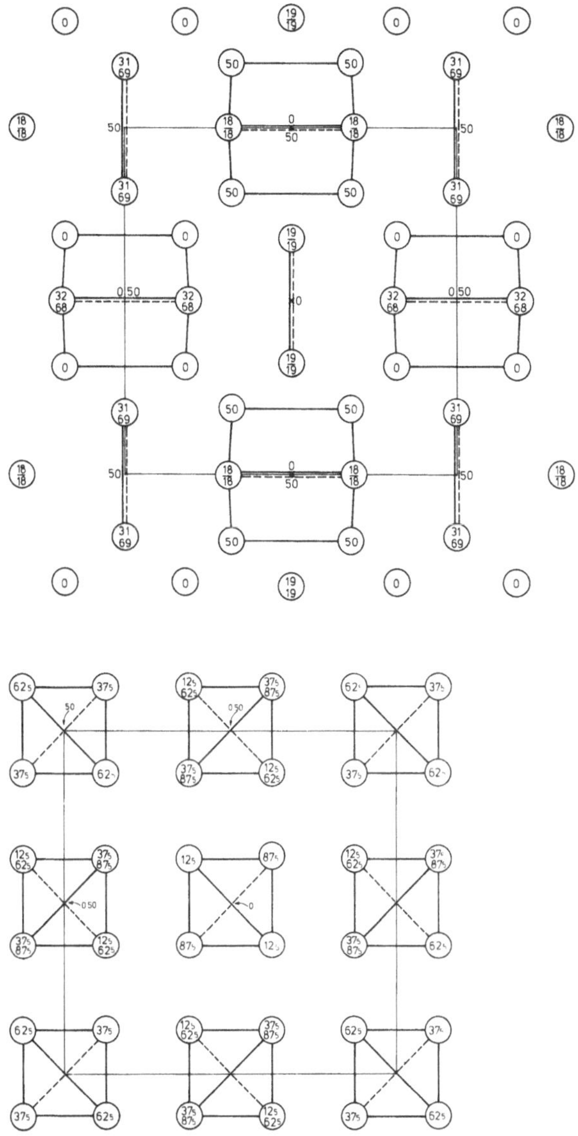

Fig. 8c. First coordination shell of points of the J* complex:

J*(12 co) in J_2 (*upper left*; Ca^{2+} position in perovskite of index 8)

J*(4 l) in I[12 i] (*upper right*; Mn^I position in $NaMn_7O_{12}$; for the second coordination shell ($4\,l^{\perp}$) is $d_1/d_2 = 0.71$)

J*(4 l) in W*[4 t_c] (*lower left*; unoccupied in sodalite)

J*(4 t) in [F_2''–I(4 t)] (*lower right*; unoccupied in tetrahedrite) [compare Fig. 8b]

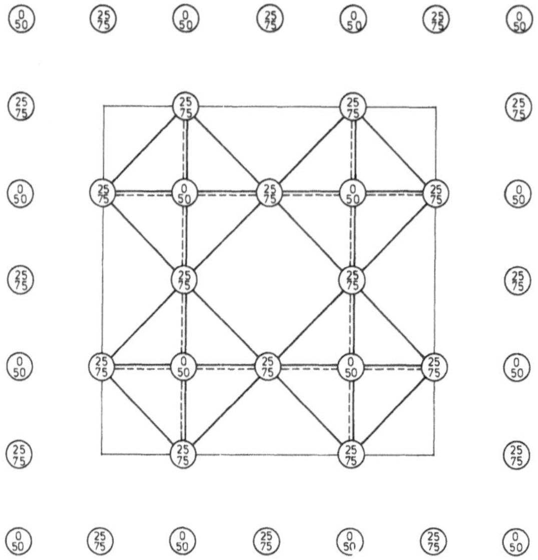

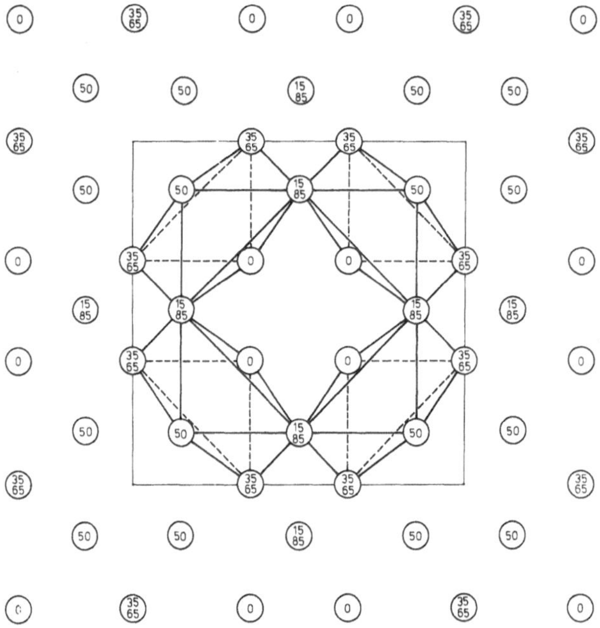

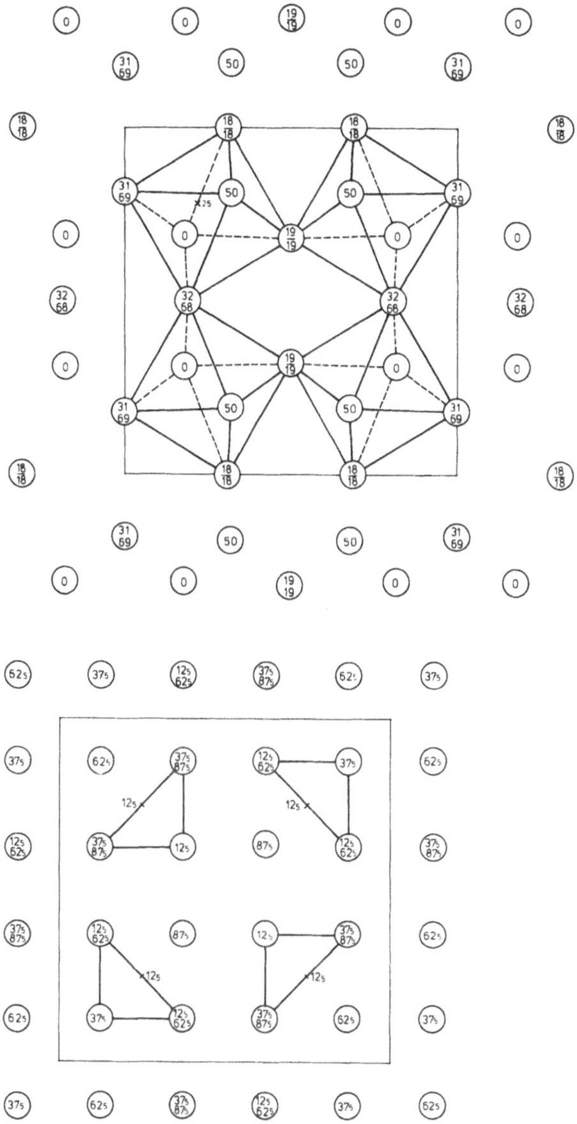

Fig. 8d. First coordination shell of the P'_2 complex:
P'_2(6 o) in J_2 (*upper left*; perovskite of index 8)
P'_2(6 o) in I[12 i] (*upper right*; Mn^{II} position in $NaMn_7O_{12}$)
P'_2(6 a) in W*[4 t_c] (*lower left*; occupied by (Na, Ca) in Hauÿne)
P_2(3 y) in [F''_2 – I(4 t)] (*lower right*; occupied by (As, Sb) in tetrahedrite) [compare Fig. 8b]

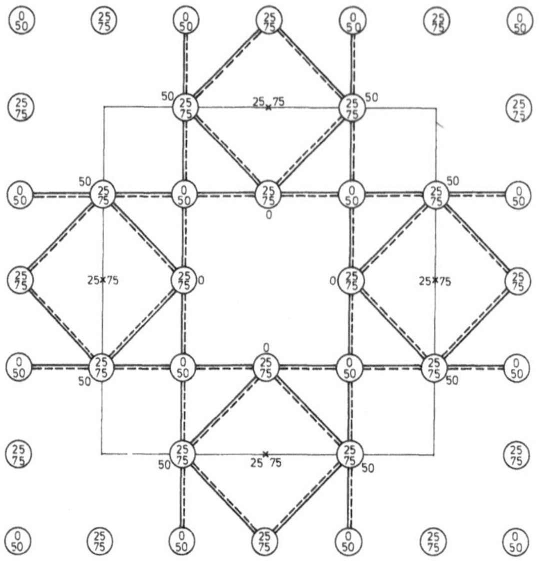
0
50
25
75
25×75

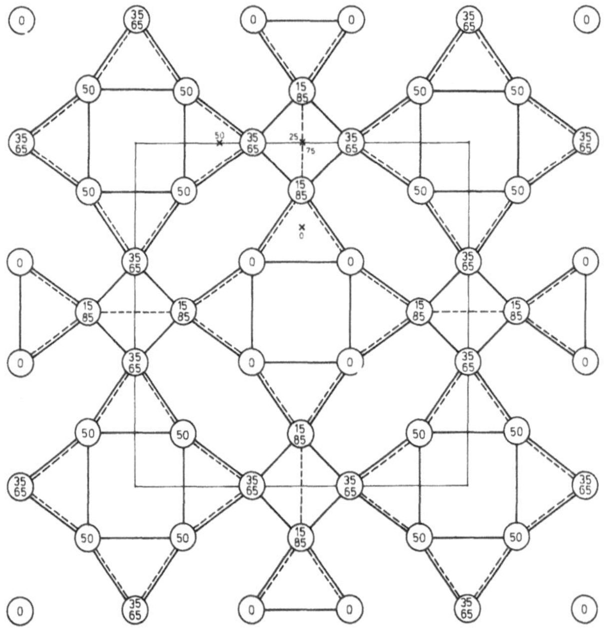
0
35
65
50
15
85
25
75

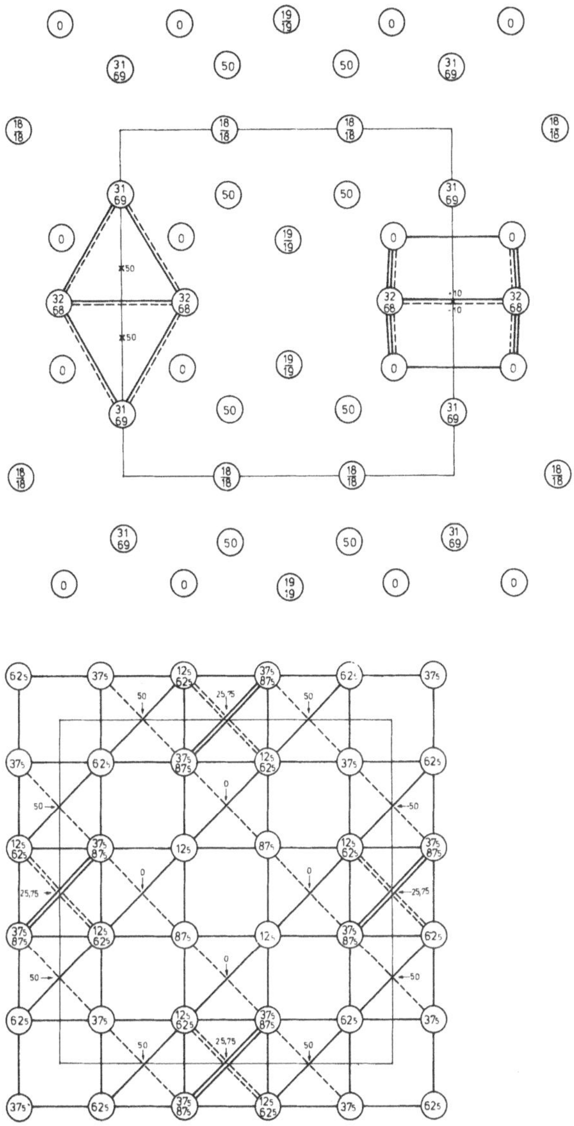

Fig. 8e. First coordination shell of the W* complex:
W*(4 l) in J_2 (*upper left*; unoccupied in perovskite of index 8)
W*(6 r) in I[12 i] (*upper right*; unoccupied in $NaMn_7O_{12}$)
W*(4 t) in W*(4 t_c) (*lower left*; occupied by (Si, Al) in sodalite and hauÿne)
W*(4 t) in [F_2''–I(4 t)] (*lower right*; occupied by (Ag, Cu) in tetrahedrite) [compare Fig. 8b]

The Heterogenous [W*(4 t) + J*(2 ℓ))] ≡ W*x[6 o]-Types

In the section I[6 o]-types it was already explained, that this "Bauverband" forms dumb-bells around J* when the parameter x becomes larger then 0.293 (the sphere-packing parameter). These dumb-bells of J*(2 ℓ) together with expanded tetrahedra around W*(4 t) (p.e., Im3, 24(g) Oyz, y = 0.35, z = 0.28) form more outstanding octahedra (Fig. 9). These octahedra share edges, consisting of the dumb-bells, mentioned above, and they share vertices for the points contributed by W*(4 t). The central points of these octahedra (0.35, $\frac{1}{2}$, 0) do not coincide with the points of W*($\frac{1}{4}$, $\frac{1}{2}$, 0). The whole arrangement may be symbolised by W*x[6 o] consisting of the following positions:

W_x^*[· 6 o]	W*[4 t]	in Im3	24(g) Oyz	y = 0.35	z = 0.28
	J*(21)	in Im3	12(d) x00	x = 0.35	
	W*x	in Im3	12(e) x0 $\frac{1}{2}$	x = 0.35	

This important octahedral "Bauverband" is built up by oxygen atoms centered by Sb, Ir or Re forming antimonates, iridiates and rhenates; these compounds can be described in the following way:

$NaSbO_3$	9.378 Im3 Z = 12	O_{24}, W*(4 t),	O_{12}, J*(21),	Sb_{12}, W*x,	$Na_{16/3}$, I(8 c)$_{stat.}$	$Na_{8/2}$, P_2' stat.	W*x[· 6 o] + P_2', I(8 c)
$KSbO_3$	9.60 Pn3 Z = 12	O_{24}, W*(4 t),	O_{12}, J*(21),	Sb_{12}, W*x,	K_8, I(4 t),	K_4, F″,	W*x[· 6 o] + F‴, I(4 t)
$Bi_3GaSb_2O_{11}$	9.49 Pn3 Z = 3	O_{24}, W*(4 t),	O_{12}, J*(21),	$(Ga + Sb_2)_{12/3}$, W*x,	O_8, I(4 t^-),	Bi_8, Bi_4, I(4 t), F″,	W*x[· 6 o] + F‴, I(4 t, 4 t)
$Ba_{0.5}IrO_3$	9.41 I23 Z = 12	O_{24}, W*(4 t),	O_{12}, J*(21),	Ir_{12}, W*x,	$Ba_{6/8}$, I(4 t)$_{stat.}$,	O_2, I,	W*x[· 6 o] + I(· 4 t)$_{stat.}$
$La_4Re_6O_{19}$	9.6316 I23 Z = 12	O_{24}, W*(4 t),	O_{12}, J*(21),	Re_{12}, W*x,	La_8, I(4 t),	O_2, I,	W*x[· 6 o] + I(· 4 t)

Besides the W*x[· 6 o] "Bauverband", clusters like (8 c), (4 t), (4 t + 4 t) are formed around I and centered clusters (· 4 t) in $Ba_{0.5}IrO_3$ and $La_4Re_6O_{19}$.

Similar – but not centered – "Bauverbände" W*x[6 o] may be found in γ-brass-like compounds, e.g. Ag_5Hg_8.

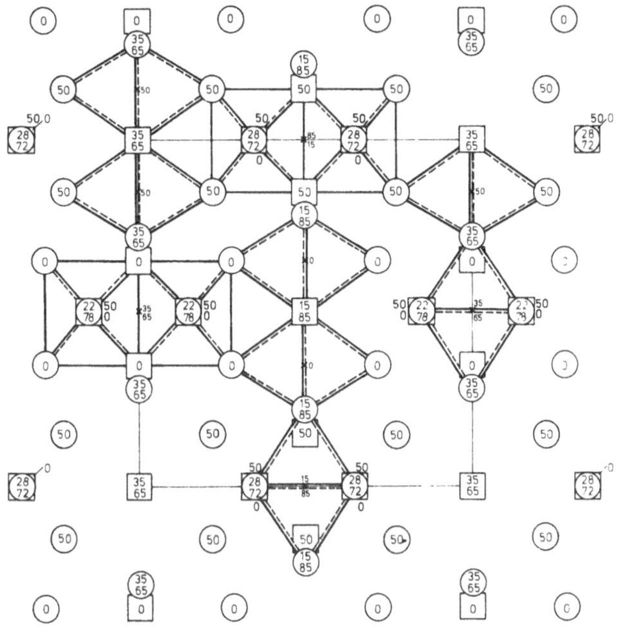

Fig. 9. The heterogeneous "Bauverband" [W*(4 t) + J*(2 l)] (abbreviated: W*x[6 o]). (It forms octahedra around W*x and describes the positions of O in $KSbO_3$; the point of the J*(2 l) are surrounded by a square □

Furthermore α-Mn, Tl_7Sb_2, Ag_8Ca_3 and the "Bauverband" of the fluorine atoms in U_2F_9 may be added.

Other homogeneous "Bauverbände" like D[4 t], [P_4' – Y**] [4 t], D[6 o], I[8 c], F[8 c], F[12 co], D[12 co], D′[12 tt], I[12 i], ..nI[12 i] and ..cP_2[12 i] are shown in Table 4 together with one representative of each "Bauverband". The symbol ..n in front of I[12 i] indicates that the icosahedron around $\frac{1}{2}\frac{1}{2}\frac{1}{2}$ is not parallel oriented to that around 000 but generated by a diagonal glide plane n. The meaning of ..c in front of P_2[12 i] is similar. In the following a few heterogeneous "Bauverbände" may be discussed.

[I + W]-Types

This heterogeneous "Bauverband" is found in Pm3n and built up by compounds of the β-W-type (A15-type) (e.g., W_3O or Cr_3Si). A great number of compounds with this structure type are superconductors, e.g., Nb_3Sn, partly with high transition temperature, and are used commercially therefore.

The density of the heterogeneous sphere packing [I + W] is 64%. A polyhedral space partition is formed in the following way:

$$W'(4\,t) \quad V_{4t} = 0.020833 \times 6 = \frac{1}{8}$$

$$P_2'(5\,by) \quad V_{5by} = 0.046875 \times 8 = \frac{3}{8}$$

$$..nI(12\,i\ (4\,t)) \quad V_{4t} = 0.020833 \times 24 = \frac{1}{2}$$

In Fig. 10 the three differently located polyhedra are illustrated.

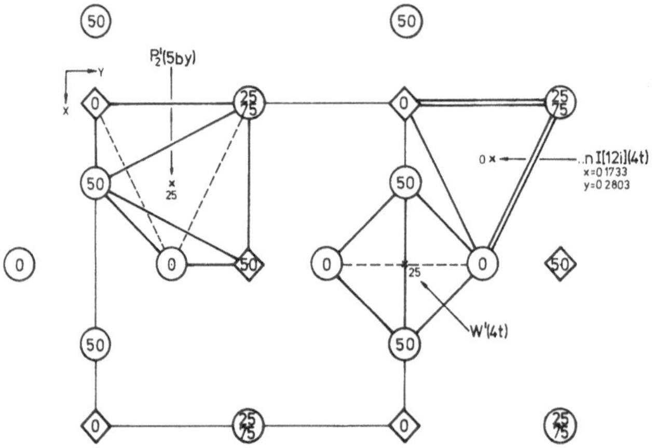

Fig. 10. Heterogeneous "Bauverband" [I + W] which describes the β-W (W_3O)-structure with its three different space-filling polyhedra W'(4 t), .. nI(12 i(4 t)) and P_2'(5 by)

Ag_3PO_4 can be considered as homeotype with the PO_4-groups around I. The structure may be symbolised by [I(·4 t) + W].

In the γ-oxygen and β-fluorine structures dumbbells rotate around all points of the heterogeneous Bauverband; the structure type may be symbolised as [I(2 ℓ) + W(2 ℓ)]; rotating dumbbells may be indicated by (2 ℓ⌒).

In this connection the structure of KGe is of special interest. The Ge form tetrahedra around I and W'; therefore the arrangement can be considered as homeotypic and symbolised by [I(4 t) + W'(4 t)]. The K-atoms form a similar arrangement, but all the tetrahedra are larger and those around I are oriented in a different manner I(4 t^-). The full symbol for KGe may be formulated as I(4 t, 4 t^-), W'(4 t), W(4 t) = I(8 c), W*(4 t).

Finally Geller (1959) pointed out that the position of the cations in the garnet structure $Ca_3Al_2(SiO_4)_3$ can be considered as a β-W arrangement of index 8:

$$\begin{array}{llll} & Al_{16}, Si_{24}, & Ca_{24} & \\ & I_2, \quad S^*, & V^* & = I_2, W_2^{S*} \\ \text{with} & S^* + V^* & = W_2 \text{ (see Fig. 3a).} & \end{array}$$

$[I_2 Y^{**}_{\underline{\underline{x}}} + {}^+S]$-types

A representative for this type is $Th_3P_4(I\bar{4}3d)$:

12 Th	12(a) $\frac{3}{8}$ 0 $\frac{1}{4}$	
16 P	16(e) xxx.	

There exist I_2 and Y^{**} as limiting forms with $x = 0$ and $x = \frac{1}{8}$, respectively. Y^{**} consists of two interpenetrating sphere packings ${}^+Y^*$ and ${}^-Y^*$ with three nearest neighbors (Fig. 3b and Fischer and Koch (1976)). The parameter $x = 1/12 = 0.083\bar{3}$ gives rise to 8 nearest neighbors with equivalent distances for the Th-atoms and 6 nearest neighbors for the P-atoms. The density is 60.9%. Recently Bärnighausen (1978) has determined $x = 0.0772$ for Sn_3S_4. With this parameter the distances split up in 4 + 4 and 3 + 3, respectively; in the paper of Bärnighausen the valence problem is discussed; it may be worthwhile to include the possibility of polarisation too, which can be made plausible by the homeotypes of the structure. The simplest homeotype is Rb_4O_6 or rather $Rb_4(O_2)_3$ for comparison. There are dumbbells of O_2 at ${}^+S$. The parameter for Ru has been determined as $x = 0.054$. $Rb_4(O_2)_3$ can be symbolised as

$$[I_2 Y^{**}_{\underline{\underline{x}}} + {}^+S(21)].$$

In eulytite ($Bi_4(SiO_4)_3$) centered SiO_4 tetrahedra are located at ${}^+S$. The x-coordinate for Bi is $x_{Bi} = 0.0857$. The symbol for this type is therefore:

$$[I_2 Y^{**}_{\underline{\underline{x}}} + {}^+S(\cdot 4\,t)].$$

The mineral langbeinite $K_2Mg_2(SO_4)_3$ is described in the subgroup $P2_13$ of $I\bar{4}3d$ where $I_2 Y^{**}_{\underline{\underline{x}}}$ splits into four parts $FY_{\underline{x}}$ and ${}^+S$ is a limiting form of the general position. The coordinates for the sulfur position with $x = 0.625$, $y = 0.467$, $z = 0.268$ are close to those for ${}^+S$ with $x = \frac{5}{8}$, $y = \frac{1}{2}$, $z = \frac{1}{4}$. The symbol of langbeinite, therefore, contains a splitting term for the different cations and indications for the additional degrees of freedom:

$$[I_2^F Y_{\underline{x}} + {}^+Sxyz(\cdot 4\,t)]$$

it should be mentioned that the O-positions are close to those of W_2' and therefore the description $[I_2 Y^{**}_{\underline{\underline{x}}} + W_2'] + {}^+S$ permits to classify eulytite as an addition structure of index 8 of $[I + W]$ (see preceeding paragraph).

[I(20 pd) + W]-Types

The clathrates 46 $H_2O \cdot 8$ M and 46 $H_2O \cdot 6$ M with M = Cl, SO_2, CH_3Cl, H_2S ... belong to this group. The heterogeneous Bauverband consists of the three parts I(12 i), I(8 c) and W in the space group Pm3n. The first two build up the pentagon-dodekahedron (20 pd) around I, and together with W a void around W′ with a large coordination polyhedron of 24 points, called truncated hexagonal trapezohedron (24 tht). In both cases around I and W′ the M atoms or molecules are located. $Ge_{38}P_8Br_8$($P\bar{4}3n$) is homeotypic to this structure type. The Br atoms replace the M-groups at I and W′; I(8 c), splitted up into I(4 t^+) and I(4 t^-), is occupied by phosphor and germanium, I(12 i) shows three degrees of freedom in this space group.

Pentagondodecahedra (20 pd) around I also exist in the structure type of skutterudite, $CoAs_3$, but as a combination of I(12 i) and P_2' in Im3. These pentagondodecahedra share those vertices, which belong to P_2'.

Deformed Structure Types

For structures not belonging to the I-, P- and F-family a classification has not been proposed because the problem is not yet been solved, how to define families in these cases. In connection with this problem two examples may be discussed namely $CaCu_3Ti_4O_{12}$ or $NaMn_7O_{12}$ in relation to the ideal perovskite and $RbNiCrF_6$ and pyrochlore in relation to the F, P_2'-type of index 8 (CaF_2). The derivations of the $NaMn_7O_{12}$ structure from the perovskite structure (index 8) is shown; in addition for each position the coordination polyhedron is given:

	Im3	Perovskite Index 8	$NaMn_7O_{12}$
(Ca)Na	2(a) 000	I (12 co)	I (12 i)
(Cu)MnI	6(b) 0 $\frac{1}{2}\frac{1}{2}$	J* (12 co)	J* (41 + 41)
(Ti)MnII	8(c) $\frac{1}{4}\frac{1}{4}\frac{1}{4}$	P_2' (6 o)	P_2' (6 o)
O	24(g) 0yz	J_2 y = z = 0.250 (8 r)	J_2 y = 0.3132 z = 0.1828 (9 + 1)

The perovskite with the ideal parameter y = z = 1/4 differs very much from the structures of $CaCu_3Ti_4O_{12}$ and $NaMn_7O_{12}$, the anion parameters of which are near by those of two invariant sphere packings.

$$6/3/c10 \text{ with } y = 0.3170; \ z = 0.1830$$

and

$$9/3/c3 \text{ with } y = 0.3010; \ z = 0.1873.$$

Especially the surrounding of the J* position occupied by Cu^{2+} and Mn^{3+} changes drastically; Fig. 11 shows that the coordination number 12 of the first coordination shell splits into 4 + 4 + 4. The sphere-packing parameters of 6/3/c10 are closest to the real parameters. With these parameters, there are 4 nearest neighbors and 4 further neighbors in the second coordination shell of oxygen. The third coordination shell is formed by 8 points of P_2'(Mn^{II}-position) and the fourth by the missing last 4 neighbors as found for the 8/3/c2-J_2 of the ideal perovskite.

Similar changes may be seen at the surrounding of the atoms of the "Bauverband" I_2 and I[12 i] with the two sphere-packing conditions (Fig. 11). The reason for the considerable distortion within the perovskite framework is the strong Jahn-Teller instability of the orbital degenerate E_g-groundstates of Cu^{2+} and Mn^{3+}, which lead to a square planar coordination of the nearest oxygen neighbours. The pyrochlore structure $R_2Q_2X_7$ (R = Ca, Ce ..., Q = Nb, Fe, Ta ..., X = 0, OH, F) may be described as a F, P_2'-type (CaF_2) (index 8) with vacancies in the anion Bauverband. For this purpose the CaF_2 type is described with a splitting of the fluorine arrangement (index 8):

CaF_2					Pyrochlore	
Index 1		Index 8			$R_2Q_2X_7$	$RbNiCrF_6$
Fm3m	Pn3m	Fd3m				
4(a) 000	F 4(b) $\frac{1}{4}\frac{1}{4}\frac{1}{4}$	F''	16(c) $\frac{1}{8}\frac{1}{8}\frac{1}{8}$	T	16R	16(Ni, Cr)
			16(d) $\frac{5}{8}\frac{5}{8}\frac{5}{8}$	T'	16Q	Vacant
8(c) $\frac{1}{4}\frac{1}{4}\frac{1}{4}$	P_2' 2(a) 000	I	8(a) 000	D	Vacant	Vacant
			8(b) $\frac{1}{2}\frac{1}{2}\frac{1}{2}$	D'	8X	8Rb
	6(d) 0 $\frac{1}{2}\frac{1}{2}$	J*	48(f) 000	J_2^*	48X	48F
			x = 0.250		0.180 < x < 0.230	x = 0.1875

The splitting $F_2'' = T + T'$ and $I_2 = D + D'$ is shown in Fig. 2. In Fd3m J_2^* only occurs as limiting form in an univariant point position. The sphere-packing symbol of J_2^* is 4/6/c2 and the density is $\frac{3}{4}$ of that of a P-lattice namely $52.3 \cdot 3/4 = 39.2\%$. Within the parameter range $\frac{3}{16} < x < \frac{1}{4}$ the sphere-packing type do not change.

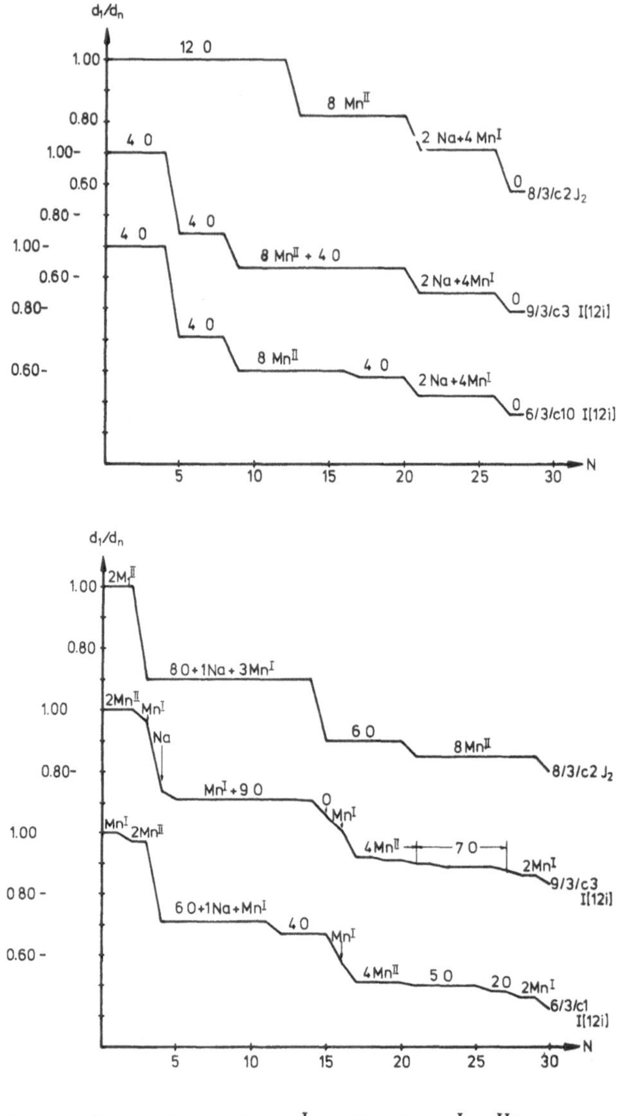

Fig. 11. The distance relations diagram for Mn^I and O in $NaMn_3^IMn_4^{II}O_{12}$ (= $NaMn_7O_{12}$) in an ideal perovskite structure with the 24 oxygen in 8/3/c2 J_2 "Bauverband" and in I[12 i], with two sphere-packing conditions 9/3/c3 and 6/3/c10.
[Parameters are found in Fig. 8a; the d_1 are the shortest distances between the central ion (*upper part* for Mn^I, *lower part* for O) and those atoms forming the first coordination shell; the d_n are the distances between the central atom and atom N of the coordination shell; the closest relation exists between the observed parameters (y = 0.3132, z = 0.1828) and I[12 i] as 6/3/c10 with y = 0.3170; z = 0.1830 in Im3 24(g) 0yz]

With x = 3/16 there exist the sphere packing 8/3/c3-D[6 o] listed in Table 4 with a density of 46.9%. This sphere packing consists of octahedra around T with common vertices similar to perovskite, but the "Bauverbände" are quite different.

If one regards the structures of $CaCu_3Ti_4O_{12}$ and $NaMn_7O_{12}$ as a distorted perovskite (index 8) and pyrochlore and $RbNiCrF_6$ (Babel 1974) as distorted and defect structures of the CaF_2-type, the former belongs to the F-family and pyrochlore and $RbNiCrF_6$ to the P-family with respect to their anion "Bauverbände". This would help to reduce the number of classes and families and give the chance to assign deformed and defect structure types with non-cubic space-groups symmetry to the cubic families, main- and subclasses, because it gives an information about possible degrees of deformation.

Symbolism for Net Structures

Symbols of structure types which may be considered as sequences of 2-dimensional hexagonal nets can be partly derived from those of cubic structure types, if the nets are arranged perpendicular [111]. According to Hermann a hexagonal closed packed layer may be symbolised by H,
a graphite layer by G,
and a Kagomé net by N.

In Fig. 12 these three different nets are shown with the P points in the voids of the G-net and the N-net marked by squares. The axis of the unit cell has to be enlarged by the factor $\sqrt{3}$ for the G-net and by a factor of 2 for the N-net with respect to the primitive lattice, formed by centering voids. The first case is indicated by the subscript V, the second case by the subscript 2, actually it should be 22 for a_1, a_2. Therefore the following equations can be written:

$$G + P = H_V = H^G$$
$$N + P = H_{22} = H^N$$

The cubic sequence of H-layers (F-type, Cu) is symbolised by (c); (h) stands for a hexagonal sequence. Other sequences are symbolised by combinations of (c) and (h) (Jagodzinski, 1949).

Cu	F =	(c) H	3-layer type
Mg		(h) H	2-layer type
Am		(hc) H	4-layer type
Sm		(hhc) H	9-layer type

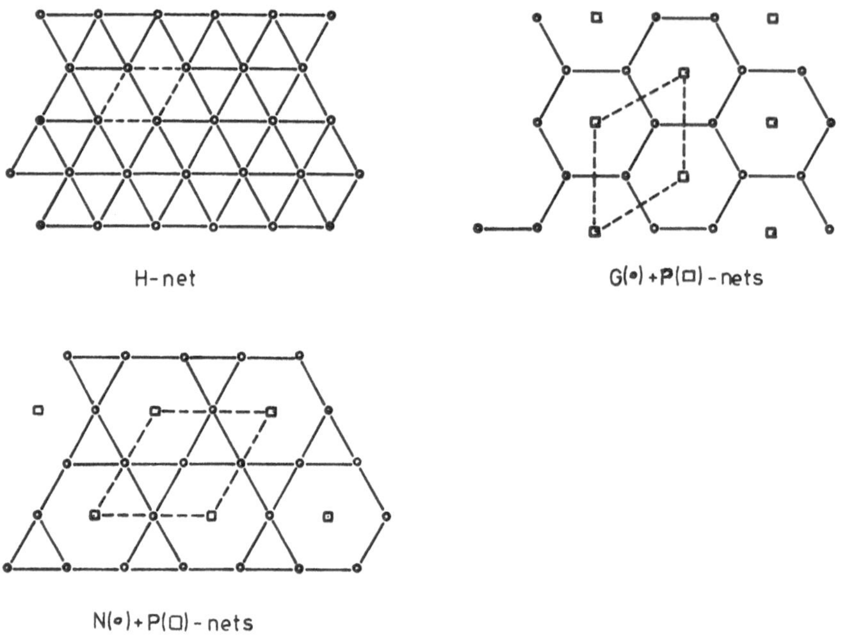

Fig. 12. A hexagonal closed packet net (*upper left*) and two splittings in a graphite layer G (*upper-right*) and a Kagomé net N (*lower left*).
[The axis of the unit cells (*dotted lines*) have the ratios 1: $\sqrt{3}$: 2]

This symbolism holds also for AB_3 compounds in an ordered arrangement:

Cu_3Au (ord.)	$J + P = F^P = (c)[N + P] = (c)\ H^N$	3-layer type
Ni_3Sn	$(h)\ H^N$	2-layer type
Ni_3Ti	$(hc)\ H^N$	4-layer type
Al_3Pu	$(hcc)\ H^N$	6-layer type

One or two primes may indicate whether the octahedral or the tetrahedral voids are occupied. In this way the "hexagonal" analogs of NaCl, ZnS, perovskite etc. may be described. Symbols for 2-layer types analogous to D and T are possible as well as for the structures of micas and clay minerals with the Si_4O_{10}-layer. In the structure of elements and intermetallic compounds other sequences of closed packed layers are found, where the atoms of the next layer do not lie above the middle of trigons but above the middle of edges between two atoms. This is the case in a tetragonal deformed F with $a_1 : c : a_2 = \sqrt{3} : 1 : \sqrt{3}$ forming a 2-layer type; an orthorhombic deformation of the cubic D in Fddd with $a : b : c = \sqrt{3} : 1 : 2 \cdot \sqrt{3}$ results in a 4-layer type and the ^+Q complex in $P6_222$ in orthohexagonal description with $a\sqrt{3}:a:c = \sqrt{3}:1:\frac{3}{2}\sqrt{3}$ gives a 3-layer type. The small letters f, d, and q in parentheses are used to characterise these different sequences.

Examples are:

Pa	(f)H	2-layer type
γ-Pu	(d)H	4-layer type

and with the ordered arrangement AB_2 and the graphite-layer splitting:

$MoSi_2$	$(f)H^G$	2-layer type
$CrSi_2$	$(q)H^G$	3-layer type
$TiSi_2$	$(d)H^G$	4-layer type

Combinations of this sequences and 1-dimensional disorder may be found too.

Summary and Outlook

Most of these homogeneous cubic connection patterns or "Bauverbände" have been described by the sphere-packing symbol (Fischer, 1973/1974), the centers of the voids within the sphere packing and the coordination polyhedra. They are designated by capital letters (symbols of lattic complexes) and by symbols for coordination polyhedra. Heterogeneous "Bauverbände" are symbolised by combinations of symbols for homogeneous Bauverbände.

By means of the coordination shells of the I lattice complex the structure types of the I-, P- and F-families can be derived. A list of main- and subclasses of the I-family of index 1, 8, 27, 64 and 512, which are realised in crystal structure are given. Some distinct structure types are described as subtraction structures of the P or addition structures of the F-family.

In addition representatives of other main- and subclasses of homogeneous and heterogeneous "Bauverbände" are mentioned with respect to the degrees of freedom for their coordinate distortions. The consequences of such distortions with respect to the change of coordination polyhedra in the "Bauverbände" have been discussed.

Starting from the rhombohedral description of simple cubic structure types with the aid of 2-dimensional nets it is possible to symbolise also the trigonal and hexagonal variants of structure types. In the future it will be tried to find the distorted structure types with non-cubic symmetry, which are homeotypic to the cubic ones down to the triclinic crystal system, where the distortion and splitting of the cubic "Bauverbände" may be of very different kind. Besides this search of homeotypes, new "Bauverbände" have to be studied which are typical for the tetragonal, hexagonal etc. crystal system. The knowledge of these homogeneous "Bauverbände" will be of some help for further arrangements of families, main- and subclasses.

Acknowledgement. This paper is a consequence of work, teaching and cooperation with Laves and Hermann. Detailed discussions over a long period with Burzlaff and Fischer and their work on theoretical crystallography, in the last decennium together with Koch, were a further step forward. Fischer and Koch wrote the first programs to analyse the coordination polyhedra in structure types via the "Wirkungsbereiche" (or Wigner-Seitz cell).

Gerlich continued to complete the program for the calculation of the coordination polyhedra and built up the possibility for a dialogue between a graphical display and the computer IBM 370/145, which could be acquired by the financing of the government of Hessen with the support of the president of the University, Zingel. The federal government via the "Institut für Dokumentationswesen" (Dr. Dr. Cremer) guaranteed a budget for several years. Finally I have to thank Prof. Dr. v. Schnering for inviting me to join a weekend seminar in Nov. 1977 with several experts to whom I am indebted for several stimulating remarks and advises.

References

Babel, D.: Z. allg. anorg. Chem. *387*, 161–178 (1972)
Bärnighausen, H.: Acta Cryst. *31*, 3 (1975)
Bokii, G.B., Smirnova, N.L.: J. Structural Chem. *4*, 683–694 (1963)
Burzlaff, H., Fischer, W., Hellner, E., Niggli, A.: Z. Krist. *139*, 246–251 (1974)
Donnay, J.D.H., Hellner, E., Niggli, A.: Z. Krist. *120*, 364–374 (1964)
Donnay, J.D.H., Hellner, E., Niggli, A.: Z. Krist. *123*, 255–262 (1966)
Fischer, W.: Z. Krist. *133*, 18–42 (1971)
Fischer, W.: Z. Krist. *138*, 129–146 (1973)
Fischer, W.: Z. Krist. *140*, 50–74 (1974)
Fischer, W., Burzlaff, H., Hellner, E., Donnay, J.D.H.: Space groups and lattice complexes. Monogr. 134. Washington: National Bureau of Standards 1973
Fischer, W., Koch, E.: Z. Krist. *140*, 324–330 (1974)
Fischer, W., Koch, E.: Acta Cryst. *A32*, 225–232 (1976)
Frank, F.C., Kasper, J.S.: Acta Cryst. *11*, 184–190 (1958)
Frank, F.C., Kasper, J.S.: Acta Cryst. *12*, 483–499 (1959)
Geller, S.: Acta Cryst. *12*, 944–945 (1959)
Göttinger Isomorphiebesprechung, Die Chemie (Angewandte Chemie) N.F. *57*, 29–56 (1944)
Hellner, E.: Acta Cryst. *19*, 703–712 (1965)
Hellner, E.: Z. anorg. allg. Chem. *421*, 37–40, 41–48, 49–60 (1976a, b, c)
Hellner, E.: Z. anorg. allg. Chem. *437*, 60–72 (1977)
Hermann, C.: Z. Kristallogr. *113*, 142–154 (1960)
Jagodzinski, H.: Acta Cryst. *2*, 201–207 (1949)
Koch, E.: Z. Krist. *140*, 75–86 (1974)
Kripyakevich, P.I.: J. Structural Chemistry *4*, 1–35 (1963)
Laves, F.: Z. Krist. *23*, 203–265 (1930)
Laves, F.: Die Chemie (Angewandte Chemie) N.F. *57*, 30–33 (1944)
Laves, F.: Crystal structure and atomic size. In: Theory of alloys. Cleveland, Ohio: Amer. Soc. Met. 1956
Laves, F.: Kristallstruktur und Kristallchemie von Elementen und metallischen Verbindungen. In: Beiträge zur Physik und Chemie des 20. Jahrhunderts. Braunschweig: F. Vieweg u. Sohn 1959

Lima-de-Faria, J.: Z. Kristallogr. *122*, 359–374 (1965)
Lima-de-Faria, J., Figueiredo, M.O.: J. Solid State Chem. *16*, 7–20 (1976)
Loeb, A.L.: J. Solid State Chem. *1*, 237–267 (1970)
Moore, P.B.: Classification and retrieval of inorganic structures, 3. Europ. Crystallogr. Meeting, Zürich 1976
Morral, F.R., Westgren, A.: Svensk. Kim. Tidskr. *46*, 153–156 (1934)
Niggli, A.: Z. Krist. *133*, 473–490 (1972)
Niggli, P.: Geometrische Kristallographie des Diskontinuums. Leipzig: Gebr. Bornträger 1919
O'Keeffe, M., Anderson, St.: Acta Cryst. *A33*, 914–923 (1977)
Parthe, E.: Crystal chemistry of tetrahedral structures. New York: Gordon and Breach Science Publishers 1964
Pearson, W.B.: Handbook of lattice spacings and structures of metals. Vol. II. Oxford: Pergamon Press 1967
Pearson, W.B.: The crystal chemistry of metals and alloys. New York: Wiley-Interscience 1972
Pearson, W.B., Shoemaker, C.B.: Acta Cryst. *B25*, 1178–1183 (1969)
Pólya, G.: Acta Math *68*, 146–254.
Propach, V.: Z. anorg. allgem. Chem. *435*, 161–171 (1977)
Reinen, D., Propach, V.: Inorg. Nucl. Chem. Letters 7, 569–572 (1971)
Sakamoto, Y., Takahasi, U.: J. Sci. Hiroshima Univ. *A35*, 1–51 (1971)
Samson, St.: Acta Cryst. *23*, 586–600 (1967)
Samson, St.: San Francisco and London: W.H. Freeman and Company 1968
Samson, St.: Acta Cryst. *B28*, 930–935 (1972a)
Samson, St.: Acta Cryst. *B28*, 936–945 (1972b)
Schubert, K.: Berlin, Göttingen, Heidelberg, New York: Springer-Verlag 1964
Schubert, K.: Structure & Bonding *33*, 139 (1977)
Shoemaker, C.B., Shoemaker, D.P.: Monatshefte f. Chemie *102*, 1643–1666 (1971)
Smirnova, N.L., Mezhueva, L.S.: J. Struct. Chem. *7*, 533–541 (1966)
Smirnova, N.L.: Sov. Phys. Cryst. *16*, 46–52 (1971)
Smirnova, N.L.: Sov. Phys. Cryst. *20*, 777–782 (1975a)
Smirnova, N.L.: Sov. Phys. Cryst. *20*, 320–325 (1975b)
Smirnova, N.L.: Sov. Phys. Cryst. *21*, 468–471 (1976)
Smirnova, N.L.: Sov. Phys. Cryst. *22*, 310–312 (1977)
Smirnova, N.L., Kurazhkovskaya, V.S., Below, N.V.: Sov. Phys. Crystallogr. *22*, 48–51 (1977a); Sov. Phys. Crystallogr. *22*, 169–171 (1977b); Sov. Phys. Crystallogr. *22*, 172–175 (1977c)
Wells, A.F.: Acta Cryst. *7*, 535–544, 545–554, 842–848, 849–853 (1954)
Wells, A.F.: V. Acta Cryst. *8*, 32–36 (1955)
Wells, A.F.: Acta Cryst. *9*, 23–28 (1956)
Wells, A.F., Sharpe, R.R.: Acta Cryst. *16*, 857–871 (1963)
Wells, A.F.: Acta Cryst. *18*, 1965, 894–900 (1965)
Wells, A.F.: Acta Cryst. *B24*, 50–57 (1968)
Wells, A.F.: Acta Cryst. *B25*, 711–719 (1969)
Wells, A.F.: *B28*, 711–713 (1972)
Wells, A.F.: Oxford: Clarendon Press 1975
Wondratschek, H.: Z. Krist. *143*, 160–170 (1976)
Zimmermann, H., Burzlaff, H.: Z. Krist. *139*, 252–267 (1974)
Zvesdinskaya, L.V., Smirnova, N.L., Belov, N.V.: Sov. Phys. Cryst. *22*, 439–442 (1977)

Formula Index

Formula	Page	References in Strukturberichte resp. Structure Reports
Ag	83	*1*, 14; *18*, 282
Ag_3AuTl_2, petzite	98	*23*, 154
Ag_8Ca_3	123	*29*, 33
Ag_5Hg_8	122	
Ag_3PO_4	124	*1*, 395; *4*, 175; *11*, 380
$AgSbO_3$	90, 95	*6*, 120
$AlAsO_4$	83	*3*, 423, 428; *4*, 166
$AlLi_3N_2$	93, 94	*11*, 20
16 $Al(OH, F)_3 \cdot 6 H_2O$	89, 100, 93, 95	*11*, 281
α-Al_2O_3, corundum	83	α-(rh) *1*, 240, 257, 777
Al_3Pu	130	*20*, 19
Am	129	*20*, 25; *24*, 42; *26*, 32; *27*, 39
Au	82	*1*, 14, 38, 748
Au_4Al	107	*8*, 13
$AuCu_3$	65, 85	*1*, 486
$AuIn_2$	82	*5*, 54
As	65	*1*, 25, 57; *34A*, 13
BPO_4	83	*3*, 92, 426; *23*, 341
Ba_2CaWO_6	102	*15*, 292
$Ba_{0.5}IrO_3$	122	*40*, 213
$Ba_2Ge_4S_{10}$	90	*39A*, 22
$Ba_4Sb_3LiO_{12}$	65, 102	*40A*, 184
$BaSiF_6$	88	*9*, 196
$BeSO_4 \cdot 4 H_2O$	88	*2*, 91; *34A*, 298
BiF_3	84, 91, 86	*2*, 22
α-BiF_3	85, 92, 86	*12*, 163
$Bi_3GaSb_2O_{11}$	122	*39A*, 215
Bi_4Si_3, eulytite	65, 98, 125	*2*, 122, 522; *31A*, 227
C, diamond	64, 65, 83, 86, 98, 85, 90	*1*, 19
CO_2	83, 84, 103, 104	*1*, 150, 226
$Ca_3Al_2(OH)_{12}$	69, 124	*3*, 129, 494; *8*, 218; *33A*, 273
$Ca_3Al_2(SiO_4)_3$, garnet	85, 69, 106, 109	*1*, 363, 411; *2*, 520; *22*, 500; *31A*, 228
CaB_6	80, 88, 106	*2*, 37; *18*, 62
$CaCu_3TiO_{12}$	101, 113, 126, 127	Reinen & Propach (1971)
CaF_2, fluorite	91, 65, 82, 100, 126, 127, 98, 86, 94	*1*, 148; *30A*, 266

Formula	Page	References in Strukturberichte bzw. Structure Reports
Ca_3PI_3	102	*40A*, 157
$CaTiO_3$, perovskite	78, 65, 101, 126, 128, 129, 115, 117, 119	*1*, 300
Cd	83	*1*, 19
$Cd(CN)_2$	97	*10*, 92
$CoAs_3$, skutterudite	65, 126	*1*, 232
CoMnSb	93, 94	*38A*, 14
Co_9S_8	93, 96	*4*, 26; *27*, 169
CoU	86, 98, 89	*13*, 94
$CrSi_2$	131	*3*, 14, 628
Cr_3Si	77, 84, 123, 86	*3*, 628; *20*, 78
$CsBe_2F_5$	77, 65, 107, 111	*38A*, 197
CsCN (kub. + rh. Mod.)	87	*9*, 138
CsCl	78, 86, 87, 65, 85, 97, 88	*1*, 74
Cu	65, 83, 84, 129	*1*, 13
Cu_9Al	122	*3*, 57, 590; *30A*, 3; *33A*, 72
Cu_3Au	130	*1*, 486
$CuCl_2 \cdot 2\,H_2O$	87, 97	*4*, 13
β-$CuFeS_2$, chalcopyrite	93, 96	*1*, 279; *2*, 48; *3*, 385; *10*, 122; *20*, 93; *23*, 141; *39A*, 51
$Cu_{18}Fe_{16}S_{22}$, thalnakite	96	*38A*, 79
Cu_2O	86, 97, 98, 92	*1*, 153
CuSbMg	86, 91	*8*, 28
Cu_3SbS_2, tetrahedrite	80, 85, 93, 96, 115, 117, 119	*1*, 335; *29*, 15; *37A*, 4
$Cu_{41}Sn_{11}$	100, 122	Arnberg, L., Jonsson, A. and Westman, S. ACS *A30*, 187–192 (1976) Booth, M. H. et al. A.C.*B33*, 30–36 (1977)
Cu_3VS_4	86, 98, 92	*3*, 94
Cu_5Zn_8 (γ-brass-type)	99, 122	*1*, 499, 533; *2*, 693; *33A*, 72
β-F	124	*29*, 229
α-Fe	65, 83	*1*, 16
$FeCO_3$	83	*1*, 295
$Fe_{13}Ge_3$	86, 92	*27*, 209
Fe_2PO_4		*20*, 303
FeS_2, pyrite	82, 84, 103, 104	*1*, 150; *34A*, 97
FeSi	80, 103	*2*, 13; *10*, 63; *11*, 146; *28*, 28
Fe_3W_3C	80, 89, 93, 94	*3*, 71; *22*, 889; *32A*, 45

Formula	Page	References in Strukturberichte resp. Structure Reports
$Ge_{38}P_8Br_8$	126	*39A*, 186
$46\,H_2O \cdot 6M$, $46\,H_2O \cdot 8M$, ($M = Cl$, SO_2, CH_3Cl, H_2S ...) clathrates	65, 126	*16*, 376
$(Hg_2)_3Cl_3O_2H$, eglestonite	93, 94	TMPM *23*, 1976, 105
$Hg(MO \cdot Cl_8)Cl_6$	71	*37A*, 201
Hg_4Pt	86, 91	*17*, 227
Hg_3TeO_6	98, 69, 106, 109	*27*, 635
InCl	101, 102	*31A*, 96
Ir_3Sn_7	79, 81	*11*, 136
$K[AlSi_2O_6]$, leucite, HT	80	*33A*, 458
$KAl(SO_4)_2 \cdot 12\,H_2O$	70, 84, 91	*3*, 108; *30A*, 376
$KBrO_3$	88	*2*, 68
KGe	124	*24*, 145; *26*, 84
$K_2Mg_2(SO_4)_3$, langbeinite	65, 98, 125	*21*, 362
α-KPF_6	102	*15*, 164
K_2PtCl_6	102	*1*, 429; *39A*, 178
$KSbF_6$	65	*13*, 285; *15*, 155
$KSbO_3$	65, 77, 122, 123	*11*, 443
K_3TlF_6	101, 103	*21*, 267
$K_2Zn(CN_4)$	95	*1*, 424; *31A*, 117
α-La	65	*2*, 171; *20*, 173
LaBi	82	*5*, 45
LaOF-β-	91	*15*, 167
$La_4Re_6O_{19}$	77, 102	*33A*, 353
$LiFe_5O_8$	93, 94	*13*, 245; *16*, 244
$LiGeN_3$	94	*17*, 169
Li_2O	78	*3*, 20, 283
Li_3PO_4	83	*24*, 397
$Li_{22}Pb_5$	85, 100	*22*, 158
Li_2TiO_3	101, 102	*9*, 177
Li_7VN_4	93, 96	*23*, 174
Mg	65, 83, 129, 78	*1*, 16
$MgAlO_4$, spinel	93, 83, 100, 95	*1*, 350; *2*, 474; *16*, 244; *33A*, 272
$MgCa(CO_3)_2$	83	*1*, 303; *23*, 419
$MgCd_3$	65	*16*, 38
$Mg_3Cr_2Al_{18}$	81, 90, 98	*18*, 14; *22*, 8
$MgCu_2$	78, 83, 100	*1*, 490

Formula	Page	References in Strukturberichte resp. Structure Reports
Mg_6MnO_8	102	*17*, 391; *18*, 524
$MgNi_2$	83	*3*, 31
Mg_3P_2	78, 86, 98, 92	*2*, 40
Mg_2SiO_4	83	*1*, 352; *2*, 121; *15*, 306; *39A*, 348
$MgZn_2$	83	*1*, 180, 228, 564; *3*, 311
α-Mn	77, 123	*1*, 65, 756; *2*, 192; *20*, 149; *32A*, 98; *34A*, 101
β-Mn	69, 107, 111	*2*, 3, 192; *3*, 221; *20*, 150
γ-Mn	83	*1*, 24, 63
$(MnMg)_9[AsMnO_{18}]Cl$, magnussonite	100	P. Moore, private communication
Mn_2O_3	93, 94	*2*, 38; *27*, 497
$Mn_{23}Th_6$	96, 99, 100	*16*, 113
$MoSi_2$	131	*1*, 219, 740, 783
NH_4CN	87	*9*, 137
$N_2H_6Cl_2$	91	*11*, 269
NH_4NO_3 (Pm3m)		*2*, 69
$(P\bar{4}2_1m)$	87	*2*, 69; *24*, 418
$NaAlSiO_4$, α-carnegieite	83, 93, 94	*2*, 158
$Na(Ca, K)(AlSiO_4)_3$, hauyne	84, 113, 77, 119, 121	*3*, 166; *33A*, 483; *26*, 537
NaCl, halite	64, 65, 82, 130, 88, 84, 85, 101, 78	*1*, 72
$Na_4Cl[AlSiO_4]_3$, sodalithe	80, 113, 115, 117, 121	*2*, 150, 562; *32A*, 490
Na_3MgCl, northurpite	81, 90, 100, 95	*2*, 80
NaK_2AlF_6, elpasolite	100, 102	*26*, 309; *38A*, 203
$Na_6Mg_2SO_4(CO_3)_4$, (tychite)	90, 93, 95	*2*, 98
$NaMn_7O_{12}$	128, 101, 113, 126, 127, 129, 115, 117, 119	*39A*, 247
Na_6PbO_4	93, 96	*39A*, 215
$2\,Na_2SO_4 \cdot NaCl \cdot NaF$, sulfohalite	93, 96	*3*, 188, 470; *33A*, 377
$NaSbO_3$	77, 122	*11*, 443
NaSeH	82	*7*, 4
NaTl	79, 86, 85, 98, 89	*2*, 733; *3*, 19
$NaZn_{13}$	81	
$NbF_{2,5}(Nb_6F_{15})$	102	*30A*, 261
NbO	101	*8*, 123; *21*, 222; *31A*, 122
Nb_3Sn	123	*19*, 243; *38A*, 134
Ne	82	*2*, 201, 22, 210
NiAl	83	*1*, 76, 147, 488; *12*, 13
$NiCu_3Sn$	100	
$NiCu_9Sn_3$	122	*17*, 156

Formula	Page	References in Strukturberichte bzw. Structure Reports
Ni_3Sn	130	*20*, 29
$NiSnCl_6 \cdot 6\,H_2O$	88	*2*, 102
NiTe	83	*1*, 87, 138, 765
$NiTe_2$	83	*6*, 166; *30A*, 76
Ni_3Ti	130	7, 14, 98; *13*, 341
$NiTi_2$	83, 89, 93, 95	*22*, 889; *27*, 296; *28*, 20
γ-O	124	*39A*, 125
Pa	87, 97, 131	*16*, 133; *18*, 269; *23*, 212
$PbMg_2$	82	*1*, 150, 229, 567; *3*, 20, 314
$Pb(NO_3)_2$	91	*1*, 304; *2*, 73
PbS	82	*1*, 74, 125, 131
α-Po	65, 84	*31A*, 61
PtCu	85	*1*, 485, 517; *9*, 76
β-$PtHg_4$	97	*17*, 227
$Pt(NH_3)_2Br_2Pt(NH_3)_2Br$	89	*11*, 483
Pt_3O_4	86, 98, 91	*8*, 147; *33A*, 270
γ-Pu	65, 131	
Pu_2C_3	93	*26*, 92
$R_2O_2X_7$,	127	Koppite *2*, 58
koppite-pyrochlore	89, 95, 93	Pyrochlore *33A*, 190
R = Ca, Ce, Na, K, Th, Y		
Q = Nb, Fe, Ti, Ta		
X = O, OH, F		
$RbNiCrF_6$	81, 89, 100, 126, 127, 95	*38A*, 205
$RbNO_3$	91	*15*, 277
$Rb_4O_6[Rb_4(O_2)_3]$	125	*8*, 150
ReO_3	85, 101	*2*, 31
α-$RhBi_4$	69, 106, 109	*20*, 54
SbO_3, senarmontite	93, 95	*1*, 245; *9*, 164
Sb_2O_4	95, 100	*3*, 54, 358; lt. *7*, 15 zu streichen
SiF_4	97, 104, 106	*2*, 37, 305; *18*, 353
SiO_2, cristobalite	83, 85, 93, 95	*1*, 169, 202
SiO_2, α-quartz	83	*1*, 166
SiO_2, β-quartz	83	*1*, 166
SiO_2, tridymite	83	*1*, 171, 203; *28*, 120; *3*, 25; *24*, 462 (low. tridymite)
Sm	129	*17*, 252; *18*, 273; *20*, 173
$(Sn_2F_3)Cl$	80, 107, 111	*40A*, 132
Sn_3S_4	125	Bärnighausen, Acta Cryst., in press

Formula	Page	References in Strukturberichte resp. Structure Reports
β-$SnWO_4$	65	*38A*, 271
$Sr_3Al_2O_6$	103	v. Schnering, private communication
Th_6Mn_{23}	99, 100, 96	*16*, 113
Th_3P_4	98, 125	*7*, 15
TiC	82	*1*, 74, 144; *4*, 6; *8*, 49; *12*, 50
$TiSi_2$	65, 131	*12*, 95
TlOF	93, 96	*38A*, 221
$TlPd_3O_4$	84, 81	
Tl_7Sb_2	80, 77, 65, 123, 84, 85, 106, 88, 99	*1*, 488; *3*, 175
$TlTa_2O_6$	95	*41*, 243
Tl_3VS_4	80, 104, 106	*29*, 91
UB_{12}	81, 84	*12*, 34; *30A*, 30
U_2F_9	123	*11*, 290; *32A*, 157
UH_3	81	*11*, 128; *15*, 81
U_4S_3	101	*8*, 113, 151
α-W	78, 84, 86, 87	*1*, 15; *23*, 262
β-W (W_3O)	77, 65, 123, 84, 86, 97, 124	*2*, 6, 191; *3*, 219; *18*, 259
WAl_{12}	81, 97, 112, 87	*18*, 30
W_3O (β-W)	77, 65, 123, 84, 86, 97, 124	*2*, 6, 191; *3*, 219; *18*, 259
Zn	83	*1*, 19, 41; *2*, 169
$Zn(BrO_3)_2 \cdot 6\,H_2O$	91	*4*, 63
$Zn_4O(BO_2)_6$	85, 113	*26*, 490
ZnS, sphalerite	64, 83, 130, 86	*1*, 76, 127
ZnS, wurtzite	86	*1*, 78, 128
ZrOS	103	*11*, 302
ZrO_2, baddelyite	91	*4*, 9; *23*, 341; *30A*, 307
Alum, $KAl(SO_4)_2 \cdot 12\,H_2O$	70, 84, 91	*3*, 108; *30A*, 376
Baddelyite, ZrO_2	91	*4*, 9; *23*, 341; *30A*, 307
α-Carnegieite, $NaAlSiO_4$	86, 93, 94	*2*, 158
Chalcopyrite, β-$CuFeS_2$	93, 96	*1*, 279; *2*, 48; *3*, 385; *10*, 122; *23*, 141; *39A*, 51
Clathrates, $46\,H_2O \cdot 6$ M or $46\,H_2O \cdot 8$ M M = Cl, So_2, CH_3, Cl, H_2O ...	65, 126	*16*, 376
Corundum (α-Al_2O_3)	83	*2*, 43
Cristobalite, SiO_2	78, 83, 85	*1*, 169

Formula	Page	References in Strukturberichte resp. Structure Reports
Diamond, C	64, 65, 83, 85, 86, 90	*1*, 19
Eglestonite $(Hg_2)_3Cl_3O_2H$	93, 94	TMPM *23*, 1976, 105
Elpasolithe, NaK_2AlF_6	102	*26*, 309; *38A*, 203
Eulytite, $Bi_4Si_3O_{12}$	65, 98, 125	*2*, 122, 522; *31A*, 227
Fluorite, CaF_2	65, 82, 100, 127, 86, 91, 94	*1*, 148; *30A*, 266
Garnet, $Ca_3Al_2(SiO_4)_3$	65, 69, 85, 124, 106, 109	*1*, 363, 411; *2*, 520; *22*, 500; *31A*, 228
Halite, NaCl	64, 65, 82, 130, 84, 101, 78	*1*, 72
Hauyne, Na(Ca, K) $(AlSiO_4)_3$	84, 113, 119, 121	*3*, 166; *33A*, 483; *26*, 537
Koppite-pyrochlore	89, 93, 95	Koppite *2*, 58
$R_2O_2X_7$	127	Pyrochlore, *33A*, 190
R = Ca, Ce, Na, K, Th, Y		
Q = Nb, Fe, Ti, Ta		
X = O, OH, F		
Langbeinite, $K_2Mg_2(SO_4)_3$	65, 98, 125	*21*, 362
Leucite (HT), $K[AlSi_2O_6]$	80	*33A*, 458
Magnussonite $(Mn, Mg)_9[As_6MnO_{18}]Cl$	100	P. Moore, private communication
Northurpite, $Na_3MgCl(CO_3)_2$	90, 100, 95, 81	*2*, 80
Petzite, Ag_3AuTl_2	98	*23*, 154
Pyrite, FeS_2	84, 82, 103	*1*, 150; *34A*, 97
Pyrochlore s. koppite		
α-Quartz, SiO_2	83	*1*, 166
β-Quartz, SiO_2	83	*1*, 166
Senarmonite	93, 95	*1*, 245; *9*, 164
Skutterudite, $CoAs_3$	65, 126	*1*, 232
Sodalite, $Na_4[AlSiO_4]_3Cl$	80, 84, 113, 77, 115, 117, 121	*2*, 150; *32A*, 490
Sphalerite, ZnS	64, 83, 86	*1*, 76
Spinel, $MgAl_2O_4$	83, 100, 93 95	*1*, 350, 417; *33A*, 272
Sulfohalite, $2\,Na_2SO_4 \cdot NaCl \cdot NaF$	93, 96	*3*, 118, 470; *33A*, 377
Thalnakite, $Cu_{18}Fe_{16}S_{22}$	96	*38A*, 79
Tetrahedrite, Cu_3SbS_2	93, 85, 96, 115, 117, 119	*29*, 15; *37A*, 4
α-Tridymite, SiO_2	83	*1*, 171; *28*, 120
Tychite, $Na_6Mg_2SO_4(CO_3)_4$	90, 93, 100, 95	*2*, 98
Wurtzite, ZnS	83	*1*, 78

Polyhalogen Cations

Jacob Shamir

Department of Inorganic and Analytical Chemistry, The Hebrew University of Jerusalem, Jerusalem, Israel

Table of Contents

1 Introduction 143
1.1 Syntheses 144
1.1.1 Acid-base Reactions 144
1.1.2 Oxidative Reactions 144
1.1.3 Substitution Reactions 145
1.2 Structure Determination 145
1.2.1 Single Crystal X-Rays 145
1.2.2 Vibrational Spectra 146
1.2.3 ^{19}F–NMR 147
1.2.4 NQR and Mössbauer Measurements 148
2 Iso-Polyhalogen Cations 148
2.1 X_3^+, X_5^+ and X_7^+ Cations 148
2.1.1 Vibrational Spectra 150
2.1.2 Absorption Spectra 151
2.1.3 NQR Measurements 151
2.1.4 Cl_3^+ 153
2.1.5 Br_3^+ 153
2.1.6 I_3^+ 154
2.1.7 I_5^+ 155
2.1.8 I_7^+ 155
2.2 X_2^+ and X_4^{+2} Cations 155
2.2.1 Single Crystal X-Rays 157
2.2.2 Vibrational Spectra 158
2.2.3 Absorption Spectra 159
2.2.4 Magnetic Moments 160
2.2.5 Br_2^+ 160
2.2.6 I_2^+ 160
2.2.7 I_4^{+2} 161
3 Hetero-Polyhalogen Cations 162
3.1 XY_2^+ Symmetric Cations 163
3.1.1 Single Crystal X-Rays 165
3.1.2 Vibrational Spectra 167
3.1.3 Absorption Spectra 173
3.1.4 ^{19}F–NMR 174
3.1.5 NQR Measurements 174

3.1.6 Mössbauer Effect 176
3.1.7 ClF_2^+ 176
3.1.8 BrF_2^+ 177
3.1.9 IF_2^+ 178
3.1.10. ICl_2^+ 178
3.1.11 IBr_2^+ 179
3.2 X_2Y^+ and YXZ^+ Cations 180
3.2.1 Vibrational Spectra 181
3.2.2 Absorption Spectra 184
3.2.3 NQR Measurements 184
3.2.4 Cl_2F^+ 186
3.2.5 I_2Cl^+, I_2Br^+ and $BrICl^+$ 186
3.3 XY_4^+ Cations 188
3.3.1 Single Crystal X-Rays 191
3.3.2 Vibrational Spectra 192
3.3.3 ^{19}F–NMR 194
3.3.4 NQR Measurements 195
3.3.5 Mössbauer Effect 195
3.3.6 ClF_4^+ 195
3.3.7 BrF_4^+ 196
3.3.8 IF_4^+ 196
3.4 XY_6^+ 197
3.4.1 X-Rays 198
3.4.2 Vibrational Spectra 199
3.4.3 ^{19}F–NMR 201
3.4.4 Mössbauer Effect 203
3.4.5 ClF_6^+ 204
3.4.6 BrF_6^+ 205
3.4.7 IF_6^+ 206
4 References 207

The existence of polyhalogen cations seems well established. Some cation-anion interaction takes place, but nevertheless their structure is basically ionic. In this review, the preparations, properties and structural data of these compounds are summarized and discussed.

Most of these cations include fluorine and can be considered as derivatives of halogen fluorides. The number of such cations not containing fluorine is fewer. This is a reflection of the fact that less inter-halogen compounds without fluorine are known, compared to the numerous halogen fluorides being in existence.

Various physical methods have been used for structure determination, the most useful ones being Raman spectroscopy, single crystal X-rays and F–NMR.

Most possible cations have indeed been synthesized and identified. It seems that this part of the subject has been studied rather extensively. On the other hand, less is known about the chemistry of these compounds, their chemical reactions and possible uses as reagents. This kind of research, as well as completion of various still missing data, will probably be of future interest.

1 Introduction

Cation formation is not restricted, as normally observed, to association with metallic species only, but it can also occur among non-metals and even in most typical ones, such as halogens, which are more noted for polyhalogen anion formation.

The most recent review dealing with this subject is that of Gillespie and Morton[70]. Since then more compounds of this type have been isolated and additional information has been obtained, justifying a new review.

Polyhalogen cations can be subdivided into two major types. The one type, *iso-polyhalogen* cations, containing several atoms of only one particular halogen, were called in the previous review[70] halogen cations. The second type, *hetero-polyhalogen* cations, contain several atoms of at least two different halogens, and were called inter-halogen cations[70].

It is rather convenient to treat these cations as being derived from self-ionic dissociation of their parent molecules, namely neutral halogens or inter-halogen compounds. These ionic dissociations can be expressed in some of the following formulae. In case of inter-halogens

$$2\,XY_n \rightleftharpoons XY_{n-1}^+ + XY_{n+1}^-$$

and if n = 1, then

$$3\,XY \rightleftharpoons X_2Y^+ + XY_2^-$$

or

$$4\,XY \rightleftharpoons X_3Y_2^+ + XY_2^-$$

In case of elemental halogens, several ionic equilibria can be assumed, according to:

$$3\,X_2 \rightleftharpoons X_3^+ + X_3^-$$

or

$$4\,X_2 \rightleftharpoons X_5^+ + X_3^-$$

or

$$5\,X_2 \rightleftharpoons 2\,X_2^+ + 2\,X_3^-$$

Polyhalogen cations of all these types have indeed been prepared and studied. All these compounds are usually highly hygroscopic and hydrolyze easily, and must therefore be handled and kept under dry conditions.

Some of these polyhalogen cations were found to react as halogenation agents.

1.1 Syntheses

1.1.1 Acid-base Reactions

The equilibria described in the previous paragraph can easily indicate that the addition of a Lewis type acid, halogen acceptor, to the parent compound would shift the self-ionization and bring about an increase in the formation of the hetero-polyhalogen cation, whereas the addition of a base would increase the formation of the counter ion, the polyhalogen anion.

This, indeed, is one of the major procedures for synthesizing these compounds. Reactions of this kind are, however, limited to those in which the formal oxidation state of the central halogen does not change being equal both in the neutral parent compound and in the polyhalogen cation. It is therefore applicable to the formation of hetero-polyhalogen cations derived from neutral inter-halogen compounds, according to the general formula

$$XY_n + MY_m \rightarrow XY_{n-1}^+ + MY_{m+1}^-$$

(XY_n = interhalogen, MY_m = Lewis acid, e.g. halides of B, P, As, Sb, etc.).

Such reactions were performed by direct interaction of the reactants. It is best to use the more volatile reactant in excess, which can then easily be pumped off, after completion of the reaction, leaving behind the pure product.

Sometimes it is preferable to perform such reactions in solution and pumping off the solvent at the end of the reaction.

1.1.2 Oxidative Reactions

Another synthetic procedure, applied mostly to the formation of iso-polyhalogen cations from the elemental halogens or from diatomic interhalogens require an oxidizer in addition to the Lewis acid, such as: I_2 + oxidizer $\rightarrow I_2^+$ or I_3^+. In some cases the Lewis acid itself being in excess can also act as an oxidizer.

An oxidative reaction is also used in direct formation of polyhalogen cations, in solutions of strong acidic media. However, solutions of this kind can also be formed by dissolution of the appropriate ionic solid in the particular solvent, being dissociated into the proper polyhalogen cation.

Related oxidative syntheses were also used in the preparations of cations, whose parent compounds have not been isolated due to their instability, such as ClF_6^+, BrF_6^+ or IBr_2^+.

1.1.3 Substitution Reactions

Some polyhalogen cations have been formed from already existing related cations by halogen substitution according to

$$2\,X_2Y^+ + Z_2 \rightarrow 2\,X_2Z^+ + Y_2\,.$$

1.2 Structure Determination

It is now generally accepted that these compounds are basically of ionic structure, although there are evidences that in most of them some cation-anion interaction of various degrees does exist.

The early structural studies were mostly limited to solutions performing such measurements as conductivity, cryoscopy and magnetic moments. These have shown that ionic species are present, thus influencing the number of species in solutions and their conductances. Based on such measurements, reactions were formulated and the composition of the species involved elucidated.

Electronic absorption spectra in the UV and visible regions were also a useful tool in identifying the presence of different, yet closely related, species.

In recent years most useful structural information was obtained from additional physical measurements, many of them performed on solids, such as single crystal X-rays, vibrational spectra, F–NMR, NQR and Mössbauer data.

1.2.1 Single Crystal X-Rays

The X-ray data, providing information regarding bond lengths and bond angles, are thus extremely valuable in deciding whether a particular compound is indeed ionic. In addition, the structure and symmetry of the species involved can be determined. In $XY^+_{n-1}\,MY^-_{m+1}$ solids, several determined distances shed light on the structural problem. In general the closest X–Y bond-length is limited to within the cation. The next closest X ... Y contact is that of a halogen of the anion bridging to the cation, forming some kind of cation-anion interaction. The ratio of the X ... Y contact to the X–Y bond-length serves as a good indicator to the extent of cation-anion interaction.

Similarly, the M–Y distances in the anion itself do also differ, although to a smaller extent. Usually the M–Y distance in the bridging position is slightly larger than the M–Y bond-length in the terminal positions. Since these differences in the anion are much smaller than in the cation, their ratio is not as significant to indicate the extent of cation-anion interaction.

Based on these atomic distances, it is possible to establish the coordination numbers in the cation and anion, which in most cases support an ionic model.

As expected, this method is limited to those compounds of which single crystals can be grown.

1.2.2 Vibrational Spectra

Vibrational spectra are also highly important and have become very useful, especially since laser Raman instrumentation has become available. Prior to that, obtaining Raman spectra of solids was more a kind of art than a standard scientific procedure. The availability of different lasers also offers a variety of exciting lines to be used in Raman studies enabling also the study of colored and even very dark, deeply colored materials. In some cases very useful data was obtained from Resonance Raman spectra.

On the whole, Raman spectra are more frequent in the literature than IR spectra.

The fundamental vibrations of many of these compounds are in the low frequency region, which is more easily accessible and observed in the Raman spectra than in the far IR. In addition, the sampling of these highly reactive and easily hydrolyzed materials is easier in Raman studies than in far IR ones. Glass and transparent fluoropolymers can serve as containers, both for solids and solutions, for Raman studies. On the other hand, the sampling for IR studies is more complicated, requiring window materials which are both transparent in the particular IR region and at the same time are not attacked by these reactive materials and thus destroying the sample.

Silver chloride windows are useful, being transparent until about 400 cm^{-1}, and plastic materials, such as polyethylene are used for the lower frequency region. The usual accepted techniques for obtaining spectra of solids, namely mulling with an organic oil or pressing into an alkali halide disc, are not very useful since they react with these reactive samples.

The interpretation of the observed vibrational spectra is not always a simple straight-forward matter. The difficulties stem mostly from the fact that both the cation and anion in such compounds are complexed and thus each ionic part has its own particular vibrations. However, both cationic and anionic frequencies occur usually in the very same spectral region, making the assignment of separate cationic and anionic vibrations quite difficult.

This is a general problem when observing a common property for both ionic parts and then attempting to assign the features for the cation and anion separately.

It is therefore necessary to assign the frequencies of one ion, and then by elimination to assign those of the counter ion. It seems best to attempt first the assignment of the anionic vibrations, which can be compared with the frequencies for such anions observed in compounds with simple cations.

In these related compounds with simple cations such as alkali metals, NO^+, NO_2^+ or even tetra-alkyl ammonium cations, no cationic spectra exist in case of a monoatomic cation, whereas a non-interfering spectra with that of the anionic spectrum is observed, in a different spectral region at much higher frequencies, in case of polyatomic cation.

In such compounds, most likely, pure ionic structures prevail allowing the anion to be undistorted and of high symmetry. Frequently the anion involved is of MX_6^- composition with octahedral $-O_h$ symmetry, thus rather simple spectra with few lines are expected. Unfortunately, this is not the case in compounds with polyhalogen cations, as far more numerous lines than expected for the anion are actually observed. This increase in number of observed lines can be a result of various factors, causing the distortion of the anions and lowering their symmetry. These are due to cation-anion interactions, halogen bridging, site symmetry lowering and factor group splitting. All these effects cause the lifting of degeneracies of vibrations and thus increase the number of observed lines.

Spectral studies can sometimes be performed in solution in which simpler and independent, non-interacting species may be expected. These are then of higher symmetry than the related ones in the solid resulting in simpler spectra of fewer lines, thus assisting in the proper assignments.

In solutions, polarization measurements of their Raman spectra are possible, thus enabling the assignment of polarized symmetric vibrations and depolarized asymmetric ones.

On the other hand, cationic species existing in the solid do not always exist as independent species in solutions, but do solvate with some solvent molecules. This then causes spectral shifts in the observed frequencies, which are in some cases larger than expected by the mere change of phase and thus in turn complicate the identification and assignment of the shifted frequencies, both in the original solid or in solution.

In many cases, force constants and force fields were calculated, which were then used in turn to support the proposed vibrational assignments. In some cases, only approximate calculations could be performed due to the lack of sufficient experimental data, as the number of observed lines is less than the number of parameters required for complete solution of the proper equations. Therefore, such results have to be dealt with carefully and in some cases these calculations failed to give meaningful results, as was shown by later studies[28,29,69,81].

1.2.3 ^{19}F–NMR

F–NMR data provided very important contributions for structural determinations of halogen fluoride cations, which were mostly studied in solution.

The observed results include chemical shifts, spin-spin coupling constants and isotope splittings.

All fluorine chemical shifts mentioned in this review have been standardized to an external reference of $CFCl_3$.

1.2.4 NQR and Mössbauer Measurements

Some useful information also enabling the calculation of charge distributions was obtained from NQR measurements of other halogen nuclei, except fluorine.

Mössbauer effect studies were also performed, contributing to structural considerations.

The experimental data will be discussed in groups of similar structures with increasing oxidation state, and within each group with increasing mass of the central atom. In each group, general and common physical properties of the various species will be discussed together and then the non-common properties of each cation.

Tables are presented in each section to summarize most of the experimental data.

2 Iso-Polyhalogen Cations

In the halogen group, like in other groups of the periodic table, the heavier elements present more metallic and electropositive properties. It is therefore no wonder that the early observed species which were easiest to be prepared and identified were those containing iodine only. Iodine most readily releases an electron thus stabilizing iodine cations. In the early studies it was assumed that the iodine cation exists as a monoatomic one $-I^+$. But in later studies this assumption could not be confirmed and it was proven that no monoatomic halogen cations exist; instead only polyatomic ones, such as I_3^+, I_2^+ were identified. These earlier studies were reviewed by Arotsky and Symons[2], whereas the later critics were included in the review by Gillespie and Morton[70] and therefore will not be repeated here.

2.1 X_3^+, X_5^+ and X_7^+ Cations

All the halogens except fluorine form X_3^+ cations in the solid state, whereas in solution only Br_3^+ and I_3^+ have been identified. The I_5^+ cation has been isolated in the solid and observed in solutions. The existence of I_7^+ cation was derived only from a thermal study. Preparations of the compounds and their properties are summarized in Table 1.

The X_3^+ cations seem to be of a bent structure of C_{2v} symmetry. Although no single crystal X-ray data has as yet been reported, other physical measurements support this proposed structure for these cationic species.

Powder X-ray data were reported for $I_3^+AlCl_4^-$ and $I_5^+AlCl_4^-$, but not indexed[97].

Various possible structures for I_5^+ have been mentioned[61].

According to the VSEPR, Gillespie-Nyholm rules, the triatomic iso- and heteropolyhalogen cations, X_3^+, XY_2^+, X_2Y^+ and YXZ^+ are expected to be of bent struc-

Table 1. X_3^+, X_5^+ and X_7^+ compounds and properties

	M.P.	Preparations and properties	Physical measurements
Cl_3^+			
AsF_6^-		$Cl_2F^+AsF_6^- + Cl_2 \xrightarrow{-78°} Cl_3^+AsF_6^-$ [69]; $Cl_2 + ClF + AsF_5 \xrightarrow{-78°} Cl_3^+AsF_6^-$ [71]. Yellow solid.	Raman [69]
SbF_6^-		$Cl_2 + ClF$ in HF/SbF_5 [70].	Raman [70]
Br_3^+			
AsF_6^-	Sublime 30–50° [76]	$O_2^+AsF_6^- + \frac{3}{2}Br_2 \rightarrow Br_3^+AsF_6^-$ [76]; $Br_2 + BrF_3$ (or BrF_5) + $AsF_5 \xrightarrow{-50°} Br_3^+AsF_6^-$ [76]. Brown-chocolate solid, decompose 70 °C[76].	IR, UV, F–NMR, diamagnetic [76]
SO_3F^-		$Br_2 + BrOSO_2F$ condensed on cold window [144];	IR [144];
		Br_3^+ in solutions of HSO_3F [76], and super acids [64,72], brown color.	Raman [64], UV [64,72,76], conductivity [64,72]
I_3^+			
$AlCl_4^-$	45 ± 1° [94,97]	$I_2 + ICl + AlCl_3 \rightarrow I_3^+AlCl_4^-$ [94,97]; $7I_2 + KIO_3 + 8AlCl_3 \rightarrow 5I_3^+AlCl_4^-$.[97] Shiny black platelets, dark brown in thin sections.	UV [97], NQR [94,97]; thermal analysis [97], powder X-rays [97]
AsF_6^-		$3I_2 + 3AsF_5 \xrightarrow{\text{in } AsF_3} 2I_3^+AsF_6^- + AsF_3$ [100], brown black.	
SbF_6^-		$3I_2 + 3SbF_5 \xrightarrow{\text{in } SO_2} 2I_3^+SbF_6^- + SbF_3$ [71], dark black.	
SO_3F^-	92° [5) 101.5° [43]	$3.33I_2 + S_2O_6F_2 \rightarrow 2I_3^+SO_3F^-$ [5,43], dark brown.	diamagnetic[43]
		I_3^+ in solutions of H_2SO_4 [3,5,61,71,78,91], oleum [3] and HSO_3F [5,63]. Deep red brown.	Raman [71]; UV[63,78]; cryoscopy, conductivity [61,63]
I_5^+			
$AlCl_4^-$	50–50.5° [94,97]	$2I_2 + ICl + AlCl_3 \rightarrow I_5^+AlCl_4^-$ [94,97], greenish black needles, dark brown-red in thin sections [97].	UV, NQR, thermal analysis, powder X-rays [97]
		I_5^+ in solutions of H_2SO_4 [61] and HSO_3F [63]	UV [63]; cryoscopy, conductivity [61,63]
I_7^+			
SO_3F^-	90.5° [43]	$I_2 + S_2O_6F_2 \rightarrow I_7^+SO_3F^-$ [43]	

tures; the central atom with an sp^3 hybridization forming a tetrahedral symmetry $- AX_2E_2$. On the other hand, the triatomic polyhalogen anions are expected to be linear; the central atom with an sp^3d hybridization forming a trigonal bipyramid and the two halogens occupying the axial positions $- AX_2E_3$ (X-halogen; E-lone pair of electrons).

2.1.1 *Vibrational Spectra*

Raman spectra have been recorded for all the three X_3^+ cations, whereas an IR bond was reported only for Br_3^+. All these are summarized in Table 2.

Table 2. Vibrational frequencies in cm^{-1} and force constants in mdyn/Å for X_3^+ cations

	ν_1 (A_1) sym. str.	ν_2 (A_1) bending	ν_3 (B_1) asym. str.	f (stretch)	d (bend)	Ref.
Cl_3^+	493, 485	225	508	2.5	0.36	69)
Br_3^+	290	–	288	–	–	72, 144)
I_3^+	207 P	114 P	233	1.7	0.31	71)

P = polarized. All frequencies are Raman lines except the ν_3 of Br_3^+ observed in the IR.

In the Raman spectra of Cl_3^+ and I_3^+ compounds, three lines were assigned to the cation. This eliminates a symmetrical linear structure $C_{\infty v}$, having a center of symmetry. In this case the mutually exclusive selection rules for IR and Raman activities would apply and only one Raman line would be expected. The observed spectra are in agreement with a bent structure of C_{2v} symmetry. In this case, all the three expected vibrations are both Raman and IR active. Force constants were calculated using a simple valence force field and a bond angle of 100° for Cl_3^+ and 114° for I_3^+ are in good agreement with the observed frequencies[69,71]. A splitting was observed in the ν_1 of Cl_3^+ at 493 and 485 cm^{-1}.

In a solution of ClF in excess SbF_5 and HF, only one broad line at 500 cm^{-1} was observed in addition to the complicated spectrum of polymeric $SbF_6(SbF_5)_n^-$ anion. This line can be assigned to ν_1 of Cl_3^+, whereas no lines could be assigned to Cl_2F^+. The possible equilibrium, $2\,Cl_2F^+ \rightarrow ClF_2^+ + Cl_3^+$ may explain the observation of a Cl_3^+ vibration[69].

The observed Raman spectrum of I_3^+ in H_2SO_4 solution was found to have the 207 and 114 cm^{-1} lines polarized and these are therefore assigned respectively as ν_1, the symmetrical stretch and ν_3, the bending mode[71].

In the case of Br_3^+, only one Raman line at 290 cm^{-1} was observed in solution of Br_3^+ in super acid[72].

On condensing vapors of a solution of Br_2 in $BrOSO_2F$ onto a cold window of a low temperature IR cell, an IR spectrum was observed, all of whose lines were assigned to the SO_3F^- anion, and the only additional extra line at 288 cm^{-1} was assigned to Br_3^+[144]. It seems therefore that the separation between the symmetric and asymmetric stretching frequencies of Br_3^+ is very small and these vibrations are not resolved.

2.1.2 Absorption Spectra

The absorption spectra of these colored materials were recorded and serve as an important tool in identifying the different species present. The I_3^+ solutions are deep red-brown, whereas the related I_2^+ solutions are of intense blue. The Br_3^+ solutions are brown, whereas the related Br_2^+ solutions are deep cherry-red.

Spectra were observed in various solvents, such as H_2SO_4, oleum, HSO_3F, HSO_3CF_3 and super-acid, as well as in solids mulled in fluorolube oil. In all of these, only slight changes were noticed by changing environments as summarized in Table 3.

Table 3. Absorption bands in nm of X_3^+ and I_5^+ cations in solutions or mulls (letters in parentheses are relative intensities)

					Ref.
I_3^+					
Oil mull	490 (m)	410 (w) (sh)	375 (s)		97)
H_2SO_4	462–459 (m)			290 (s)	4,5,78,142)
HSO_3F	472–467 (m)			305–297 (s)	5,63,142)
HSO_3CF_3	462 (m)			295 (s)	142)
Br_3^+					
HSO_3F			375, 310		76)
Super-acid			375 (sh), 300		64,72)
I_5^+					
Oil mull	485 (m)		350 (m)	245 (s)	97)
H_2SO_4	450 (m)		330		3)
HSO_3F	450 (m)		345 (s), 270 (m)	240 (s)	63)

2.1.3 NQR Measurements

I_3^+ and I_5^+ compounds of $AlCl_4^-$ were studied by NQR of the halogens[94,97]. The observed frequencies and their assignments are listed in Table 4. The ^{35}Cl resonances indicate that $I_3^+AlCl_4^-$ shows dimorphism. The α-form, metastable at room temperature, was formed when the compound was melted and then thermally shocked by quenching it in liquid nitrogen. However, on allowing the compound to stand at room temperature for about one year, a harder β-form was formed, showing a different NQR spectrum.

The ^{35}Cl resonances for $AlCl_4^-$ observed in the region of 10.74–11.16 MHz appear to be complete and do support a relative ionic formulation. It is known from other ionic $AlCl_4^-$ compounds that their average resonances are in the region of 10.6–11.3 MHz, whereas in compounds with more coordinated structure or dimeric $Al_2Cl_7^-$ anion, the resonances shift to higher frequencies of the 12.6–13.0 MHz region.

Table 4. NQR data in $I_n^+AlCl_4^-$ [97]

Compound	ν_Q MHz	Assignment	Average (MHz)	Range (MHz)
β-$I_3^+AlCl_4^-$	9.960 10.874 11.046 11.091	^{35}Cl in $AlCl_4^-$	10.74	1.13
α-$I_3^+AlCl_4^-$	10.129 10.590 11.090 11.452	^{35}Cl in $AlCl_4^-$	10.82	1.32
	308.6	^{127}I central (ν_1)		
	415 428	^{127}I terminal (ν_1)		
	527	^{127}I central (ν_2)		
$I_5^+AlCl_4^-$	10.988 11.086 11.124 11.449	^{35}Cl in $AlCl_4^-$	11.16	0.46
	400 ± 5	^{127}I		

Iodine resonances were observed in the α-I_3^+ $AlCl_4^-$ form, but none were observed in the β-I_3^+ $AlCl_4^-$ form. Both observed transitions of the central iodine allowed the computation of the asymmetry parameter $\eta = 0.373$. This parameter is related to the bonding angle according to $\eta = -3 \cos\phi$ and therefore the bonding angle in I_3^+ was found to be equal to 97.1°.

A single bonding model for a C_{2v} symmetry was used. Two equivalent bonding orbitals on the central iodine, generated from the s, p_x and p_y atomic orbitals, overlap pure p orbitals of the terminal atoms. The remaining orbitals of the central iodine as well as all other atomic orbitals were considered as being filled and non-bonding. From the observed frequencies for the central atom and each terminal atom, the σ-bond population sum was calculated as 1.89 electrons per bond, compared to the expected value of 2.0 electrons.

Charge distributions in I_3^+ were calculated for each atom separately. In addition, the charge of the central iodine was also calculated by deducing the calculated charges of the terminal atoms from the total cationic charge of I_3^+, being equal to + 1. This latter value is listed in parentheses.

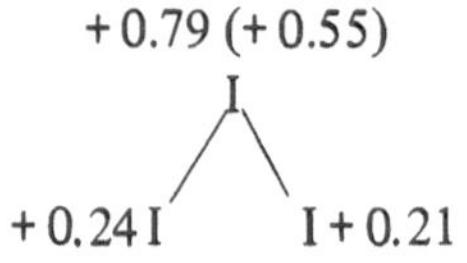

This difference in the values of the charge of the central iodine, which were computed by two different methods and which neglected any contribution to the bonding of s orbitals of the terminal atoms as well as some anion coordination to the central iodine, reflects the limitations of the model used.

The single resonance of ^{127}I observed in $I_5^+AlCl_4^-$ seems more characteristic of the terminal iodine, but no detailed analysis was possible due to the incompleteness of the observed data.

2.1.4 Cl_3^+

The Cl_3^+ cation was obtained either by a substitution reaction[69] according to

$$2\,Cl_2F^+AsF_6^- + Cl_2 \rightarrow 2\,Cl_3^+AsF_6^- + F_2$$

or by a three-body reaction, which might be explained to include a polarization step[71], which has also been observed in other preparations, namely:

$$Cl_2 + \overset{+\delta}{Cl}—\overset{-\delta}{F} + AsF_5 \rightarrow Cl_3^+AsF_6^-$$

2.1.5 Br_3^+

The Br_3^+ cation was obtained either by the reaction of a dioxygenyl compound with bromine

$$2\,O_2^+AsF_6^- + 3\,Br_2 \rightarrow 2\,Br_3^+AsF_6^- + 2\,O_2$$

or by direct reaction of $Br_2 + AsF_5$ in excess BrF_3 or BrF_5 [76], acting as oxidizers.

In the IR spectrum of $Br_3^+AsF_6^-$, a line was observed[76] at 709 cm^{-1} which was assigned to ν_3 of AsF_6^- anion.

The F–NMR of the same compound dissolved in excess BrF_5 shows a broad signal at 62 ppm, which was also assigned to the AsF_6^- anion[76]. Thus both these measurements indicate that the compound is of ionic structure.

The IR spectrum of the product obtained by condensation of a solution of Br_2 in $BrOSO_2F$ was interpreted[144] as that of $Br_3^+SO_3F^-$.

Solutions containing the Br_3^+ cation were obtained in super-acid by dissolving elemental bromine and adding $S_2O_6F_2$ as oxidizer[64,72]. Care must be taken to have the concentrations of bromine and the oxidizer in the correct ratio of $Br_2/S_2O_6F_2 = 3$. Only in this ratio will the reaction

$$3\,Br_2 + S_2O_6F_2 + 2\,H_2SO_3F^+ \rightarrow 2\,Br_3^+ + 4\,HSO_3F$$

yield the Br_3^+ cation. Other cations will be formed if different ratios will be involved.

Dissolving the solid $Br_3^+AsF_6^-$ in fluorosulphuric acid also yielded Br_3^+ in solution[76]. The presence of Br_3^+ in solutions was supported by conductivity data and Raman spectra[64,72]. The UV spectrum was also recorded and indicated that Br_3^+ was also formed in solution of $BrOSO_2F$ in super-acid[64]. This seems to result from a disproportionation reaction according to

$$4\,BrOSO_2F = Br_3^+ + Br(OSO_2F)_3 + SO_3F^-$$

2.1.6 *I_3^+*

$I_3^+AlCl_4^-$ was prepared by reacting directly equimolar quantities of I_2, ICl and $AlCl_3$ in an evacuated tube. In this case, too, a three-body reaction including a polarization step seems to follow according to

$$I_2 + \overset{+\delta}{I}—\overset{-\delta}{Cl} + AlCl_3 \rightarrow I_3^+AlCl_4^-$$

However, no reaction takes place when refluxing these reactants in solutions of CCl_4, $CHCl_3$ and CH_3CN[97].

The preparation of $I_3^+SO_3F^-$ by direct reaction of iodine with $S_2O_6F_2$ was performed in dry air at a pressure of 1 atm., in order to reduce the evaporation of iodine from the reaction mixture. The formed solid lumps were broken up into fine powder by cooling them to $-190°$. The excess iodine was pumped off at room temperature[5].

Salts of AsF_6^- and SbF_6^- were also prepared by direct reactions of iodine with MF_5 (M = As, Sb), without any additional oxidizer[71,100]. In these cases, the Lewis acids – MF_5 also act as oxidizers according to

$$3\,I_2 + 3\,MF_5 \rightarrow 2\,I_3^+MF_6^- + MF_3\ .$$

Solutions of I_3^+ were prepared by dissolving solid ionic compounds[5,71], or by direct preparations in solutions, such as: dissolving iodosyl compounds[91] or N-iodo organic compounds[78] in sulphuric acid; dissolving elemental iodine with an oxidizer such as HIO_3[61], I_2O_5, KIO_3 or $IOSO_4$ in sulphuric acid or oleum[3]. The oxidation by HIO_3 proceeds according to

$$HIO_3 + 7\,I_2 + 8\,H_2SO_4 \rightarrow 5\,I_3^+ + 3\,H_3O^+ + 8\,HSO_4^-$$

I_3^+ can also be prepared in solution, similarly to Br_3^+, by oxidation with $S_2O_6F_2$[63]. However, in this case no super-acid is necessarily required as a solvent and the reaction can already take place in a less acidic medium of fluoro-sulphuric acid, in accordance with

$$3\,I_2 + S_2O_6F_2 \rightarrow 2\,I_3^+ + 2\,SO_3F^-$$

The stabilization of Br_3^+ in solution requires most acidic solutions such as superacid and fluorosulphuric acid, whereas I_3^+ can already be stabilized in milder acidic media, such as sulphuric acid. This is explained by the fact that Br_3^+ is more electrophilic than I_3^+, as it is of smaller size and has a more highly concentrated charge. Solutions of I_3^+ were used to iodinate organic compounds[78,91]. $I_3^+AsF_6^-$ was found to react[100] as follows:

$$[Se(C_2F_5)]_2 + 2\,I_3^+AsF_6^- \rightarrow I_2 + 2\,[Se(C_2F_5)I_2]^+\,AsF_6^-$$

2.1.7 I_5^+

$I_5^+AlCl_4^-$ was prepared[97] similarly to $I_3^+AlCl_4^-$ according to

$$2\,I_2 + ICl + AlCl_3 \rightarrow I_5^+AlCl_4^-$$

The existence of I_5^+ in solution was supported by conductivity and cryoscopic measurements[61,63]. These solutions are prepared similarly to those of I_3^+ solutions using HIO_3 as oxidizer, but in the presence of larger quantities of iodine, allowing the reaction $I_3^+ + I_2 \rightarrow I_5^+$ to take place. Similarly the ratio of $I_2/S_2O_6F_2 = 5:1$ is decisive[63] in the formation of I_5^+ according to

$$5\,I_2 + S_2O_6F_2 \rightarrow 2\,I_5^+ + 2\,SO_3F^-$$

Different ratios of these reactants result in the formation of I_3^+ or I_2^+. Solutions of I_5^+ were also used in organic iodinations[78,91].

2.1.8 I_7^+

The $I_7^+SO_3F^-$ is the only reported compound with an I_7^+ cation. It has been identified in a thermal study of the I_2–$S_2O_6F_2$ system. No properties except the melting point were reported[43].

2.2 X_2^+ and X_4^{+2} Cations

Only bromine and iodine have been shown to form diatomic cations, both in the solid state and in solutions. Preparations of these compounds and their properties are summarized in Table 5.

Table 5. X_2^+, X_4^+ and X_2^{+2} compounds and properties

	M.P.	Preparations and properties	Physical measurements
Br_2^+			
$Sb_3F_{16}^-$	69° [48]	$9\,Br_2 + 2\,BrF_5 + 30\,SbF_5 \rightarrow 10\,Br_2^+Sb_3F_{16}^-$ [48]; scarlet solid.	Single crystal X-rays[48,51], Raman[51], Magnetic[48].
		Br_2^+ in solution of super-acid[64,72]; cherry red.	Raman[17,64,72]; UV[51,64,72]; conductivity, magnetic[72].
I_2^+			
$Sb_2F_{11}^-$	122°–123° [44,84] 110°–115° [108]	$I_2 + SbF_5(ex)$ or $I_2 + 2\,SbF_5 \xrightarrow{\text{in } IF_5} I_2^+Sb_2F_{11}^-$ [84]; $2\,I_2 + 5\,SbF_5 \xrightarrow{\text{in } SO_2} 2\,I_2^+Sb_2F_{11}^- + SbF_3$ [44]; blue crystals.	Single crystal X-rays, Raman, UV[44], IR, magnetic[84];
$Ta_2F_{11}^-$	120° [84]	$I_2 + 2\,TaF_5 \xrightarrow{\text{in } IF_5} I_2^+Ta_2F_{11}^-$ [84], blue solid	IR, UV [84]
		I_2^+ in solutions of: HSO_3F [3,62,63,67], SbF_5 [84], IF_5 [6,84], $H_2S_2O_7$ and oleum [3,8,66,78], blue	Raman[67]; UV[6,62,63,66,84]; conductivity[6,8,63,78,84,101,103]; cryoscopy[63,66], magnetic[6,62,63].
Cl_4^+		Cl_2 in $SbF_5 \xrightarrow{77\,°K} Cl_4^+$ [46]	ESR[46]
I_4^{+2}		in solution of HSO_3F[65], deep red-brown.	UV, conductivity, cryoscopy, magnetic[65].

Based on ESR data, it was claimed that Cl_2^+ and ClF^+ species exist in solutions of ClF or ClF_3 in SbF_5[98,99]. In later studies[31,73] these data could not be confirmed and it was assumed that the original observed ESR signals were due to chlorine oxides or oxyfluorides which were formed by reaction of the highly reactive fluorinating compounds, attacking the glass or quartz apparatus. No vibrational line around $640\ cm^{-1}$, as expected for Cl_2^+, was observed. Instead a new line at $774\ cm^{-1}$ was observed and assigned to Cl–O bond[73].

It was claimed that distilling Cl_2 onto SbF_5 at 77 °K caused the observation of an ESR signal only in the solid state. This was interpreted to result from the Cl_4^+ cation[46].

In general, it is more difficult to stabilize X_2^+ cations compared to X_3^+ ones. The stabilization of diatomic cations requires stronger acidic environments. Whereas I_3^+ is already stable in sulphuric acid, the I_2^+ cation is almost completely disproportionated in this solvent to I_3^+ and I^{+3} species[61,62,63]. The I_2^+ cation can, however, already be stabilized in oleum and fluorosulphuric acid. If the concentration of iodine in such solutions is increased, some disproportionation will also take place even in these stronger acidic media, as indicated by the appearance of the I_3^+ absorption band at 305 nm and that of $I(OSO_2F)_3$ at 210 nm[63,66].

The bromine cations are even more electrophilic than the iodine ones. Therefore, still stronger acidic solutions are required for their stabilization. The Br_2^+ cation, similar to I_2^+, is less stable than the Br_3^+ cation. Unlike I_3^+, the Br_3^+ is not stable in sulphuric acid, but is stable in fluorosulphuric acid, whereas the Br_2^+ can only be stabilized in super-acid[64,72].

2.2.1 Single Crystal X-Rays

Single crystal X-ray studies were performed on both $Br_2^+Sb_3F_{16}^-$ [48,51] and $I_2^+Sb_2F_{11}^-$ [44]. Some of the data are summarized in Table 6.

Table 6. Single crystal X-rays data of X_2^+ compounds

	$Br_2^+Sb_3F_{16}^-$ [51]	$I_2^+Sb_2F_{11}^-$ [44]
	Monoclinic	*Monoclinic*
a	13.58 ± 0.02 Å	13.283(5) Å
b	7.71 ± 0.01 Å	8.314(3) Å
c	14.33 ± 0.02 Å	5.571(2) Å
β	93.7 ± 0.2°	103.75(2)°
V	1497 Å^3	597.5 Å^3
D_{calc}	3.68 g cm^{-3}	3.92 g cm^{-3}
z	4	2
X–X, bond length	2.15 Å	2.557(4) Å
X ... F, closest contact	2.86 Å	2.89 Å
$r_{Sb-F(terminal)}$	1.83 Å	1.85 Å
$r_{Sb-F(bridging)}$	2.10; 1.97 Å	2.00 Å
$R = r_{bridg}/r_{ter}$	1.11	1.08

The data support an ionic formulation since the cation-anion contacts are rather long, although shorter than the sum of van der Waals radii, indicating very small cation-anion interaction. Actually, the shortest I ... F contact is almost the largest cation-anion contact observed in a series of $Sb_2F_{11}^-$ compounds with the following cations of decreasing contacts: $SbCl_4^+$, I_2^+, XeF_3^+, Br_2^+, BrF_4^+. This decreasing distance indicates increase of the acidity of the cations[44].

The terminal fluorine in the anion closest to the cation is in *trans* position to the internal bridging fluorine in the anion. The Sb–F (terminal) distance is similar to those observed in SbF_6^- anions, like in $ClF_2^+SbF_6^-$ or $XeF^+SbF_6^-$, namely 1.84 and 1.83 Å, respectively.

The $Sb_2F_{11}^-$ anion itself is symmetric, namely the distances between both antimony atoms and the bridging fluorine are the same: 2.00 Å. This is in contrast to many other salts with this dimeric anion in which these distances are non-equal. The cation-anion interaction causes a movement of the closest fluorine to the cation towards the central atom in it, as well as some charge donation to the cation. This in turn involves some loss of electron density at the Sb atom of that half closer to the cation thus becoming more positive and resulting in stronger attraction of the internal bridging fluorine to this antimony than to the antimony in the second half of the $Sb_2F_{11}^-$. As a result, the two internal Sb–F (bridging) distances become different, bringing about an asymmetric dimeric anion.

2.2.2 *Vibrational Spectra*

Most of the reported spectra are Raman and resonance Raman data of the deeply colored Br_2^+ and I_2^+ cations. In spite of the fact that no stable Cl_2^+ cation has been observed, its fundamental vibration is known from the electronic spectrum of the gas phase. The fundamental frequencies of X_2^+, force constants and bond lengths are summarized in Table 7. The stretching frequencies of all the X_2^+ are higher than those of the neutral X_2. Similarly, their force constants are larger and their bond lengths are shorter. All these are consistent with the removal of an electron from an anti-bonding π^* orbital in the formation of the cation[44,71] with a formal bond order of 1.5. This also explains the fact that the stretching frequencies of X_2^+ are higher than the average stretching frequencies of X_3^+ (Table 2), which have a formal bond order of only 1.0.

Table 7. Vibrational stretching frequencies in cm^{-1}, force constants in mdyn/Å and bond lengths in Å in X_2 and X_2^+ species (based on Ref. [71])

	Cl_2	Cl_2^+	Br_2	Br_2^+	I_2	I_2^+
Stretching frequency	554	646	317	368	213	238
Force constant	3.16	4.29	2.38	3.05	1.70	2.12
Bond length	1.98	1.89	2.27	2.13	2.66	2.56
Ref.	71)	44,71)	48,71)	48,51,72)	71)	44,67)

The vibration frequency in solid $Br_2^+Sb_3F_{16}^-$ was observed[48] at 368 cm^{-1} as expected, slightly higher than in solution at 360 cm^{-1} [48,51]. A resonance Raman spectrum was observed in solution with the overtones at: 716, 1070, 1430, 1780 and 2130 cm^{-1} [17,72]. The intensity of the fundamental of Br_2^+ at 360 cm^{-1} was compared to the intensity of a solvent peak at 400 cm^{-1}. It was noticed that varying the exciting lines changed the relative intensity of the Br_2^+ fundamental. This indicates that the observed spectrum is indeed a resonance Raman spectrum. It was found that this intensity decreases in the following order:

$$I_{514.5\,nm} > I_{488\,nm} > I_{632.8\,nm}$$

The electronic absorption band of Br_2^+ is at 510 nm, and therefore the highest resonance Raman intensity is obtained using as excitation line the 514.5 nm in the absorption band.

In the solid $I_2^+Sb_2F_{11}^-$, the fundamental stretching frequency is the same as in solution at 238 cm^{-1} [44,67]. As resonance Raman takes place, two overtones were observed in the solid[44] and five in dilute solution in HSO_3F[67]. The electronic absorption of I_2^+ at 640 nm has a molar extinction of $\epsilon = 2600$ and its value at 632.8 nm is $\epsilon = 2580$, thus the resonance Raman was observed using the 632.8 nm and 611.8 nm lines of an He–Ne laser.

The IR reported spectra[84] for $I_2^+Sb_2F_{11}^-$ and $I_2^+TaF_6^-$ include only lines of SbF_5 and SbF_6^- as well as TaF_5 and TaF_6^-, but no cationic vibrations were observed, which is as expected of a symmetric diatomic species, to have only a Raman active line, and no IR active one.

2.2.3 Absorption Spectra

The cherry-red solution of Br_2^+ in super-acid has a strong absorption at 510 nm[51,64,72]. In SbF_5 solution the band is reported to be at 500 nm[51].

The absorption spectra of the blue I_2^+ have been studied in different solvents such as $H_2S_2O_7$, oleum, IF_5, HSO_3F and SbF_5. A reflectance spectrum from the solid was also reported[84]. The data is listed in Table 8, showing only slight changes in different environments.

The deep red-brown solution of I_4^{+2} formed from dimerization of I_2^+ has an entirely different spectrum[65]. The very intense absorption band at 357 nm is not found in I_2^+, I_3^+ or I_5^+ spectra and is therefore indicative of the presence of an entirely different species.

Table 8. Absorption bands in nm of I_2^+ and I_4^{+2} (letters in parentheses are relative intensities)

					Ref.
I_2^+					
$H_2S_2O_7$	640 (s)	510–470 (w)	420–410 (w)		8,66,103)
Oleum	648–637 (vs)	507–490 (s)	417–406 (s)		4,6,62,78)
HSO_3F	647–640 (s)	500–490	410–406		4,62,63,78)
IF_5	641 (s)	508	418		6)
SbF_5	633 (s)	510 (w)	408 (w)		84)
$I(SbF_5)_2$ (solid)	635 (s)	510 (w)	412 (w)		84)
I_4^{+2}					
HSO_3F		470 (m)		357 (vs), 290 (s)	65)
Br_2^+					
Super-acid		510			64,72)
SbF_5		500			51)

2.2.4 Magnetic Moments

X_2^+ cations have an odd electron and are therefore expected to be paramagnetic. The magnetic moment of $Br_2^+Sb_3F_{11}^-$ solid at room temperature is: 1.6 B.M.[48] and that of Br_2^+ in super-acid is about 2.0 B.M.[72]. No ESR signal was observed in solution down to – 100 °C and in the solid down to – 196 °C.

Magnetic moments observed for I_2^+ in IF_5 and HSO_3F solutions are about 2.0 ± 0.1 B.M.[6,62,63]. This value decreases with increasing concentration of iodine, due to disproportionation to I_3^+ and I^{+3}, which both have an even number of electrons. No ESR signal was observed down to 4.2 °K[84]. The I_4^{+2} was observed to be diamagnetic[65].

2.2.5 Br_2^+

The $Br_2^+Sb_3F_{16}^-$ compound was formed by reacting bromine with BrF_5 and SbF_5[48] according to

$$9\,Br_2 + 2\,BrF_5 + 30\,SbF_5 \rightarrow 10\,Br_2^+Sb_3F_{16}^-$$

Br_2^+ can be formed in solution of super-acid[64,72], being oxidized by $S_2O_6F_2$ in the correct concentration according to

$$2\,Br_2 + S_2O_6F_2 + 2\,H_2SO_3F^+ \rightarrow 2\,Br_2^+ + 4\,HSO_3F$$

Changing this ratio of $Br_2/S_2O_6F_2$ to equal 3 : 1 would form the Br_3^+ cation. Br_2^+ was also formed[64] by dissolving $BrOSO_2F$ in super-acid as disproportionation takes place:

$$5\,BrOSO_2F \rightleftharpoons 2\,Br_2^+ + Br(OSO_2F)_3 + 2\,SO_3F^-$$

No dimerization into Br_4^{+2}, as in I_4^{2+}, was observed[72].

2.2.6 I_2^+

The earliest report[108] of the existence of I_2^+ included its preparation by direct reaction of iodine with excess SbF_5, which probably serves also as oxidant, being reduced to SbF_3. Such a reaction was also reported to take place in SO_2 solution[44], from which SbF_3 precipitated out and the filtered solution was evaporated, leaving behind solid $I_2^+Sb_2F_{11}^-$. Iodine was also found to react with SbF_5 or TaF_5 in IF_5 solution[84], forming $I_2^+M_2F_{11}^-$ (M = Sb or Ta).

The I_2^+ cation was prepared in various solutions by dissolving I_2, ICl, KI, IBr, I-acetate, $I(py)_2NO_3$, $IOSO_2F$, $IOSO_2Cl$ and N-iodo organic compounds in 65% oleum or 44.9% oleum ($H_2S_2O_7$)[4,8,62,66,101,103]. In these acidic solutions no additional oxidizer to the oxidizing solvent itself was necessary. Solutions in HSO_3F can be prepared by dissolving iodine with an additional oxidizer such as $S_2O_6F_2$, HIO_3 or $K_2S_2O_8$ [3,62,63,67]. The ratio of oxidizer to iodine is very important since otherwise different iodine cations, such as I_3^+ and I_5^+, are formed[63]. The reaction is according to

$$2\,I_2 + S_2O_6F_2 \rightarrow 2\,I_2^+ + 2\,SO_3F^-$$

Exposing the brown solution of I_2 in IF_5 to moisture turns its color to blue, being indicative of I_2^+ [6]. Reacting brown solutions of I_2 in IF_5 with pentafluorides of P, As, Sb, Nb and Ta formed the blue paramagnetic solutions with the same well-known absorption spectrum of I_2^+ [84]. Such solutions are also obtained by dissolving I_2 or ICl directly in SbF_5 [84].

$I_2^+Sb_2F_{11}^-$ was found[100] to yield similar products as $I_3^+AsF_6^-$ does in the following reaction:

$$[Se(C_2F_5)]_2 + 2\,I_2^+Sb_2F_{11}^- \rightarrow 2\,[Se(C_2F_5)I_2]^+Sb_2F_{11}^-$$

2.2.7 I_4^{+2}

The I_4^{+2} cation was formed on cooling a solution of I_2^+ in fluorosulphuric acid below – 60 °C, when a dimerization process takes place[65] according to $2\,I_2^+ \rightleftharpoons I_4^{+2}$. This dimerization was noticed by a dramatic change of the blue color of I_2^+ to deep red. This color change is rapid and reversible. The optical densities were measured at 640 nm, absorption of the blue I_2^+ and at 357 nm, specific absorption of the redbrown I_4^{+2}. Plotting log D_{640} versus log D_{357}, which were measured as a function of concentration, resulted in a straight line with a slope of 2 at – 70 °C and 1.9 at – 86.5 °C. This indicated that 2 mole of I_2^+ form one mole of the new species. The calculated dimerization constant was

$$K_D = 23 \pm 2\ mol^{-1}\ kg \text{ at } -70\,°C \text{ and } 170 \pm 20\ mol^{-1}\ kg \text{ at } -86.5\,°C\,.$$

From these constants $\Delta H = 10 \pm 2$ kcal was calculated.

The magnetic moment of I_2^+, which equals 2 B.M., decreased with decreasing temperature, becoming finally zero. Since I_4^{+2} contains an even number of electrons, it is expected to be diamagnetic. Cryoscopic and conductivity measurements also supported a dimerization process. The structure of I_4^{+2} is not known, a tetrahedral, square planar or acyclic chain all being reasonable.

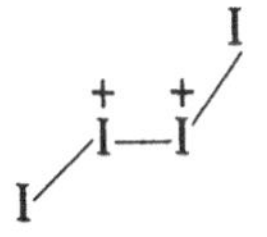

3 Hetero-Polyhalogen Cations

Numerous hetero-polyhalogen cations such as XY_2^+, X_2Y^+, XY_4^+, XY_6^+ and possibly $X_3Y_2^+$ have been prepared and studied. In all of them the more electronegative halogen is in a terminal position and the least electronegative one in the central position. Therefore, any halogen can be in the terminal position and any one, except fluorine, can be in the central position.

Although it has been claimed[29)] that the $(ClFCl)^+$ cation is of a symmetric structure with the fluorine in the center position, this has been shown to be erroneous[69)]. Instead it was shown to be of asymmetric structure – $(ClClF)^+$ – the fluorine in the terminal position and will be discussed later in detail under the X_2Y^+ Section 3.2.4.

According to the charge distribution, the positive charge is mostly located on the central atom. Its formal oxidation state is similar to that in the parent compound, whereas the oxidation state of the terminal halogen equals – 1.

Table 9. ClF_2^+ compounds and properties

Anion	M.P.	Preparations and properties	Physical measurements
BF_4^-	30° [115)]	$ClF_3 + BF_3$ [115)]. Dissociation pressure: $\log p_{mm} = (12 \pm 0.6) - (2576 \pm 16)/T$ [115)]. $\Delta H_d^0 = 23.6$ kcal mole^{-1} [25a)]; $\Delta H_{f,298}^0 = -333.3$ kcal mole^{-1} [25b)].	Raman [26,68)]; IR [25,26,115)]; conductivity [25a,115)]
PF_6^-		$ClF_3 + PF_5$ [25a)]; Dissociation pressure: $\log p_{mm} = 10.53 - 1798/T$ [25a)]; $\Delta H_d^0 = 16.4$ kcal mole^{-1} [25a)]; $\Delta H_{f,298}^0 = -436.7$ kcal/mole^{-1} [25b)].	
AsF_6^-		$ClF_3 + (AsF_5, AsCl_3, AsF_3)$ [25a,114)]; vapor pressure at 20°: 5 torr [114)].	Single crystal X-rays [90)]; Raman [26,68,133)]; IR [25a,26)]; F–NMR [20,36)]; conductivity, cryoscopy, powder X-rays [25a)]
SbF_6^-	75°–78° [114)] 285° [130)], 135° [50)]	$ClF_3 + (SbF_5, SbCl_5, SbCl_3, SbF_3)$ [25a,113,114)].	Single crystal X-rays [50)]; Raman [23,68,135)]; IR [23)]; F–NMR [20,36)]; NQR [86)]; conductivity [25a,113,114)]; DTA [130)]
$Sb_2F_{11}^-$	11° [130)]		DTA [130)]
$Sb_4F_{21}^-$	25° [130)]		DTA [130)]
PtF_6^-	171° [14)]	$ClF_3 + (PtF_5, PtF_4, PtF_6, O_2^+PtF_6^-)$ [13,14,77,106)]; $ClF_5 + PtF_6$ [35,37)]. $F_2/He + (K_2PtCl_4, K_2PtCl_6$ in melt$)$ [105)]; $ClF_2^+PtF_6^- \xrightarrow{350°} PtF_4 + ClF_3 + F_2$ [105)].	Raman [35,37)]; IR [35,37,77,106)]; EPR [77)]; magnetic [14)]
SnF_6^{2-}			Mössbauer [45,127,128)]

3.1 XY_2^+ Symmetric Cations

The XY_2^+ cations were prepared both in the solid state and in solutions, details of which are summarized in Table 9 for ClF_2^+ compounds, in Table 10 for BrF_2^+ compounds and in Table 11 for IF_2^+, ICl_2^+ and IBr_2^+ compounds.

The XY_2^+ cations can be considered as derivatives of the XY_3 interhalogen parent compounds. No ICl_3, however, is known but only its dimer, I_2Cl_6, forming the ICl_2^+ cation. The supposed parent compound of IBr_2^+, namely IBr_3, has not been isolated or identified.

Various physical measurements such as single crystal X-rays, vibrational spectra, F–NMR, NQR and Mössbauer studies support an ionic structural model, with the XY_2^+ cation being of a C_{2v} symmetry. However, some cation-anion interaction does

Table 10. BrF_2^+ compounds and properties

Anion	M.P.	Preparations and properties	Physical measurements
AsF_6^-		$BrF_3 + AsF_5$ [33]; vapor pressure at 23°: 2 torr [33].	Raman [33,131,133]; IR [33,131]
SbF_6^-	128° [131]	BrF_3 + (SbF_5, SbF_3, Sb_2O_5, SbOCl) [33,124,145]; yellow solid;	Single crystal X-rays [47,49];
	129.8° [59]	Dissociation pressure: log $p_{mm} = 7.66 - 3030/T$ [124];	Raman [23,33,131,135]; IR [23,33]; NQR [86]
	130° [124]	Heat of formation: – 1,773 KJ mole^{-1} [104].	
$Sb_3F_{16}^-$	33.5° [59]		
TaF_6^-		BrF_3 + (Ta, Ta_2O_5) [80]; yellow solid; Heat of formation: – 2,251 KJ mole^{-1} [104]	
NbF_6^-		BrF_3 + (Nb, Nb_2O_5) [80]; yellow; decompose at 150–180° [80].	Solubility [123]
BiF_6^-		$BrF_3 + BiF_5$ [80]	
GeF_6^{2-}	Sublime 20° [18]	2 $BrF_3 + GeF_4$ [33]; vapor pressure at 23.6°: 4 torr [33].	Single crystal X-rays [53]; Raman [33]; IR [18,33]; powder X-rays [33]
SnF_6^{2-}		BrF_3 + (Sn, $SnCl_2$, $SnCl_4$) [145]; Heat of formation: – 1,505 KJ mole^{-1} [104].	NQR [86]; Mössbauer [45,127,128]; conductivity [145]
AuF_4^-		BrF_3 + Au [55,121]; decompose at 50° [55], 180° [121]	Mössbauer [55]
SO_3F^-		$BrF_3 + SO_3$; $BrF_5 + SO_3$; BrF_3 + $Br(OSO_2F)_3$ [79]	UV, F–NMR [79]
PtF_6^{2-}	136° [14]	BrF_3 + ($PtCl_4$, $PtBr_4$) [122]; red solid, decompose at 180–200° [122]	
PdF_4^-		BrF_3 + ($PdCl_2$, $PdBr_2$) [122].	
BF_4^-	180° [137]	$BrF_3 + BF_3$ [137]; denial of existence [32]	

Table 11. IF_2^+, ICl_2^+ and IBr_2^+ compounds and properties

	M.P.	Preparations and properties	Physical measurements
IF_2^+			
BF_4^-		$IF_3 + BF_3$, stable only in solution of CH_3CN [111,112]	
AsF_6^-		$IF_3 + AsF_5$, decompose $-22°$ [111,112]	
SbF_6^-		$IF_3 + SbF_5$, decompose $45°$ [111,112]	F–NMR [112]
ICl_2^+			
$SbCl_6^-$	83.5° [140]	$ICl_3 + SbCl_5$ [140]; $I_2 + 2\,SbCl_5 + Cl_2$ (ex. liquid) $\xrightarrow[\text{temp.}]{\text{room}}$ $ICl_2^+SbCl_6^-$ [120], wine red.	Single crystal X-rays [141]; Raman [60,71,119]; UV [120]
$AlCl_4^-$	105° [140]	$ICl_3 + AlCl_3$ (in CCl_4, $CHCl_3$) [140]; $I_2 + 2\,AlCl_3 + Cl_2 \xrightarrow{-70°}$ $ICl_2^+AlCl_4^-$ [54], wine red	Single crystal X-rays [141]; Raman [60,119]; NQR [54,94,96,97]
$Sb_2F_{11}^-$	62° [143]	$ICl_3 + SbF_5$ [143], orange-red	Raman [143]
SO_3F^-	41°–42° [142,143] 40° [148] 34°–35° [5]	$IOSO_2F + Cl_2$ [5,142,148]; $IBr_2^+SO_3F^- + Cl_2$ [143], bright orange	Raman [142,143,148]; UV, NQR [142]
SO_3Cl^-	8° [101]	$ICl_3 + SO_3$ [101]	IR, conductivity [101]
		ICl_2^+ in solutions of: H_2SO_4 [116,142,143], $H_2S_2O_7$ and oleum [66,8], HSO_3F [142,148] and HSO_3Cl [101]	Raman [71]; UV [116,142,143]; conductivity [66,101,116,142,148]; cryoscopy [66,101,116,142]
IBr_2^+			
$Sb_2F_{11}^-$	64°–65° [143]	$IBr_2^+SO_3F^- + SbF_5$ [143]	
SO_3F^-	95°–97° [142]	$IOSO_2F + Br_2$ [142,148]; $I_3^+SO_3F^- + Br_2$ [143]	Raman [142,143]
$SO_3CF_3^-$	72°–75° [143]	$ISO_3CF_3 + Br_2$ [143]	Raman [143]
		IBr_2^+ in solutions of: H_2SO_4 [116,142,143], $H_2S_2O_7$ [8], HSO_3F and HSO_3CF_3 [143,148]	UV [116,142,143], conductivity, cryoscopy [116,142,148]

exist, mostly through two covalently halogen bridgings of two separate anions towards the cation, such as

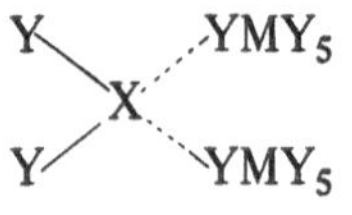

Thus the central halogen in the cation can be considered with a coordination number 4, the four ligands forming a distorted square plane. In this square, two positions

are occupied by two terminal halogens of the XY_2^+ cation and two others by bridging halogens of two separate anions. Such a geometry is in accordance with the valence shell electron pair repulsion theory for AX_4E_2 coordination, the two lone pairs of non-bonding electrons being in axial positions, forming an overall octahedral structure. A similar structure prevails around each iodine atom in the planar I_2Cl_6 molecule, the parent compound of ICl_2^+.

Several bond models are possible:

a) A purely ionic model in which only localized p orbitals of the central halogen are involved forming p-σ bonds. A bond angle of 90° would then be expected, which could slightly increase, as a result of the mutual repulsion of the two terminal halogens;

b) a covalent bridging model with d^2sp^3 hybridization of the central halogen with an octahedral structure. A bond angle of 90° would be expected, in this case, as well. However, since the two bridging halogens form weaker bonds than the two terminal halogens, a slight increase in the bonding angle may be expected;

c) an ionic model with sp^3 hybridization with a tetrahedral structure of AX_2E_2, where a bond angle of 109° 27′ would be expected.

On the basis of actually observed X-ray data, it seems that no exclusive model is appropriate[90)]. On the whole, a basically ionic model of sp^3 hybridization seems to exist with a minor contribution of d^2sp^3 hybridization bridging model.

Theoretical calculations for the geometry and electronic structure of ClF_2^+ were reported[85)]. Using an NDDO approximation for a MO–LCAO–SCF method and considering the role of d electrons, it was concluded that ClF_2^+ is of C_{2v} symmetry. In another study[139)] based on self-consistent field theory and minimum basis set (MBS) or double zeta (DZ) it was found that the bond angle in ClF_2^+ is 98.2° or 97.4°, respectively, which is in good agreement with the experimental values obtained from X-rays (3.1.1).

3.1.1 Single Crystal X-Rays

Single crystal X-ray data of compounds containing the ClF_2^+, BrF_2^+ and ICl_2^+ cations are summarized in Table 12. Considerable differences in X–Y distances are noticed. The shortest distance is within the XY_2^+ cation, whereas the "next closest" one is between X and a bridging Y, which actually is a part of the anion. On the other hand, the differences between the M–Y distances in the anion itself between and a bridging halogen and a terminal one are rather minor. Therefore it is quite clear that the bridging halogen is indeed a part of the anion, thus indicating that the structures of these compounds are basically ionic. This conclusion has also been derived[53)] for the structure of $(BrF_2^+)_2GeF_6^{2-}$, a case which has been discussed for a long time and thought, on the basis of IR data, ot be of covalent structure.

However, even the larger X...Y contact is yet shorter than the sum of van der Waals radii, indicating some cation-anion interaction. The extent of this interaction

Table 12. Single crystal X-ray data of XY_2^+ containing solids

	$ClF_2^+AsF_6^-$ [90]	$ClF_2^+SbF_6^-$ [50]	$BrF_2^+SbF_6^-$ [47,49]	$(BrF_2^+)_2GeF_6^{2-}$ [53]	$ICl_2^+SbCl_6^-$ [141]	$ICl_2^+AlCl_4^-$ [141]
	Monoclinic	*Triclinic*	*Orthorhombic*	*Monoclinic*	*Tetragonal*	*Monoclinic*
a	10.676(9) Å	5.60 ± 0.01 Å	10.12 ± 0.01 Å	5.07(1) Å	6.98 ± 0.03 Å	6.92 ± 0.03 Å
b	7.673(7) Å	10.55 ± 0.01 Å	5.81 ± 0.01 Å	13.83(2) Å	6.98 ± 0.03 Å	11.02 ± 0.05 Å
c	8.064(7) Å	5.30 ± 0.01 Å	10.95 ± 0.01 Å	6.45(1) Å	24.2 ± 0.1 Å	6.11 ± 0.03 Å
α		92.1 ± 0.2°				
β	113.40(5)°	91.8 ± 0.2°		116.6(3)°		99.1°
γ		91.5 ± 0.2°				
V	606.3 $Å^3$	313 $Å^3$	644 $Å^3$	404 $Å^3$		
V per F	19 $Å^3$	19.5 $Å^3$	20 $Å^3$	20 $Å^3$		
D_{calc}	2.874 g cm^{-3}	3.28 g cm^{-3}	3.65 g cm^{-3}	3.46 g cm^{-3}	3.00 g cm^{-3}	2.64 g cm^{-3}
D_{obs}					3.0 ± 0.2 g cm^{-3}	2.4 ± 0.3 g cm^{-3}
Z	4	2	4	2	4	2
Y–X–Y	103.17°	95.9°	93.5°	90.7(9)°	92.5°	96.7°
$r_{X-Y(terminal)}$	1.541(14) Å	1.58 Å	1.69 Å	1.71 Å	2.31 Å	2.275 Å
$r_{X...Y(bridging)}$	2.339(14) Å	2.38 Å	2.29 Å	2.21 Å	2.925 Å	2.87 Å
$R = r_{bridg}/r_{term}$	1.52	1.51	1.36	1.29	1.27	1.26
$r_{M-Y(terminal)}$	1.67 Å	1.84 Å	1.835 Å	1.73 Å	2.32 Å	2.065 Å
$r_{M...Y(bridging)}$	1.74 Å	1.93 Å	1.91 Å	1.825 Å	2.425 Å	2.15 Å
$R = r_{bridg}/r_{term}$	1.04	1.05	1.04	1.05	1.05	1.04

can be judged from the ratio of the bridging contact X...Y to that of the terminal bond length X–Y in the cation. This ratio $R = r_{bridging}/r_{terminal}$ decreases with increasing cation-anion interaction. This trend of decreasing R's is noticed in the series of compounds of $ClF_2^+ > BrF_2^+ > ICl_2^+$, as well as in the couples $ClF_2^+AsF_6^- > ClF_2^+SbF_6^-$, $BrF_2^+SbF_6^- > (BrF_2^+)_2GeF_6^{2-}$ and $ICl_2^+SbCl_6^- > ICl_2^+AlCl_4^-$ compounds.

The bond angles of the cation may vary between 90.7° and 103.17° and are in rather good agreement with the theoretically calculated values for ClF_2^+, being 97.4° and 98.2°[139]. It seems that the change in bonding angle too is related to the cation-anion interaction. Decrease in this interaction, as noticed also from changes in the R values, is coupled with an increase in the bond angle. This is noticed in the various series and couples as mentioned above, with regard to the R's,

The two bridging fluorines in the ClF_2^+ compounds are in the *trans* position within the anion, whereas in both BrF_2^+ compounds and in $ICl_2^+SbCl_6^-$, the two bridging halogens are in the *cis* position. This indicates a weaker cation-anion interaction in the ClF_2^+ compounds.

In $(BrF_2^+)_2\,GeF_6^2$ two pairs of *cis* bridging fluorines to each cation exist, and they are in *trans* position to each other[53].

3.1.2 Vibrational Spectra

Both Raman and IR spectra were recorded for the various XY_2^+ cations. The interpretations of these spectra are quite complicated and the reasons for this have already been discussed in the introduction of structural determination (1.2.2).

Eliminating the anionic frequencies, several lines can then be assigned to two stretching vibrations and one bending mode of the XY_2^+ cations. If a bent structure of C_{2v} symmetry is assumed for these cations, altogether three vibrations are expected, namely ν_1-symmetric stretch, ν_2-bending mode and ν_3-asymmetric stretch. All of these are active both in the Raman and in the IR.

Another possible structure is that of a linear one with $D_{\infty h}$ symmetry. In this case the selection rules for IR and Raman activity are mutually exclusive. Only one Raman line not coinciding with two IR lines would be expected. However, as a result of some crystal field effects, the mutual exclusive selection rules might break down. As a result, even in a linear structure all three lines might be observed both in the Raman and IR.

Thus, merely the observed number of IR and Raman bands would not be decisive whether an XY_2^+ cation is bent or linear. In addition to this information, the closeness between ν_1 and ν_3 could be indicative of a bent structure of C_{2v} symmetry with a bonding angle of around 90°, whereas in a linear structure these two vibrations would be far apart[26].

The spectral data are summarized in Tables 13, 14 and 15 for ClF_2^+, BrF_2^+ and ICl_2^+, IBr_2^+ compounds. It is noticed that in general all three expected vibrations have been observed, and in many cases in both the Raman and IR. Therefore the vibrational spectra support a bent structure with C_{2v} symmetry.

Table 13. Vibrational frequencies in cm^{-1} and force constants in mdyn/Å of ClF_2^+ cations

Ref.	25a)		26)			68)			23)		135)		131)	37)		77)	33)
Anion	BF_4^-	AsF_6^-	BF_4^-	AsF_6^-		BF_4^-	AsF_6^-	SbF_6^-	SbF_6^-		SbF_6^- in HF	HF/ClF_3	AsF_6^- in ClF_3	PtF_6^-			
Spectra	IR	IR	IR	IR	R	R	R	R	IR	R	R	R	R	IR	R	IR	
ν_1 (A_1)	518	519	798	810	811	798 788	809 806	809 805	803	810	810 P	785 P	782	789	788 784	529	
ν_2 (A_1)	–	–	537	558 520	544	396 373	384	387	375	385	384	384	–	381 376	381	–	
ν_3 (B_1)	536	558	813	818	–	808	821	830	830	833	810(?)	802	–	799	799	505	
f_r (stretch)			4.77 ± 0.07			4.6	4.7	4.8									4.88
d (bend)			1.18 ± 0.1			0.61	0.61	0.62									0.63
α (degree)			90°–120°			95°–100°	95°–100°										95°

P = polarized.

Table 14. Vibrational frequencies in cm^{-1} and force constants in mdyn/Å of BrF_2^+ cations

Ref.	23)		33)						131)				135)		60)
Anion	SbF_6^-		SbF_6^-		AsF_6^-		GeF_6^{2-}		AsF_6^-	SbF_6^-	AsF_6^- or SbF_6^- in BrF_3		SbF_6^- in HF	HF/BrF_3	
Spectra	IR	R	IR	R	IR	R	IR	R	R	R	IR	R	R	R	
ν_1 (A_1)	–	706	705	705	713	706	688	690	709	704	–	625	707 P	700 P	
ν_2 (A_1)	370	367	–	362	–	360	–	344	307 294	308	292	–	361	362	
ν_3 (B_1)	715	–	692	702	698	703	(661?)	(657?)	(699)	(705)	635	–	695	689	
f_r (stretch)					4.60										4.44
f_α (bend)					0.47										0.61
α (degree)					95°										95°

P = polarized.

Table 15. Raman frequencies in cm^{-1} and force constants in mdyn/Å of ICl_2^+ and IBr_2^+ cations

Cation	ICl_2^+									IBr_2^+			
Anion	$SbCl_6^-$			$AlCl_4^-$		$Sb_2F_{11}^-$	SO_3F^-	in H_2SO_4	in HSO_3F	SO_3F^-			$SO_3CF_3^-$
Ref	119)	71)	60)	119)	60)	143)	148)	71)	142)	143)	142)	148)	143)
ν_1 (A_1)	366	370 362 359	371	365	366	394	360	387	400 380	256	256	236	248
ν_2 (A_1)	149	148	147	151	149 143	–	161	148	–	127	124	–	–
ν_3 (B_1)	372	–	364	–	350	398	380	387	–	256	256	236	258
f_r (stretch)	2.15	2.20	2.19		2.09								
d (bend)	0.25	0.19	0.19		0.18								
α (degree)	92.5°		92.5°		96.7°								

Some force constant calculations were performed[26,33,60,68,71,119] using a bond angle of about 95–100°.

Examining the literature regarding the ClF_2^+ vibrations can serve as a good example for the difficulties involved in proper assignments. In spite of the fact that the ClF_2^+ compounds show much clearer and relatively simpler spectra than those of BrF_2^+ compounds, quite diversified assignments were reported. The early IR data for $ClF_2^+BF_4^-$ and $ClF_2^+AsF_6^-$ [25a] turned out to be only partial and led to erroneous assignments. Probably based on this information, similar erroneous assignments were made for $ClF_2^+PtF_6^-$ [77].

Reexamined spectra of ClF_2^+ compounds of BF_4^- and AsF_6^- anions[26], which seem indeed to present the correct frequencies, brought about a correction of the assignment of the stretching vibrations. However, it seems, as was shown later[68], that the assignment of the bending mode of the cation was exchanged with some anionic vibration. The force constant calculation based on these assignments did not seem to prevent this error.

The best spectral assignments of ClF_2^+ compounds with BF_4^-, AsF_6^- and SbF_6^- anions, including critical discussion of the data in the literature, were later reported[68]. A splitting in ν_1 of ClF_2^+ was observed in all three compounds studied. These splittings were correlated with factor group splitting[68].

Similar results, which are in agreement with these data, were reported for $ClF_2^+PtF_6^-$ [37], including the split of ν_1 and for $ClF_2^+SbF_6^-$, without such splitting[23].

In the series of ClF_2^+ compounds with SbF_6^-, AsF_6^-, BF_4^- and PtF_6^- anions, a slight decrease of both stretchings and bending frequencies is noticed. This is also reflected in the force constants, thus indicating some increase in cation-anion interactions[26].

The spectra of BrF_2^+ compounds with SbF_6^- and AsF_6^- are more complicated than those of similar compounds of ClF_2^+. From X-ray data (3.1.1) it is known that whereas the two bridging fluorines of MF_6^- anion in ClF_2^+ compounds are in *trans* position, these are in BrF_2^+ compounds in *cis* position. As a result the octahedral structure of the anion is distorted and its symmetry reduced from O_h to D_{4h} in the case of ClF_2^+ compounds and to C_{2v} or even to D_2 in the case of BrF_2^+ compounds[33]. This causes a considerable increase of the number of anionic vibrations, which are in the BrF_2^+ compounds both IR and Raman active, whereas in the ClF_2^+ compounds the anion still has a center of symmetry and activity in the Raman and IR is mutually exclusive. Yet, in BrF_2^+ compounds too, in spite of the fact that the data are less complete, assignments have been made for all three cationic vibrations. Whereas in ClF_2^+ compounds $\nu_1 < \nu_3$, the opposite is true in BrF_2^+, except for one report[23] in which the assignment is different.

The most difficult spectra in this group are those of $(BrF_2^+)_2GeF_6^{2-}$. In an early study[18] it was claimed that this is not an ionic compound, since the strong IR vibration of the GeF_6^{2-} – ν_3 – found in ionic compounds at 600 cm^{-1}, has not been observed in the IR spectrum of $(BrF_2^+)_2GeF_6^{2-}$. More detailed vibrational data[33] were, however, interpreted to be in agreement with an ionic structure claiming that the decrease of BrF_2^+ frequencies to 690 and 344 cm^{-1} is similar to the observed decrease in ClF_2^+ frequencies, in the series of $SbF_6^- > AsF_6^- > BF_4 > PtF_6^-$, although to a lesser extent. In this series, it was interpreted to reflect increasing cation-anion interaction, which could also be the case in the changes observed in $(BrF_2^+)_2GeF_6^{2-}$. One also has to consider the different composition ratio which in all other cases is 1:1 and in the case of germanium 2:1. This may lead to significant structural changes. The strongest evidence for the ionic structure came later from the X-ray data reported for this compound[53)]. Indeed the structure of the GeF_6^{2-} is distorted from an octahedral one to a D_{4h} symmetry.

Some of the Raman spectra of XF_2^+ compounds were studied not only in the solid state but also in different solutions. Without giving details, it was stated[36)] that the Raman spectra of HF solutions of $ClF_2^+BF_4^-$ and $ClF_2^+SbF_6^-$ show all the expected bands for the individual ions and that their frequencies did not deviate by more than 10 cm^{-1}, at the most, from those observed in the solid. This suggests that in both phases the same discrete ions are present. Splittings of various bands which were observed in the solid were missing in the HF solutions, indicating that in the solution, independent non-interacting species of both the cation and anion were present and thus show simpler spectra[135)].

It was noticed that on dissolving $ClF_2^+SbF_6^-$ in HF, the Raman active symmetric stretching frequency of ClF_2^+ at 810 cm^{-1} was indeed polarized, but the asymmetric stretch was not observed. On the other hand, on dissolving $ClF_2^+SbF_6^-$ in HF, including an excess of the base ClF_3, the symmetric polarized stretching vibration was observed at 785 cm^{-1} and the asymmetrical depolarized one at 802 cm^{-1}, indicating some decreasing shift in frequencies[135)]. Similarly, on dissolving AsF_5 in excess base ClF_3, the ν_1 of ClF_2^+ was also observed at lower frequency, at 782 cm^{-1}, close to that mentioned in the former case[133)]. Also in a mixture of HF–ClF_3, where the following equilibrium exists:

$$ClF_3 + HF \rightleftharpoons ClF_2^+ + HF_2^- ,$$

a shoulder at 785 cm^{-1} is observed in its Raman spectrum[135)]. On dissolution of $ClF_2^+SbF_6^-$ in excess acid SbF_5, the lines of the SbF_6^- anion seem to change as to indicate the formation of the dimeric anion $Sb_2F_{11}^-$. The ClF_2^+ stretching vibrations seem to collapse into one line at a higher frequency at a higher frequency at 820 cm^{-1} [68)].

The lowering of the XF_2^+ frequencies when excess of the parent compound XF_3 is present seems to result from solvation of the cation with its parent compound[135)].

Similar observations have also been noticed in BrF_2^+ compounds[135)]. Dissolving $BrF_2^+SbF_6^-$ in HF lowers the stretching vibrations and in the presence of excess BrF_3 it is further lowered to 700 and 689 cm^{-1}. Polarization measurements of the Raman spectra proved clearly that the polarized symmetric stretch ν_1 is at higher frequency than the asymmetric one – ν_3. This is in contrast to polarization results in ClF_2^+, in which $\nu_3 > \nu_1$ [135)].

The Raman and IR spectra of solutions of AsF_5 or SbF_5 in excess BrF_3 show BrF_2^+ lines at 635 (IR), 625 (R) and 292 (IR) cm^{-1}. In pure BrF_3, where self-ionization is known to exist, two lines at 625 (R) and 635 (IR) cm^{-1} are associated with the stretching frequencies[131)] of the cation. These decreases in frequencies of ClF_2^+ and BrF_2^+ in solutions indicate that the force constants of Cl–F and Br–F are lower in solution compared to the solid state. The observed decrease, which is smaller in ClF_2^+ solutions than in BrF_2^+ solutions, indicates that the cation-solvent interaction in the former case is weaker than in the latter case [133].

Raman spectra of ClF_3–BrF_3 mixtures were also studied [133], in which it was noticed that the BrF_4^- bands already observed in pure BrF_3 decreased in intensity. However, this intensity decrease was to a lesser extent than could be expected on the basis of simple dilution of BrF_3 caused by the addition of ClF_3. This was explained by the fact that the dilution indeed caused a decrease in the concentration of BrF_4^- resulting from self-ionization, but actually an additional dissociation process takes place which increases the concentration of BrF_4^- according to equilibrium:

$$ClF_3 + BrF_3 \rightleftharpoons ClF_2^+ + BrF_4^-$$

which reflects that BrF_3 is more acidic than ClF_3. From the intensity changes the equilibrium constant was calculated:

$$K = \frac{[ClF_2^+][BrF_4^-]}{[ClF_3][BrF_3]} = (1 \pm 0.4) \times 10^{-4}$$

The ν_1 of BrF_2^+ as observed in the Raman spectrum of pure BrF_3 itself is at 625 cm^{-1}. Its intensity changes were studied as correlation of temperature and addition of acid or base [136]. It was assumed that increase of temperature increases the depolymerization of BrF_3. The Raman spectrum of the BrF_3–HF system was also studied [132]. The following equilibrium was assumed to exist:

$$BrF_3 + HF \rightleftharpoons BrF_2^+ + HF_2^-$$

No lines attributable to HF_2^- were observed, but the ν_1 of BrF_2^+ was observed at 625 cm^{-1}. From its intensity changes with correlation to concentration changes the equilibrium constant was calculated:

$$K = \frac{[BrF_2^+][HF_2^-]}{[BrF_3][HF]} = 2.62 \times 10^{-3}$$

The Raman data of ICl_2^+ and IBr_2^+ compounds are summarized in Table 15. The number of observed lines of $SbCl_6^-$ anion, which is of an expected O_h symmetry, indicates the lifting of the degeneracy and thus the ν_2 and ν_5 are split. This is due to the bridging chlorines between the cation and anion, thus distorting the expected octahedral structure of $SbCl_6^-$. This is similar to the observations in fluorine compounds. In one study [119], polarization measurements were performed in a transparent single crystal, which indicated clearly that the 366 cm^{-1} line is polarized and the 372 cm^{-1} line is not. Thus these two frequencies were assigned as the symmetric and asymmetric stretchings respectively. The dissimilarity towards polarization of these two closely spaced lines excludes their being interpreted as resulting from chlorine isotope splitting of the very same vibration[71]. If this were so, then both lines, being of the same vibration, would behave similarly towards polarization measurements, both being polarized or depolarized.

Some increases of frequencies of ICl_2^+ were observed in compounds with $Sb_2F_{11}^-$ [143] and SO_3F^- [148]. In solution of the latter one in fluorosulphuric acid, a broad band is observed at 380–400 cm^{-1} apart from the solvent bands[142]. This could be due to the collapse of ν_1 and ν_3 into one line.

Similarly, the ICl_2^+ vibrations were observed in sulphuric acid solution at 387 and 148 cm^{-1}, the first one probably being the collapsed stretchings ν_1 and ν_3 and the latter one the bending ν_2 [71].

Only one frequency was observed in the $IBr_2^+SO_3F^-$ compound which was assigned to the coinciding ν_1 and ν_3 vibrations of the cation[142,143,148]. On the other hand, in the compound with $SO_3CF_3^-$ anion two separate frequencies were observed[143].

Mean amplitudes were computed for ClF_2^+, BrF_2^+ and ICl_2^+, which are 0.0408, 0.0395 and 0.0406 Å, respectively[11].

3.1.3 Absorption Spectra

In BrF_2^+ solution an absorption band was observed at 402 nm[79]. The absorption spectra of ICl_2^+ and IBr_2^+ vary slightly in different solvents and are summarized in Table 16. The spectra are useful in identifying the presence of different, yet closely related, species. The spectrum of ICl_2^+ has also been observed in a transparent crystal[120].

Table 16. Absorption bands in nm of ICl_2^+ and IBr_2^+ cations in solutions and solid (letters indicate relative intensities)

						Ref.
ICl_2^+						
Solid	590 (sh)	470				120)
H_2SO_4		452–448 vw	360–355 w	318 s		116,142)
$H_2S_2O_7$		452	360			8)
HSO_3F		486 vw	395 w		275 (sh) s	142)
HSO_3CF_3		471 vw	374 w	310 s		142)
IBr_2^+						
H_2SO_4	540–535 vw		358–355 m		260 s	116,142)
$H_2S_2O_7$	560				258	8)
HSO_3F	560 (sh) vw	455 w	361 m		232 vs	142)
HSO_3CF_3	545 vw	453 (sh) w	360 m		260	142)

3.1.4 ¹⁹F–NMR

The F–NMR spectra were observed for several XF_2^+ cations. Lowering the temperature of the solutions of $ClF_2^+SbF_6^-$ and $ClF_2^+AsF_6^-$ in HF to 0 °C allowed the observation of the separate NMR signal of ClF_2^+ besides that of the anion and solvent. As the temperature was further reduced to – 30 °C and – 60 °C, the width of this signal was narrowed. Acidification of these solutions by adding SbF_5 or AsF_5 respectively enabled the observation of the separate NMR signal of ClF_2^+ even at room temperature [20,36]. The chemical shift of ClF_2^+ is at 23–25 ppm using $CFCl_3$ as external reference.

The single broad line observed in HF solutions at room temperature for the cation, anion and solvent indicates a rapid fluorine exchange between ClF_2^+ and the solvent according to

$$ClF_2^+ + HF_2^- \rightleftharpoons ClF_3 + HF$$

or

$$ClF_2^+ + 2\,HF \rightleftharpoons ClF_3 + H_2F^+$$

The observation of a sharp singlet resonance for the cation is in agreement with a C_{2v} symmetry in which both fluorines in this cation are equivalent. No Cl–F, spin-spin coupling splitting was observed due to rapid relaxation caused by interaction of the chlorine quadrupole moment and the asymmetric electric field gradient.

In solutions of BrF_2^+ cations obtained by reacting BrF_3 with SbF_5, SO_3 and $Br(OSO_2F)_3$ or BrF_5 with SO_3 separate signals were observed for the cation and anion. The chemical shift of BrF_2^+ is at 47–49 ppm, using $CFCl_3$ as external reference [79].

Two overlapping signals of different widths were observed in solution of $IF_2^+SbF_6^-$ in acetonitrile [112]. Their derivative recording showed an intensity ratio of 2.6, thus lending support that these are indeed to be assigned to the IF_2^+ and SbF_6^- ions, separately in which case the calculated ratio would be 3 : 1.

3.1.5 NQR Measurements

In Section 2.1.3 it was already mentioned that the ^{35}Cl NQR resonances are a good criterion for the extent of ionicity of the $AlCl_4^-$ anion. In ionic structures these resonances are in the range of 10.6–11.3 MHz [97]. In the case of $ICl_2^+AlCl_4^-$ the resonances of the anion were observed at an average of 11.09 MHz, as listed in Table 17 [94,96]. The structure of $ICl_2^+AlCl_4^-$ is therefore to be considered as ionic. The ^{35}Cl resonances of the cation in $ICl_2^+SO_3F^-$ [142] are slightly lower than those observed in $ICl_2^+AlCl_4^-$ [54]. Down frequency shift is associated with stronger coordination of the anion to the central atom of the cation [97]. This therefore indicates that ICl_2^+ is more strongly

Table 17. NQR data of ICl_2^+ compounds

Compound	ν_Q, MHz	Assignment	Average (MHz)	Range (MHz)
$ICl_2^+AlCl_4^-$	10.802 [a] 10.843 11.297 11.413	^{35}Cl in $AlCl_4^-$	11.09	0.61
	38.690 [b] 39.086	^{35}Cl in ICl_2^+	38.89	0.40
	458 [c]	^{127}I (ν_1)		
$ICl_2^+SO_3F^-$	37.902 [d] 38.343	^{35}Cl in ICl_2^+	38.12	0.44
$ICl_2^+ICl_4^-$	33.916 [e] 35.680	^{35}Cl of terminal Cl in I_2Cl_6	34.80	1.76

[a] Ref. [94,96]. [b] Ref. [54]. [c] Ref. [97]. [d] Ref. [142]. [e] Ref. [42].

coordinated with SO_3F^- than with $AlCl_4^-$. This trend is even stronger in the ^{35}Cl resonances of I_2Cl_6 which can be formulated as: $ICl_2^+ICl_4^-$. In this latter case the resonances of the terminal chlorines are calculated to be in the region of 33.35 MHz and were observed in the region of 33.916 MHz [42].

The asymmetry parameter on the iodine is related to the bonding angle according to $\eta = -3 \cos \phi$ and knowing this angle from X-ray data as 96.7°, the η was computed to equal 0.350. From calculations based on observed resonances, the σ-bond population was found to be 1.97 electrons per bond compared to the expected value of 2.00.

The charge distributions were calculated as in the case of I_3^+ (2.1.3) for each atom separately. In addition, the number in parentheses of the central atom was calculated by the difference from the terminal charges and the total ionic charge of + 1.

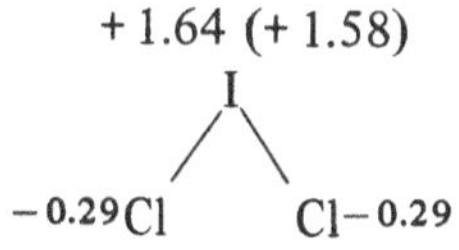

Some NQR data of ^{35}Cl, ^{121}Sb and ^{123}Sb resonances were reported for compounds of $ClF_2^+SbF_6^-$, $ClF_2^+Sb_2F_{11}^-$, $ClF_2^+Sb_4F_{21}^-$ and $BrF_2^+SbF_6^-$ [86]. The results indicate that the donor capability of ClF_3 is larger than that of BrF_3.

The ^{127}I resonance is the highest frequency observed in the series of $AlCl_4^-$ compounds with $I_3^+ < I_2Cl^+ < ICl_2^+$ and will be discussed under the section of NQR measurements of X_2Y^+ cations (3.2.3).

3.1.6 Mössbauer Effect

Complexes of ClF_3 and BrF_3 with SnF_4 were studied by Mössbauer effect. The presence of quadrupole splitting rules out the possibility of an octahedral symmetry for SnF_6^{2-}. It is more likely of a structure of a tetragonal bipyramid, D_{4h}, in which two fluorines in *trans* positions are coordinated to two separate cations[45,127,128].

Another study reported some data of ^{197}Au Mössbauer study of $BrF_2^+AuF_4^-$ [55].

3.1.7 ClF_2^+

Most of the solid compounds containing the ClF_2^+ cation were prepared by direct reaction between the parent compound ClF_3 and the proper Lewis acids, fluoride acceptors such as BF_3 and pentafluorides of P, As and Sb, forming 1:1 compounds. In reaction with SnF_4, a compound 2:1 is formed with an SnF_6^{2-} anion.

Some fluorinations can take place prior to the complex formation. It has been shown that excess ClF_3 can react with chlorides or low valency fluorides[25a,113,114] to form first the Lewis acid and then the ClF_2^+ compounds. A decreasing stability has been noticed[25a,113,114] in the series of ClF_2^+ compounds with the $SbF_6^- > AsF_6^- > BF_4^- > PF_6^-$ anions.

DTA data were reported for the ClF_3–SbF_5 system. It was shown that various ionic compounds exist with different melting points, in contrast to what had been reported earlier[130].

The following compositions were identified for congruently melting compounds $ClF_2^+SbF_6^-$, $ClF_2^+Sb_2F_{11}^-$ and $ClF_2^+Sb_4F_{21}^-$, melting at 285 °C, 11° and 25 °C, respectively. An additional covalent compound 3 $ClF_3 \cdot 2\,SbF_5$ was also identified, melting incongruently between – 71 °C and – 73 °C.

It was claimed that the reported[114] melting point of 75–78 °C for $ClF_2^+SbF_6^-$ seemed to have been of an impure material.

The $ClF_3 \cdot PtF_5$ compound was reported[13,14] to be formed by reaction of ClF_3 with $O_2^+PtF_6^-$ and was thought to be a covalent, fluorine bridged compound. It is more likely that it is an ionic compound as shown in reactions of ClF_3 with PtF_6. In this reaction, $ClF_2^+PtF_6^-$ was formed in addition to the oxidation of ClF_3 to ClF_5, which in turn forms $ClF_4^+PtF_6^-$ [77,106]. In the reaction between ClF_5 and PtF_6 under UV irradiation, a mixture of $ClF_2^+PtF_6^-$ and $ClF_6^+PtF_6^-$ is formed. This is explained on the basis that the UV radiation decomposes the $ClF_5 \rightarrow ClF_3 + F_2$ [35,37]. Another reaction is fluorinating $PtCl_4^{2-}$ or $PtCl_6^{2-}$ in a melt of LiF–NaF–KF at 500 °C with an F_2/He mixture[105].

In the liquid phase, whether in melt or solutions in excess ClF_3, the ClF_2^+ compounds are conductors, indicating ionic dissociation[25a,113–115].

Solutions of various ClF_2^+ compounds in HF, ClF_3, IF_5 and SbF_5 were reported and served for various physical measurements such as Raman spectra, F–NMR, conductivity and cryoscopy, all indicating the presence of the ClF_2^+ cation.

In addition, the presence of ClF_2^+ was identified in solutions of ClF_3 itself in HF or BrF_3 without the addition of any outside Lewis acid [133,135]. In HF solution an ionization process takes place according to

$$ClF_3 + HF \rightleftharpoons ClF_2^+ + HF_2^- .$$

The dissolution of ClF_3 in HF reduces its Hammet acidity from – 11 in pure HF to – 8.2 in solution and – 6.0 in pure ClF_3 [135]. In BrF_3 solution a similar ionization [133] process takes place:

$$ClF_3 + BrF_3 \rightleftharpoons ClF_2^+ + BrF_4^- .$$

3.1.8 BrF_2^+

Some solid compounds containing the BrF_2^+ cation were prepared by the general synthetic method of direct reaction between the parent compound BrF_3 and the proper Lewis acids [33,80,124]. In some other preparations excess BrF_3 was used to fluorinate elements, oxides and halides. The fluorides formed then reacted to form the BrF_2^+ compounds [55,80,121,122,145].

Reacting BrF_3 and SO_3 forms $BrF_2^+SO_3F^-$ [79]. Similarly, it turned out that this compound was also obtained as the final product of reacting $BrF_5 + SO_3$.

In the system BrF_3–SbF_5, two congruently melting compounds were found, namely $BrF_2^+SbF_6^-$ and one with a polymeric anion $BrF_2^+Sb_3F_{16}^-$. Two additional compounds $3\,BrF_3 \cdot SbF_5$ and $3\,BrF_3 \cdot 2\,SbF_5$, melting incongruently, were identified [59].

The compound $(BrF_2^+)_2GeF_6^{2-}$ has been subject to controversy as to whether it is indeed ionic like the other compounds or a covalent one. Single crystal X-ray data (3.1.1) and vibrational spectra (3.1.2) support an ionic structure of this compound. The existence of $BrF_2^+BF_4^-$ [137] with a melting point at 180 °C was claimed. It was to be formed by bubbling BF_3 through BrF_3 or by condensing both reactants at – 196 °C, warming up to room temperature and then vacuum evaporating to dryness.

In a later study [32] this could not be confirmed as the solid formed at – 80 °C could be pumped off at 23 °C, after having melted to a yellow liquid at 0 °C.

Comparing the stability of various BrF_2^+ compounds, several series of decreasing stability were noticed:

$$SbF_6^- > AsF_6^- > GeF_6^{2-}$$ [33]

$$SbF_6^- > TaF_6^- > SnF_6^{2-}$$ [104]

and

$$TaF_6^- > NbF_6^-$$ [80]

The BrF_2^+ cation has also been identified in solutions formed by dissolving BrF_2^+ compounds in excess BrF_3 [123,131] or HF [132,135]. The high conductivity of pure BrF_3 was explained as resulting from self ionic dissociation:

$$2\,BrF_3 \rightleftharpoons BrF_2^+ + BrF_4^-$$

Indeed, in the Raman spectrum of BrF_3 itself, BrF_2^+ lines were identified [136].

Dissolving BrF_3 in HF also involves an ionic dissociation according to

$$BrF_3 + HF \rightleftharpoons BrF_2^+ + HF_2^-$$ [132]

The dissolution of BrF_3 in HF reduces its Hammet acidity from − 11 for pure HF to − 8.0 in solution and − 7.0 in pure BrF_3 [132].

$BrF_2^+SbF_6^-$ can be used to fluorinate many inorganic compounds, including refractory oxides, feldspars and others, which resist fluorination by BrF_3 itself. These compounds are readily converted to fluorides, releasing elemental oxygen which can be used for its quantitative analysis [124]. This fluorinating agent allows fluorinations to be performed at temperatures up to 500 °C.

3.1.9 IF_2^+

This is the least stable XF_2^+ cation in similarity to the instability of IF_3 itself. The IF_2^+ cation was identified by F–NMR [112]. The stability of its compounds is in decreasing order in the series with $SbF_6^- > AsF_6^- > BF_4^-$ anions. The compounds with AsF_6^- and SbF_6^- were prepared by direct reaction of IF_3 with the proper Lewis acid in liquid AsF_5 and that of BF_4^- is stable only in acetonitrile solution [111,112].

3.1.10 ICl_2^+

Most of the compounds containing the ICl_2^+ cation were prepared by direct reaction between I_2Cl_6 and the proper Lewis acid, such as $AlCl_3$, $SbCl_3$, SbF_5 and SO_3 [101,102,140,143]. In contrast to XF_2^+ cations which are mostly prepared from their parent compounds XF_3, the ICl_2^+ can also be prepared without first preparing the parent compound I_2Cl_6. The $ICl_2^+AlCl_4^-$ and $ICl_2^+SbCl_6^-$ were also prepared by reacting an excess of chlorine with stoichiometric mixtures of $I_2 + 2\,AlCl_3$ or $I_2 + 2\,SbCl_5$. The first reaction [54] takes place at − 70 °C, whereas the second one takes place at high pressure, at room temperature, under the pressure of liquid chlorine [120].

$ICl_2^+SO_3F^-$ was prepared [5,142,148] by chlorination of iodo-fluorosulfate according to $IOSO_2F + Cl_2 \rightarrow ICl_2^+SO_3F^-$ or by halogen substitution of IBr_2^+ compound [143] according to

$$IBr_2^+SO_3F^- + Cl_2 \rightarrow ICl_2^+SO_3F^-$$

Some of these solids dissolve in H_2SO_4, HSO_3F, HSO_3Cl and nitrobenzene [101,102,120,142,148].

The ICl_2^+ cation has also been identified while prepared directly in solution, using various oxidation processes as in the following reactions[116] according to

$$HIO_3 + 2\,I_2 + 5\,Cl_2 + 8\,H_2SO_4 \rightarrow 5\,ICl_2^+ + 3\,H_3O^+ + 8\,HSO_4^-$$

or by dissolving directly in $H_2S_2O_7$ [66], which also acts as an oxidizing agent according to

$$ICl_3 + 3\,H_2S_2O_7 \rightarrow ICl_2^+ + HS_3O_{10}^- + HSO_3Cl + 2\,H_2SO_4\,.$$

On the other hand, ICl_3 itself does not dissolve in H_2SO_4 [116] unless a mixture of I_2/HIO_3 is added, reacting according to

$$I_2 + 3\,HIO_3 + 10\,ICl_3 + 24\,H_2SO_4 \rightarrow 15\,ICl_2^+ + 9\,H_3O^+ + 24\,HSO_4^-\,,$$

Iodine can also be chlorinated in $H_2S_2O_7$ [8] according to

$$(I_2^+ \text{ in } H_2S_2O_7) + Cl_2 \rightarrow ICl_2^+$$

3.1.11 IBr_2^+

The IBr_2^+ compounds cannot be prepared by the direct interaction of a parent compound since no IBr_3 or its dimer I_2Br_6 (like I_2Cl_6) has been synthesized. It was therefore prepared by other methods using oxidation reactions or halogen substitution as summarized in Table 11.

Oxidation with bromine forms the following compounds according to

$$IOSO_2F + Br_2 \rightarrow IBr_2^+SO_3F^-$$ [142,148]

and

$$IOSO_2CF_3 + Br_2 \rightarrow IBr_2^+SO_3CF_3^-$$ [143] .

A halogen substitution reaction was performed[143] according to

$$I_3^+SO_3F^- + Br_2 \rightarrow IBr_2^+SO_3F^-\,.$$

Another reaction is an exchange of anions[143] by reacting an excess of SbF_5 with $IBr_2^+SO_3F^-$ which formed $IBr_2^+Sb_2F_{11}^-$. All these mentioned IBr_2^+ compounds dissolved in H_2SO_4, HSO_3F and HSO_3CF_3. The IBr_2^+ was also prepared directly in solution of sulfuric acid[116] according to

$$HIO_3 + 2\,I_2 + 5\,Br_2 + 8\,H_2SO_4 \rightarrow 5\,IBr_2^+ + 3\,H_3O^+ + 8\,HSO_4^-\,.$$

It was also prepared in disulfuric acid[8] by bromination of I_2^+ according to

$$I_2^+ + Br_2 \rightarrow IBr_2^+.$$

3.2 X_2Y^+ and YXZ^+ Cations

Asymmetric X_2Y^+ and YXZ^+ cations have been isolated and identified both in the solid state and in solutions, as are summarized in Table 18.

Table 18. X_2Y^+ and YXZ^+ compounds and properties

	M.P.	Preparation and properties	Physical measurements
Cl_2F^+			
BF_4^-		$2\,ClF + BF_3$ [29], stable at − 127°	Raman [69]; IR [29]
AsF_6^-		$2\,ClF + AsF_5$ [29], dissociation pressure: $\log p_{mm} = 11.7124 - 2391.6/T$ [29]; H_d° = 32.83 kcal · mole^{-1}; $H^\circ_{f,298} = -355.4$ kcal · mole^{-1} [29]	Raman [69]; IR [29]
I_2Cl^+			
$SbCl_6^-$	70° [95]	$2\,ICl + SbCl_5$ [71]; $I_2 + SbCl_5 + Cl_2(ex) \xrightarrow{-78^\circ} I_2Cl^+SbCl_6^-$ [118,120]	
	63° [108]	$I_2 + 2\,SbCl_5 \rightarrow I_2Cl^+SbCl_6^- + SbCl_3$ [57,95,108]; dark brown, orange-red in thin sections	Raman [71,120]; UV [120]; NQR [95]
$AlCl_4^-$	53° [97] 110° [56]	$2\,ICl + AlCl_3$ [97]; purplish black with metallic sheen	UV, NQR [94,97]; thermal analysis [56,97]; powder X-rays [97]
SO_3F^-	39–40° [142]	$IOSO_2F + ICl$ [142]; $I_3^+SO_3F^- + ICl_2^+SO_3F^-$ [143]	Raman [142]; UV [142,143]
$GaCl_4^-$	Non-congruent melting [1]		
$TaCl_6^-$	102° [126]		
		I_2Cl^+ in solutions of H_2SO_4, HSO_3F, HSO_3CF_3 [142,143], oleum [120]	UV [120,142,143] conductivity [142,143]
I_2Br^+			
SO_3F^-	70° [142]	$IOSO_2F + IBr$ [142]; $I_3^+SO_3F^- + IBr_2^+SO_3F^-$ [143]	Raman [142]; UV [142,143]
		I_2Br^+ in solutions of H_2SO_4, HSO_3CF_3 [142,143]	conductivity [142,143]
$BrICl^+$			
$SbCl_6^-$		$IBr + SbCl_5 + Cl_2(ex) \xrightarrow{-78^\circ} BrICl^+SbCl_6^-$ [120]	Raman [120]
SO_3F^-	64–65° [143]	$ICl_2^+SO_3F^- + IBr_2^+SO_3F$ [143];	Raman, UV [143]
		$BrICl^+$ in solutions of H_2SO_4 and HSO_3F [143]	conductivity [143]
$I_3Cl_2^+$			
$SbCl_6^-$	63° [108]	$I_2 + SbCl_5$ [57,58,108]; $I_4Cl_3^+$ and $I_5Cl_4^+$ of $SbCl_4^-$ [58]	

In the central position X is the least electronegative atom. Another X is in one terminal position whereas the second one is occupied by Y, the more electronegative atom. This is the case in the Cl_2F^+, I_2Cl^+ and I_2Br^+ cations. Only one member of the YXZ^+ group is known, namely the $BrICl^+$ cation with iodine in the central position.

The X_2Y^+ cations can be treated as derivatives of XY interhalogens obtained according to

$$2\,XY + MX_5 \rightarrow X_2Y^+MX_6^-$$

No parent compound for the $BrICl^+$ cation is known, which could have been speculated to be $BrICl_2$ or Br_2ICl.

In general, less structural data are known about these compounds. No X-ray data are as yet available, and the observed vibrational spectra cannot allow a clear determination of their structure. Based on the similarity of the compositions of these cations to those of XY_2^+ cations, it may be assumed that their structures are also similar, namely bent and not linear.

Some reports have claimed[57,58,108] that polymeric cations were also prepared according to

$$nICl + SbCl_x \rightarrow (I_nCl_{n-1})^+SbCl_{x+1}^-$$
$$n = 3 - 5$$

3.2.1 Vibrational Spectra

Mostly Raman spectra were observed and only a few IR spectra were reported. Based on the vibrational spectra alone, it is impossible to elucidate the structures, whether the X_2Y^+ cations are bent with C_s symmetry or linear with $C_{\infty v}$ symmetry. In both these structures, altogether three vibrations would be expected and all of them are both Raman and IR active. In the observed spectra, as summarized in Table 19, three cationic frequencies were assigned for all the observed species. The observed frequencies can be dealt with as functional group stretching vibrations of ν(X–Y) and ν(X–X) and the bending mode δ. It is easily noticed[69,120] that both stretching frequencies in X_2Y^+ cations are just slightly lower than the similar ones observed for the vibrations in the diatomic XY and X_2 neutral molecules. This indicates that only slight coupling exists between the ν(X–Y) and ν(X–X) vibrations in the X_2Y^+ cations.

Such consideration also supported the proposed structure for the only reported YXZ^+ cation, namely $BrICl^+$, with the iodine in the central position, bromine and chlorine in the terminal positions. The observed stretching frequencies were assigned as ν(I–Cl) and ν(I–Br). The existence of another structure like $BrClI^+$ or $ClBrI^+$ with chlorine or bromine, respectively, in the central position, can be eliminated. In both such alternative structures, a functional group frequency ν(Br–Cl) should have been observed as expected at 425–435 cm^{-1} in accordance with the neutral diatomic BrCl vibration. In the case of $ClBrI^+$, no ν(I–Cl) stretching frequency at

Table 19. Vibrational frequencies in cm^{-1} of X_2Y^+ and YXZ^+ cations

Species	Cl_2F^+				I_2Cl^+			$BrICl^+$		I_2Br^+	ClF	Cl_2	ICl	IBr	I_2
Anion	AsF_6^-		BF_4^-		$SbCl_6^-$		SO_3F^-	$SbCl_6^-$	SO_3F^-	SO_3F^-					
Spectra	R	IR	R	IR	R	R	R	R	R	R					
Ref.	69)	29)	69)	29)	120)	71)	142)	120)	143)	142)	a	b	b	b	c
ν(Cl–F)	744	593 586	743	594 588							785				
ν(Cl–Cl)	535 528	535 527 520	540 516	532 528 519								558			
ν(I–Cl)					356 350	358 351	360	370 362	373 370				382		
ν(I–Br)								255	263	258				262	
ν(I–I)					190	184	197			198					205
δ (bending)	299 293	293 258	296	–	126	124	126(?)	149	166	–					
R(Cl–F)	.95		.95												
R(Cl–Cl)	.95		.95												
R(I–Cl)					.92	.93	.94	.96	.97						
R(I–Br)								.97	1.00	.98					
R(I–I)					.93	.90	.96			.97					
Ratio of I–Cl split					1.017	1.02		1.022	1.008						

[a] Gutman, V. (ed.): Halogen chemistry, Vol. 1. New York: Academic Press 1967.
[b] Siebert, H.: Anwendung der Schwingungsspektroskopie in der Anorganischen Chemie. Berlin, Heidelberg, New York: Springer 1966, p. 40.
[c] Shamir, J., Claassen, H. H.: unpublished.

R = ratio of X–Y or X–X stretching frequency in cation to that of the related diatomic molecule.

350–370 cm^{-1} would have been at all expected. In fact, a ν(I–Cl) and ν(I–Br) were indeed observed whereas no ν(Br–Cl) was observed, thus supporting a $BrICl^+$ structure with iodine in the central position.

Comparison of the frequencies of a functional group in the different cations shows that the slight changes observed might be correlated with the mass of the second terminal atom. The average ν(I–Cl) in the various $SbCl_6^-$ salts with the cations $ClICl^+$, $BrICl^+$ and $IICl^+$ are 369, 366 and 353 cm^{-1}, respectively. Thus, the (I–Cl) frequencies seem to decrease with increasing mass of the second terminal atom, which indicates yet some slight coupling between the functional group stretching vibrations.

In Table 19 some data are also included for the ratios of the particular functional group frequency as observed in an X_2Y^+ cation to that observed in the related neutral X_2 or XY molecules. It seems that within each cation this ratio $R = \nu(X-Y)_{cation}/\nu(XY)_{molecule}$ is quite constant. Thus, the relative decreases in the frequencies of both stretching vibrations in the very same cation are rather equal.

Splittings were observed in some Cl–Cl and I–Cl vibrations. Some of these were explained[71,120] as resulting from chlorine isotopes. Using a diatomic model for I–Cl vibrations, the calculated ratio of $\nu_{^{127}I-^{35}Cl}/\nu_{^{127}I-^{37}Cl} = 1.022$. The observed ratios for ν(I–Cl) frequencies are in rather good agreement with the calculated value, as following: in $I_2Cl^+SbCl_6^-$: 1.017[120] or 1.02[71], in $BrICl^+SbCl_6^-$:1.022[120] and in $BrICl^+SO_3F^-$:1.008[143].

Force constants were also calculated[120].

The ratio of the observed split in $Cl_2F^+BF_4^-$ is 1.047, much larger than the expected value for the chlorine isotopes split, and was therefore explained[69] to be due to factor group splitting.

Observed IR spectra of Cl_2F^+ compounds were at first interpreted[29] as to support a symmetric structure of C_{2v} symmetry of $Cl-F-Cl^+$. This would have been a unique case in which the most electronegative atom occupys the central position. The frequencies at 586 and 529 cm^{-1} were assigned to the asymmetric and symmetric stretching vibrations, respectively. This would be in confirmation of the observed general pattern of bent XY_2^+ species with a bonding angle around 90–100°, in which the two stretching frequencies are rather close to each other. Force constants were also calculated[29] assuming a bond angle around 110° and were to support as well the symmetric C_{2v} structure. Splittings observed in the stretching frequencies were explained to be due to the chlorine isotopes, whereas the splitting in the bending mode was explained to be due to crystal field effects.

The supposed Cl–F stretching frequencies are much lower than similar vibrations observed in ClF and ClF_2^+ around 800 cm^{-1}. This discrepancy was explained to result from a large increase in the polarity of the Cl–F bond in Cl_2F^+ compared to that in ClF or ClF_2^+. This decrease of frequencies was supposedly also reflected in low force constant.

Based on later observed Raman spectra[69] and reinterpretation of the earlier reported IR spectra, the Cl_2F^+ cation was shown to be of the general asymmetric structure with C_s symmetry, having a chlorine in the central position and the most electronegative atom, the fluorine in a terminal position. The 744 cm^{-1} line was assigned to the stretching frequency ν(Cl–F) and that at 535 cm^{-1} as ν(Cl–Cl), close to the diatomic ClF and Cl_2 frequencies. The formerly assigned cationic asymmetric stretching at 593 cm^{-1} was reassigned as ν_2 of the anion[69] and not at all of the cation.

3.2.2 Absorption Spectra

Absorption spectra were observed for the various non-containing fluorine cations, whether in solution or in solids mulled with halocarbon grease. The spectrum of $I_2Cl^+SbCl_6^-$ was obtained from a transparent crystal[120]. All these are summarized in Table 20, and it is noticed that only rather small spectral changes were observed in different solutions.

Table 20. Absorption bands in nm of X_2Y^+ and YXZ^+ cations in solutions and solid (letters indicate relative intensities)

						Ref.
I_2Cl^+						
Crystal			450			120)
Mull		475 m	450 m	310 s		97)
H_2SO_4			438–432 m	300–295 s		142,143)
Oleum	642		460			120)
HSO_3F	635 (sh)		461 m	300 s	230 s	142,143)
HSO_3CF_3	630 w		450 m	315 s	235 s	142)
I_2Br^+						
H_2SO_4		490 m	408	300 m	270–255 s	142,143)
HSO_3F	635 w		420–414 m	300 s	261–260 s	142,143)
HSO_3CF_3	620 w		418 m	300 s	260 s	142)
$BrICl^+$						
H_2SO_4			488 w	340 m	270 s	143)

3.2.3 NQR Measurements

Both $I_2Cl^+AlCl_4^-$ [97] and $I_2Cl^+SbCl_6^-$ [95] compounds were studied by NQR and the observed frequencies are summarized in Table 21. The observed signals of $I_2Cl^+SbCl_6^-$ were stronger than those of $I_2Cl^+AlCl_4^-$, and even the weaker signals of ^{37}Cl, which is of lower concentration than the ^{35}Cl, according to their natural abundance, have been observed.

Table 21. NQR data of I_2Cl^+ compounds

Compound	ν_Q, MHz	Assignment	Average (MHz)	Range (MHz)	$\frac{\nu_Q \text{ of } ^{35}Cl}{\nu_Q \text{ of } ^{37}Cl}$
$I_2Cl^+AlCl_4^-$ 97)	10.297				
	10.474	^{35}Cl in $AlCl_4^-$	10.83	0.99	
	11.265				
	11.283				
	37.912	^{35}Cl in I_2Cl^+	38.02	0.125	
	38.127				
	417	^{127}I(central) (ν_1)			
$I_2Cl^+SbCl_6^-$ 95)	15.846	^{123}Sb (ν_1)			
	16.376				1.26881
	16.554	^{37}Cl in $SbCl_6^-$	18.19	3.87	1.26924
	19.569				1.26879
	20.247				1.26878
	20.778	^{35}Cl in $SbCl_6^-$			
	21.011	^{35}Cl in $SbCl_6^-$ + ^{121}Sb (ν_1)			
	21.029	^{123}Sb (ν_2)			
	24.829	^{35}Cl in $SbCl_6^-$	23.08	4.91	
	25.689				
	30.169	^{37}Cl in I_2Cl^+			1.26888
	32.896	^{123}Sb (ν_3)			
	35.705	^{121}Sb (ν_2)			
	38.281	^{35}Cl in I_2Cl^+			
	429	^{127}I(central) (ν_1)			
	517	^{127}I(terminal) (ν_1)			

The frequencies of $AlCl_4^-$ averaged at 10.83 MHz, in good agreement with the frequency range of ionic $AlCl_4^-$ species (2.1.3 and 3.1.5). The $SbCl_6^-$ frequencies average at 23.08 MHz and agree well with other $SbCl_6^-$ salts. The fact that four Cl transitions are described for the $SbCl_6^-$ anion indicate it to be of C_s or C_{2v} symmetry. The asymmetry parameter $\eta = 0.38$ is rather large for an O_h symmetry, which is expected for $SbCl_6^-$. However, even only a 3% imbalance in p-orbital population relative to the axial symmetry would cause such an asymmetry parameter and especially if angular distortions occur.

The agreement between the observed ^{37}Cl signals with those calculated from observed ^{35}Cl signals using their isotope ratio of nuclear electric quadrupole moments of 1.26878 is very good and this ratio was indeed used in their assignment. The ratio of the averaged frequencies is: 1.26883.

The observed ^{35}Cl frequencies for the I_2Cl^+ cation are at 38.281 MHz in the $SbCl_6^-$ compound and slightly lower at 38.02 MHz in the $AlCl_4^-$ compound, thus indicating that the cation-anion coordination is slightly stronger in $AlCl_4^-$ than in the

$SbCl_6^-$ compound. The same can possibly also be the case with regard to the ^{127}I(central)-ν_1 frequencies, being at 429 and 417 MHz in $SbCl_6^-$ and $AlCl_4^-$ salts respectively.

Comparing the average ^{35}Cl resonances in various $AlCl_4^-$ salts, it is noticed that in the series of β-I_3^+, I_2Cl^+ and ICl_2^+ these are 10.74, 10.83 and 11.09 MHz, respectively. These plausible frequencies show progression with increasing acidity expected for iodine in the cation[97].

The ^{127}I(central)-ν_1 frequencies in the series of α-I_3^+, I_2Cl^+ and ICl_2^+ salts of $AlCl_4^-$ are 309, 417 and 458 MHz, respectively. This trend is explained to arise through the inductive effect of the addition of a hypothetical I^+ or Cl^+ to the iodine in ICl. Terminal iodine frequencies are in comparison 421 and 517 MHz in α-$I_3^+AlCl_4^-$ and $I_2Cl^+SbCl_6^-$, respectively.

Charge distributions were calculated for the I_2Cl^+ cation to be:

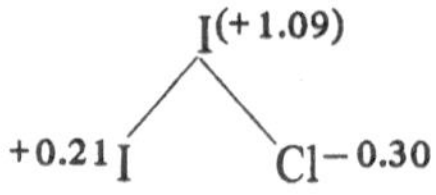

3.2.4 Cl_2F^+

The Cl_2F^+ compounds with BF_4^- and AsF_6^- were prepared in the regular direct reaction of the reactants involved, ClF and BF_3 or AsF_5, respectively[29]. The reactants were condensed at − 196 °C and after completion of the reaction the excess ClF was pumped off.

The $Cl_2F^+AsF_6^-$ was reported to react forming new compounds such as:

$$Cl_2F^+AsF_6^- + Cl_2(ex) \rightarrow Cl_3^+AsF_6^- \quad [69]$$

$$Cl_2F^+AsF_6^- + OSF_2 \rightarrow OSClF_2^+AsF_6^- \quad [87,88]$$

and

$$Cl_2F^+AsF_6^- + Xe \rightarrow XeF^+AsF_6^- + Cl_2 \quad [39]$$

It seems that Cl_2F^+ is a very strong oxidizer and the decreasing series of oxidizing power $Cl_2F^+ > XeF^+ > Xe_2F_3^+ > BrF_2^+$ was noticed.

3.2.5 I_2Cl^+, I_2Br^+ and $BrICl^+$

The $I_2Cl^+AlCl_4^-$ compound was prepared by direct reaction between the reactants, according to

$$2\,ICl + AlCl_3 \rightarrow I_2Cl^+AlCl_4^-$$

Thermal analyses of the ICl–$AlCl_3$ system were reported[56,97], but the determined melting points for the solid complex differ considerably and might have resulted from differences in the purity of the materials used[97].

Attempts to prepare $I_2Cl^+AlCl_4^-$ by refluxing the reactants in carbon tetrachloride, chloroform or acetonitrile have failed[97].

$I_2Cl^+SbCl_6^-$ was prepared by several methods. It was synthesized by direct reaction of ICl in hot solution of $SbCl_5$ [71]. It was also prepared by reacting a binary mixture of iodine with excess $SbCl_5$ which also serves as an oxidant[95,108] according to

$$I_2 + 2\,SbCl_5 \rightarrow I_2Cl^+SbCl_6^- + SbCl_3$$

It was also noticed that reactions like $nI_2 + ICl + SbCl_5$ always yielded only the $I_2Cl^+SbCl_6^-$, even if n = 1–3. This is in contrast to $AlCl_4^-$ compounds which in similar reactions yielded various compounds of the type $I_{2n+1}^+AlCl_4^-$ [95]. In another reaction, equimolar quantities of iodine and $SbCl_5$ were reacted in excess liquid chlorine at – 78 °C to form $I_2Cl^+SbCl_6^-$ [118,120]. In a related a similar reaction but of different molar ratio of $I_2 + 2\,SbCl_5$ with excess liquid chlorine, the $ICl_2^+SbCl_6^-$ was obtained but only at high pressures of liquid Cl_2. This reaction took place in a closed vessel at room temperature, keeping liquid chlorine under its own pressure. However, if this reaction was carried out similarly to the reaction of $I_2 + SbCl_5$ at a lower pressure and a temperature of – 78 °C, a mixture of approximately equal quantities of $I_2Cl^+SbCl_6^-$ and $ICl_2^+SbCl_6^-$ was obtained. This clearly points out that the $I_2Cl^+SbCl_6^-$ is easier formed, in spite of the fact that the quantities involved in the reaction are sufficient to yield $ICl_2^+SbCl_6^-$, as indeed was the case when the reaction took place at a higher temperature and pressure. It should also be mentioned that iodine trichloride (I_2Cl_6) is actually formed by direct reaction of iodine and chlorine already at – 78 °C. Therefore, if the formation of the complexed salts $ICl_2^+SbCl_6^-$ or $I_2Cl^+SbCl_6^-$ were to take place in a stepwise reaction, namely first forming ICl or ICl_3 and only then reacting with $SbCl_5$, then pure $ICl_2^+SbCl_6^-$ should be formed also at – 78 °C, similarly to the formation of pure $I_2Cl^+SbCl_6^-$. It has therefore been suggested[120] that a three-body reaction may take place according to:

$$\text{(a)}\qquad I{-}I + \overset{+\delta}{Cl}{-}\overset{-\delta}{Cl} + SbCl_5 \rightarrow I_2Cl^+SbCl_6^-$$

$$\text{(b)}\qquad Cl{-}I + \overset{+\delta}{Cl}{-}\overset{-\delta}{Cl} + SbCl_5 \rightarrow ICl_2^+SbCl_6^-$$

Whereas the first reaction (a) can take place directly the second one (b) can only occur after a prior reaction of the formation of $I_2 + Cl_2 \rightarrow 2\,ICl$ and only then in turn form $ICl_2^+SbCl_6^-$. Therefore no pure $ICl_2^+SbCl_6^-$ was formed in spite of the presence of the proper concentrations. It also seems that once the more stable ionic $I_2Cl^+SbCl_6^-$ was formed the excess chlorine present did not further oxidize or chlorinate it to $ICl_2^+SbCl_6^-$.

Polymeric cations $I_nCl_{n-1}^+$ were also formed[57,58,108] in the system $IClSbCl_5$ by varying the molar ratios such as

$$3\,ICl + SbCl_5 \rightarrow I_3Cl_2^+SbCl_6^-\,,$$

and even higher ones such as $I_4Cl_3^+$ and $I_5Cl_4^+$ [58] cations.

$IOSO_2F$ can react with ICl or IBr to form I_2X^+ according to

$$IOSO_2F + IX \rightarrow I_2X^+SO_3F^- \quad (X = Cl \text{ or } Br)$$ [142]

The solid containing $BrICl^+$ was obtained in the reaction:

$$ICl_2^+SO_3F^- + IBr_2^+SO_3F \xrightarrow{95^\circ} 2\,BrICl^+SO_3F^- .$$

Another reaction yielded[120] a $BrICl^+$ compound similar to the formation of I_2Cl^+ according to

$$IBr + SbCl_5 + Cl_2(ex) \xrightarrow{-78^\circ} BrICl^+SbCl_6^- .$$

A non-congruent compound of $2\,ICl \cdot GaCl_3$ was reported[1].

An $ICl \cdot TaCl_5$ compound was claimed[126] to exist with a melting point of 102 °C, but in a later study[97] its existence could not be confirmed.

All these cations were also found to exist in solutions which were obtained either directly or by dissolving solid compounds in H_2SO_4, HSO_3F, oleum and HSO_3CF_3[143]. Several reactions were reported to take place in solutions such as:

(a) $I_3^+ + IX_2^+ \rightarrow 2\,I_2X^+ \quad (X = Cl \text{ or } Br)$ [143]

(b) $IBr_2^+ + ICl_2^+ \rightarrow 2\,BrICl^+$ [143]

From these reactions it is noticed, as has also been observed in the synthesis of $I_2Cl^+SbCl_6^-$, that the asymmetric cations are more stable than the symmetric ones[143]. Some additional reactions were reported to form I_2X^+ using HIO_3 as oxidant

$$HIO_3 + 2\,I_2 + 5\,IX + 8\,H_2SO_4 \rightarrow 5\,I_2X^+ + 3\,H_3O^+ + 8\,HSO_4^-$$

or

$$HIO_3 + 7\,ICl + 8\,H_2SO_4 \rightarrow 3\,I_2Cl^+ + 2\,ICl_2^+ + 3\,H_3O^+ + 8\,HSO_4^-$$

A further equilibrium

$$I_2Cl^+ + ICl \rightleftharpoons I_3Cl_2^+$$

was also reported[61].

3.3 XY_4^+

The only heteropolyhalogen XY_4^+ cations identified are the XF_4^+ halogen fluorides ($X = Cl$, Br and I), as summarized in Table 22. Attempts to prepare the ICl_4^+ ca-

Table 22. XF_4^+ compounds and properties

	M.P.	Preparations and properties	Physical measurements
ClF_4^+			
AsF_6^-		$ClF_5 + AsF_5$ [30]; Dissociation pressure: $\log p_{mm} = 12.2008 - 2763.6/T$ [30]; $\Delta H_d^\circ = 25.23$ kcal · mole^{-1}; $\Delta H_{f,298}^\circ = -376.8$ kcal · mole^{-1} [30]	Raman, IR [38]
SbF_6^-	88° [38] 120° [129]	$ClF_5 + SbF_5/HF$ [30,38]	Raman [36,38,41]; IR [38]; F–NMR [36,38], NQR [86]; DTA [129]; powder X-rays [30]
$Sb_2F_{11}^-$	64° [129]		NQR [86]; DTA [129]
$Sb_4F_{21}^-$	62° [129]		NQR [86]; DTA [129]
PtF_6^-		$ClF_5 + PtF_6$ [35,37,106]; reddish; vapor pressure at 23°: 2 torr [106]	IR [35,37,106]; powder X-rays [106]
BrF_4^+			
AsF_6^-		$BrF_5 + AsF_5$ [38]	
$Sb_2F_{11}^-$	60° [110] 60–61° [93]	$BrF_5 + SbF_5$ [38,110,134]	Single crystal X-rays [89]; Raman [38,75,134,135]; F–NMR [38,15,93,127]
SnF_6^{2-}		Yellow [127]	Mössbauer [45,127,128]
IF_4^+			
SbF_6^-	103° [147]	SbF_5 in hot IF_5 [146,147]	Single crystal X-rays [7]; Raman [38,117]; IR, F–NMR [38]
$Sb_2F_{11}^-$		$IF_5 + CrO_3 + SbF_5$ [52]	Single crystal X-rays [52]
SO_3F^-		$IF_5 + SO_3$ [147]	
SnF_6^{2-}			Mössbauer [45,128]
PtF_6^-	140° [14]	$IF_5 + (PtF_5, O_2^+PtF_6^-)$ [13,14]	
$CrF_4Sb_2F_{11}^-$		$2\,O_2^+[CrF_4Sb_2F_{11}]^- + 3\,IF_5 \rightarrow 2\,IF_4^+[CrF_4Sb_2F_{11}]^- + 2\,O_2 + IF_7$ [19]. Decompose: 153° [19]	Raman, powder X-rays [19]

tion have failed [8,120,142]. As in other compounds, these compounds have also been shown by various physical methods to be of ionic structures containing XF_4^+ cations. The parent compounds, the halogen pentafluorides, are of a pseudo-octahedral symmetry, C_{4v}, in which the five fluorines occupy the five corners of a square pyramid and a non-bonding pair of electrons occupying the sixth corner of the pseudo-octahedron. This is an energetically more favored structure than the less energetically favored C_{2v} structure of the XF_4^+ cations [38]. These have a pseudo-trigonal bipyramidal structure, two fluorines being in the axial positions and two others in the equatorial triangular plane, whereas the third position in this plane is occupied by a non-bonding pair of electrons. Such a structure is in accordance with the VSEPR, Gillespie-Nyholm rules, with sp^3d hybridization[52]. In contrast, XY_4^- anions with sp^3d^2 hybridization would be of D_{4h} symmetry, square planar structure, and 2 non-bonding pairs of electrons in the axial positions. Another possible structure was assumed in which the equatorial plane bonds including the free electron pair are due to sp^2 hybridization, whereas the bonding of the two axial bonds involves mainly one delocalized p-electron pair of the central halogen for the formation of a semi-ionic three-center four-electron pp σ-bond[38]. This latter assumption was based on the relatively large difference between the covalent bond strengths of the equatorial bond and axial bond in ClF_4^+. In a later study[41] these respective force constants were found to have slightly other values but are still significantly different from each other.

Theoretical calculations using the MBS (minimum basis set) or DZ (double zeta) basis methods indicated that the symmetry of ClF_4^+ would be of a C_{4v} square pyramid structure[139], which disagrees with the proposed symmetry based on physical measurements such as X-rays, vibration and F–NMR spectra, as described in this chapter.

However, when including d-functions of the central chlorine but not of the fluorines in DZ basis calculations, a sizeable lowering of energy was obtained in a C_{2v} symmetry compared to a C_{4v} symmetry[138]. These later calculations resulted in geometrical dimensions which include two different axial and equatorial bond lengths, as well as different axial and equatorial bond angles, as listed in Table 23.

Table 23. Single crystal X-ray data of XF_4^+ compounds

	$BrF_4^+Sb_2F_{11}^-$	$IF_4^+Sb_2F_{11}^-$	$IF_4^+SbF_6^-$	ClF_4^+
Ref.	89)	52)	7)	138)
	Monoclinic	*Monoclinic*	*Tetragonal*	
a	14.19 ± 0.03 Å	8.52(1) Å	5.892 Å	
b	14.50 ± 0.03 Å	14.82(2) Å	–	
c	5.27 ± 0.01 Å	9.98(1) Å	10.255 Å	
β	90.6 ± 0.1°	112.7(3)°		
V	1085 Å^3	1163 Å^3		
V per F	18.1 Å^3	19.4 Å^3		
Z	4	4	2	
D_{calc}	3.72 g cm^{-3}	3.75 g cm^{-3}		
∢ ax	173°	160°	148°	169.6°
∢ eq	96°	92°	107°	109.7°
$r_{X-F(ax)}$	1.855 Å	1.84 Å	1.83 Å	1.63
$r_{X-F(eq)}$	1.77 Å	1.77 Å	1.79 Å	1.57
$r_{X-F(bridg)}$	2.365	2.53 Å		
$R = r_{bridg}/r_{term}$	1.30	1.40		

3.3.1 Single Crystal X-Rays

Some single-crystal X-ray data of the $BrF_4^+Sb_2F_{11}^-$ [89], $IF_4^+Sb_2F_{11}^-$ [52] and $IF_4^+SbF_6^-$ [7] are summarized in Table 23. It is clearly noticed that the compounds are basically of ionic structure, although some cation-anion interaction also exists. The ratio of the distances between the central atom of the cation and the bridging fluorine of the anion to that of the terminal fluorine in the cation itself is indicative of the degree of ionicity of the compound. Yet, the X ... F bridging contact is smaller than the van der Waals radii, indicating some cation-anion interaction.

In the XF_4^+ cations two different X–F bond lengths as well as two different F–X–F bond angles were observed. These prove that the XF_4^+ cation is of C_{2v} symmetry. The repulsion between the non-bonding electron pair and bond pairs decreases the angles between axial and equatorial bonds from their expected values of 180° and 120°, respectively.

The IF_4^+ cation in $IF_4^+Sb_2F_{11}^-$ compound [52] is bridged to two fluorines of two separate anions. Including the non-bonding electron pair, the iodine is 7-coordinated, forming a distorted pentagonal bipyramid.

In addition, there are two more bridging fluorines with longer distances than the first couple of bridging fluorines already mentioned. The short bridgings are at distances of 2.51 and 2.55 Å and the longer ones at 2.83 and 2.94 Å. Including these long bridging fluorines, a total of 8 fluorines surround the iodine forming a distorted cube. Such extra-long contacts were not observed in the $BrF_4^+Sb_2F_{11}^-$ compound, which may reflect the greater size of the iodine atom compared to that of bromine.

The bridging fluorine in the dimeric $Sb_2F_{11}^-$ anion is about half-way between the two antimony atoms, with distances of 2.03 and 2.05 Å, thus forming almost a symmetrical anion. The internal bridging fluorine in the anion is in a *cis* position to the cation-anion bridging fluorine.

The data of $IF_4^+SbF_6^-$ is less complete but also includes different axial and equatorial distances of X–F, as well as different F–X–F bonding angles [7]. This supports a C_{2v} symmetry of the cation.

The data of $BrF_4^+Sb_2F_{11}^-$ does not seem to be of sufficient accuracy [52], as is noticed from the fact that the reported Br–F bond lengths are the same as those reported for I–F ones.

The BrF_4^+ cation, too, is bridged to two separate $Sb_2F_{11}^-$ dimeric anions. Including the non-bonding electron pair, the bromine seems to be 7-coordinated. This causes a distortion of the ideal trigonal bipyramid to a puckered pentagonal bipyramid or a distorted octahedron [89].

The dimeric $Sb_2F_{11}^-$ anion is asymmetric, the internal bridging fluorine being at different distances from the two antimony atoms. The bridging fluorines are in a *cis* position to each other.

Powder X-ray data were reported for $ClF_4^+SbF_6^-$ [30], $ClF_4^+PtF_6^-$ [106] and $IF_4^+(CrF_4Sb_2F_{11})^-$ [19].

3.3.2 Vibrational Spectra

Both Raman and IR spectra of all the various XF_4^+ cations have been studied in the solid state as well as in HF solutions, and are partially summarized in Table 24. The observed spectra are very complicated with a large number of lines. As has already been discussed in earlier sections, the number of observed vibrations assigned to the anions MF_6^- is larger than expected for a highly symmetric, O_h, octahedral structure. This stems from the fact that the anion is actually distorted, causing a lowering of symmetry and increasing the number of observed lines. The vibrations of the dimeric $Sb_2F_{11}^-$ are also rather numerous.

Assuming that the XF_4^+ cations are of a C_{2v} symmetry, the number of expected Raman lines is equal to the maximum possible vibrations, in accordance with $3\,n-6 = 9$; ($n = 5$). Therefore, merely the complexity of the observed spectra and the large number of included frequencies strongly support the suggestion that the XF_4^+ cations are indeed of C_{2v} symmetry and thus in agreement with the X-ray data for BrF_4^+ and IF_4^+ cations (3.3.1) and F–NMR data for ClF_4^+ (3.3.3). The assignments of the observed frequencies were compared to those of the isoelectronic compounds of SF_4 SeF_4 and TeF_4. Some of the compounds were not studied in pure materials, but in mixtures with others like the PtF_6^- salts which included a mixture of both ClF_4^+ and ClF_6^+ cations [35,37,106]. Similarly, $Sb_2F_{11}^-$ compounds included a mixture of both BrF_4^+ and BrF_6^+ cations [75]. The fact that the additional accompanying species were XF_6^+ cations, which are of octahedral structure, thus expected to show a rather simple spectral pattern consisting of only a few lines, assisted in the selection of the proper lines to be assigned to the XF_4^+ cations. However, the selection of the proper XF_4^+ vibrations is very complicated and in no case were all the expected vibrations identified and assigned.

It was stated [38], that the assignments should be considered as tentative only. The best interpreted spectra are those of ClF_4^+. In spite of the good agreement in various reports on observed frequencies [38,41], some differences with regard to their assignments are present. In both spectra not all the expected nine vibrations were observed and the listing of nine frequencies for the ClF_4^+ includes the ν_6, which has actually been assumed to coincide with the ν_1. In a later study [41], some of the assignments have been partially changed.

The observed frequencies for $ClF_4^+SbF_6^-$ in HF solution [38,41] were in good agreement with those observed in the solid, confirming that the ClF_4^+ cation also exists in solution.

The assignments of BrF_4^+ are mostly in common in various reports, with only few differences among them. The assignments for ν_3 and ν_7 vibrations have been reversed in some reports [38,134]. Similarly, the assigned frequency as ν_8 vibration in one study [38] has been assigned by others [75,134] as a vibration of the anion and not at all as of the BrF_4^+ cation.

The spectra of BrF_4^+ in HF solution has been interpreted in one study [134] as being mostly that of the molecular BrF_5 and only two frequencies at 726 and

Table 24. Vibrational frequencies in cm^{-1} and force constants in mdyn/Å of XF_4^+ cations

Cation	ClF_4^+					BrF_4^+				IF_4^+		
Anion	AsF_6^-		SbF_6^-			$Sb_2F_{11}^-$				SbF_6^-		
Spectra (Ref.)	IR [38]	R [38]	IR [38]	R [38]	R [41]	IR [38]	R [38]	R [75]	R [134]	IR [38]	R [38]	R [117]
$A_1 - \nu_1 - \nu$ sym XF_2 eq	796	799	803	802	807 P	730	723	725	726	728	729	732
$A_1 - \nu_2 - \nu$ sym XF_2 ax	568	567	–	568	578 P	606	606	601	606	–	614	617
$A_1 - \nu_3 - \delta$ sciss XF_2, eq + ax, sym	511	519	510	515	385 P	–	385	382	424	345	341	343
$A_1 - \nu_4 - \delta$ sciss XF_2, ax + eq, asym	–	237	–	235	250 P	–	219	214	216	–	151	145
$A_2 - \nu_5 -$ XF_2 twist	–	473	–	475	478	–	–	–	–	–	–	–
$B_1 - \nu_6 - \nu$ asym XF_2 ax	–	–	–	[795]	[795]	690	704	700–710	702	–	–	–
$B_1 - \nu_7 -$ XF_2 eq wagging	536	538	535	534	515	419	–	418	385	388	385	390
$B_2 - \nu_8 - \nu$ asym XF_2 eq	827	830	825	822	829	730	736	–	–	719	720	723
$B_2 - \nu_9 - \delta$ sciss XF_2 ax out of plane	395	–	386	–	385	369	369	371	365	311	316	–
$- f_r -$ (eq)				4.58	4.78							
$- f_R -$ (ax)				3.44	3.73							

Numbers in square brackets are assumed values.

P = polarized.

606 cm^{-1} were assigned to stretching vibrations of the BrF_4^+ cation. As a result, it is assumed that the equilibrium of

$$BrF_4^+ + HF \rightleftharpoons BrF_5 + HF_2^-$$

lies far to the right. In another study[38] the interpretation of the spectrum of a similar solution was that it is due mostly to BrF_4^+ species. However, the BrF_4^+ frequencies at 726 and 606 cm^{-1}, which were mentioned in the former study[134] were not observed except as for a shoulder around 600 cm^{-1}.

The observed frequencies for the $IF_4^+SbF_6^-$ compound reported by several authors [38,117] are in good agreement.

Some vibrations of the IF_4^+ cation were also assigned in the spectrum of the complexed compound of $IF_4^+(CrF_4Sb_2F_{11})^-$. The observed frequencies at 729, 709 and 325 cm^{-1} were compared[19] and related to the 729, 720 and 341 cm^{-1} observed in $IF_4^+SbF_6^-$ [38,117].

Force constants were calculated for ClF_4^+ [38] but were slightly changed in a later study[41].

3.3.3 ^{19}F–NMR

Spectra of F–NMR of all the various XF_4^+ cations were obtained in solutions. Two separate signals of equal intensities at – 274 and – 256 ppm which were assigned to the cation were observed in a solution of $ClF_4^+SbF_6^-$ in an acidified HF–AsF_5 mixture[36]. This split resonance was observed only in the lower temperature range of – 70 °C to – 80 °C, whereas at higher temperatures (0 °C to – 60 °C) these collapse into a single average resonance at – 265 ppm. These observations strongly indicate the presence of two pairs of non-equivalent fluorines in ClF_4^+, in agreement with a C_{2v} symmetry as derived from X-ray data (3.3.1) and vibrational spectra (3.3.2).

In similarity with the assignment of the observed resonances in SF_4, it was assumed that the signals at – 274 and – 256 ppm are of the axial and equatorial fluorines, respectively.

In a solution of BrF_4^+ in HF or even in an acidified mixture of HF–AsF_5, only one resonance assigned to the cation was observed at – 197 ppm[38]. Another broad signal at 130 ppm was assigned to a solvent-anion HF–$Sb_2F_{11}^-$, averaged resonance resulting from fast exchange between them. However, this broad anionic-solvent signal was split at – 60 °C to – 80 °C into several peaks at 76, 93, 120 and 127 ppm, in similarity to the chemical shifts of $Sb_2F_{11}^-$ in HF solution.

In the $BrF_4^+Sb_2F_{11}^-$ melt a chemical shift at – 167 ppm was assigned[93] to the BrF_4^+ cation, which is quite different from the above-mentioned chemical shift in HF solution.

In a solution of BrF_5 in excess SbF_5, a single resonance assigned to the BrF_4^+ cation was observed[75] at – 180.4 ppm, which is closer to the one observed in HF solution.

In an IF_4^+ solution in HF only one line was observed at 133 ppm. Lowering the temperature to −80 °C or acidification of the solvent with excess AsF_5 did not enable the separation of the signal. This chemical shift is therefore an average of all fluorines of the cation, anion and solvent and is due to rapid exchange between all these species.

The exchange rates, whether inter- or intramolecular, decrease in the series $IF_5 \cdot SbF_5 > BrF_5 \cdot 2\,SbF_5 > ClF_5 \cdot SbF_5$, which might be explained by the decreasing size and polarizability of the corresponding halogen fluorides[38].

The absence of splittings of X–F spin-spin couplings in XF_4^+ species is similar to the observations in XF_2^+ species (3.1.4) and is due to rapid relaxation caused by interaction of the quadrupole moment of Cl or Br, with the asymmetric electric field gradient about the central halogen.

3.3.4 NQR Measurements

Resonances of ^{35}Cl, ^{121}Sb and ^{123}Sb of ClF_4^+ compounds with SbF_6^-, $Sb_2F_{11}^-$ and $Sb_4F_{21}^-$ anions were reported[86]. Comparing these results and those obtained for compounds formed with XF_3-halogen trifluorides, it seems that the donor capability decreases in the order:

$$ClF_3 > BrF_3 > ClF_5$$

3.3.5 Mössbauer Effect

Tin tetrafluoride complexes $SnF_4 \cdot 2\,L$ (L = BrF_5 or IF_5) were studied by Mössbauer effect[45,127,128]. The results were similar to those observed with halogen trifluoride compounds and have already been discussed (3.1.6). No SnF_6^{2-} species of O_h symmetry is present, but are most likely of D_{4h} symmetry, in which two *trans* fluorines are bridged to two separate cations[127].

3.3.6 ClF_4^+

Solid compounds with AsF_6^- and SbF_6^- anions were prepared by direct reaction of the reactants[30]. It was claimed[38] that good quality material could be obtained by combining an excess of ClF_5 with SbF_5 dissolved in HF and pumping off the solvent and excess reagent after completion of the reaction. An additional compound of $3\,ClF_5 \cdot 4\,SbF_5$ was also formed[38].

DTA studies of the ClF_5–SbF_5 system[129] indicated that three congruently melting compounds were formed. All these three compounds are ionic with infinite chains of discrete ClF_4^+ cation and SbF_6^-, $Sb_2F_{11}^-$ and $Sb_4F_{21}^-$ anions, connected through relatively weak fluorine bridgings and melting at 120°, 64° and 62 °C, respectively. It was claimed that the earlier reported[30] melting point of 35–88 °C for $ClF_4^+SbF_6^-$ was probably due to impure substances containing possibly an excess of SbF_5 or HF. It was also assumed[129] that a pure compound cannot be obtained in HF solution, but only by direct interaction of stoichiometric amounts.

3.3.7 BrF_4^+

A thermal unstable $BrF_5 \cdot AsF_5$ compound was formed by direct reaction[38]. In reaction with SbF_5, only a compound with a dimeric anion, $Sb_2F_{11}^-$, was formed, even if a larger excess of BrF_5 is used. It was first reported[110] to be formed by reacting SbF_5 with BrF_5 at its melting point of – 61 °C, obtaining a white solid, which melts to a red liquid at 60 °C, and of $BrF_5 \cdot 2\,SbF_5$ composition. The white solid turned reddish on exposure to air or moisture, indicating the presence of Br_2^+ [38].

In the Raman spectrum of BiF_5 solution in BrF_5, an additional line to those of BrF_5 was observed at 733 cm^{-1} and was assigned to a stretching frequency of BrF_4^+, indicating the formation of an ionic compound[134]. However, on evaporation of the solution, only BiF_5 was left behind as established from its Raman spectrum, thus indicating that no stable solid complexed compound was formed.

NbF_5 behaves similarly to BiF_5, dissolving in BrF_5 [134], but in this case no additional Raman line to be assigned to BrF_4^+ was observed, not even in the solution. Similarly, no solid complexed compound was formed on evaporation of the solution.

The chemical reactivity of $BrF_4^+Sb_2F_{11}^-$ seemed to be stronger than that of the parent compounds themselves[93]. The parent compounds are stable in acetonitrile solution, whereas the complex reacts vigorously in CH_3CN.

Reacting BrF_5 with SO_3 did not form the expected $BrF_4^+SO_3F^-$. Passing through various steps of covalent compounds an ionic compound is finally formed[79], which was shown by F–NMR to be $BrF_2^+SO_3F^-$.

3.3.8 IF_4^+

A white solid, $IF_4^+SbF_6^-$ was formed by dissolving SbF_5 in hot IF_5 [146,147]. Another compound with a dimeric $Sb_2F_{11}^-$ anion was formed by reacting SbF_5 with a mixture of $IF_5 + CrO_3$, which forms first CrO_2F_2. However, no adduct of CrO_2F_2 was formed in this liquid reaction mixture; instead $IF_4^+Sb_2F_{11}^-$ crystallized out as was shown by its X-rays[52].

A compound with a more complicated anion was formed[19] according to the reaction

$$3\,IF_5 + 2\,O_2^+(CrF_4Sb_2F_{11})^- \rightarrow 2\,IF_4^+(CrF_4Sb_2F_{11})^- + 2\,O_2 + IF_7$$

The dioxygenyl compound was first prepared by formation of the red-brown viscous liquid according to

$$CrF_5 + 2\,SbF_5 \rightarrow CrF_4Sb_2F_{11} ,$$

which is an excellent electron acceptor, which in turn reacts with oxygen, forming the dioxygenyl compound according to

$$O_2 + CrF_4Sb_2F_{11} \rightarrow O_2^+(CrF_4Sb_2F_{11})^- .$$

Other IF_4^+ compounds were prepared as well, such as $IF_4^+PtF_6^-$, which was obtained[14] by reacting $O_2^+PtF_6^-$ or PtF_5 with IF_5.

$IF_4^+SO_3F^-$ was obtained[147] by reacting IF_5 with SO_3.

3.4 XY_6^+

The only heteropolyhalogen XY_6^+ cations identified are XF_6^+ halogen fluorides (X = Cl, Br and I) as summarized in Table 25. These compounds have been shown by vibrational spectra as well as by F–NMR measurements to be of ionic structure with XF_6^+ cations.

The first one which was the only one known for a long time was the IF_6^+ cation, derived from its parent compound IF_7. The other two cations, BrF_6^+ and ClF_6^+, have been discovered only recently. Their supposed parent compounds, BrF_7 and ClF_7 have not been identified, and the cations were formed by oxidation of halogen pentafluorides with extremely strong oxidizers, such as PtF_6 or KrF^+ and $Kr_2F_3^+$ compounds, thus forming heptavalent halogen species.

These cations were shown by physical measurements to be of octahedral structure with O_h symmetry with a sp^3d^2 hybridization[27]. Another possible model involving three delocalized p-electron pairs of the central halogen forming three semi-ionic, three-center four-electron pp σ-bonds did not seem proper.

This was based on the magnitude of the calculated force constants. In the first model the force constant would be similar to that of a single covalent I–F bond, whereas in the second model the force constant would be only about half that of a covalent single I–F bond[27].

The calculated f_r in IF_6^+ and BrF_6^+ are actually the highest volumes obtained for any I–F and Br–F bonds, respectively and therefore do support the first bonding model.

Theoretical calculations using the NDDO approximation for an MO–LCAO–SCF method resulted in an octahedral symmetry for the ClF_6^+ cation with a bond length of 3.06 Å[24].

Table 25. XF_6^+ compounds and properties

	M.P.	Preparations and properties	Physical measurements
ClF_6^+			
PtF_6^-		$PtF_6 + FClO_2$ [35,37]; $ClF_5 + PtF_6$ (UV radiation)[35,37,107]; yellow-orange, canary yellow; decompose: 140–180° [107]	Raman[35,37]; IR[35,37,107]; F–NMR[36]; powder X-rays [107]
BrF_6^+			
AsF_6^-		$BrF_5 + (KrF^+, Kr_2F_3^+)$[40,74,75]	Raman[40,74,75]; IR[40]; F–NMR[74,75]; powder X-rays[40]
$Sb_2F_{11}^-$		White solid[40,74,75]	Raman[74,75]; IR[40]; F–NMR[74,75]
IF_6^+			
SbF_6^-	175–180°	$IF_6^+Sb_2F_{11}^- + IF_7 \xrightarrow{190^\circ} IF_6^+ SbF_6^-$ [82]	Raman, IR, powder X-rays[82]
Sb_2F_{11}		$IF_7 + 2\,SbF_5 \xrightarrow{100^\circ} IF_6^+Sb_2F_{11}^-$ [82]	
$IF_7 \cdot 3\,SbF_5$	92–94° [114]	$IF_7 + SbF_5$ [114]	
AsF_6^-	Sublime 120–140° [114]	$IF_7 + AsF_5$ [27,114]; Dissociation pressure: $\log p_{mm} = 16.418 - 4800/T$ [27]; $\Delta H_d^\circ = 43.9$ kcal · mole^{-1}; $\Delta H_{f,298}^\circ = -538$ kcal · mole^{-1} [27]	Raman[27,34]; IR[27,28]; F–NMR[12,21,83]; Mössbauer[22]; powder X-rays[27]
BF_4^-		$IF_7 + BF_3$ [114]; vapor pressure at −60°: 10 torr [114]	
AuF_6^-		$4\,O_2^+AuF_6^- + 3\,IF_5 \rightarrow IF_6^+AuF_6^-$ [15]	Raman, powder X-rays[15]

3.4.1 X-Rays

No single crystal X-ray data have been reported yet for this group of compounds. However, some powder X-ray data are known.

The data of $ClF_6^+PtF_6^-$ were obtained from a pure material but from a mixture containing also the $ClF_4^+PtF_6^-$ compound. Subtracting the lines of the latter compound, it seemed that the pattern for $ClF_6^+PtF_6^-$ is relatively simple and thus indicates a high symmetry[107]. This is in agreement with the more detailed information of some other XF_6^+ compounds, all of which are cubic[27,40,83] as summarized in Table 26.

Table 26. Powder X-ray data of XF_6^+ compounds

Compound	(Ref.)	Cell unit	a_0 in Å	Z	Volume per F' atom in $Å^3$	Density in $g \cdot cm^{-3}$ Calc.	Exp.
$ClF_6^+PtF_6^-$	107)	Face centered cubic	9.394	4	17.27	3.068	
$BrF_6^+AsF_6^-$	40)	Face centered cubic	9.49	4	17.8	3.33	3.28
$IF_6^+AsF_6^-$	27)	Face centered cubic					
$IF_6^+SbF_6^-$	82)	Cubic	6.069	1	18.6	3.54	3.48
$IF_6^+AuF_6^-$	15)	Cubic	9.573	4			

3.4.2 Vibrational Spectra

Both Raman and IR spectra have been observed for all the XF_6^+ cations. These spectra are rather simple ones since both the cations and MF_6^- anions are of the highly symmetrical octahedral structure, O_h, with a center of symmetry. Therefore the selection rules for Raman and IR active vibrations are mutually exclusive. Only three Raman lines, ν_1, ν_2 stretchings and ν_5 bending mode are expected, whereas two non-coinciding IR lines, ν_3 stretching and ν_4 bending mode are expected. Indeed, the reported spectral data as summarized in Table 27 clearly indicate that these compounds are ionic, both cations and anions of octahedral structure.

In case of BrF_6^+ and IF_6^+ compounds, pure compounds have been studied, whereas in the case of ClF_6^+ compound, no pure compound was available and only mixtures of ClF_2^+, ClF_4^+ and ClO_2^+ compounds were studied.

Table 27. Vibrational frequencies in cm^{-1} and force constants in mdyn/Å of XF_6^+ cations

Cation	ClF_6^+			BrF_6^+				IF_6^+				
Anion	PtF_6^-			AsF_6^-		$Sb_2F_{11}^-$		AsF_6^-		SbF_6^-		SbF_{11}^+
Spectra	IR	IR	R	IR	R	IR	R	IR	R	IR	R	R
Ref.	107)	35,37)	35,37)	40)	74,75)	40)	74,75)	27,28)	27,34)	82)	82)	82)
$\nu_1(A_{1g})$			679		658 P		656		708 P		710	705
$\nu_2(E_g)$			630		668 dp		666		732 dp		733	729
$\nu_3(F_{1u})$	890	890		775		775		797 790		797		
$\nu_4(F_{1u})$	540	582		430		430		343		344		
$\nu_5(F_{2g})$			513		405 dp		404		340		339	340
$\nu_6(F_{2u})$										250		
f_r	4.68[37]; 4.98[40]			4.9[40]				5.60[27]; 5.42[40]				

P = polarized; dp = depolarized

The vibrations of both BrF_6^+ and IF_6^+ show a rather unusual pattern compared to that usually observed in MX_6 species. In all other octahedral species except BrF_6^+ and IF_6^+, the frequencies of ν_1 are higher than those of ν_2, whereas in BrF_6^+ and IF_6^+ these are reversed, namely $\nu_1 < \nu_2$. It was first claimed [27] that based on the relative intensities of the Raman lines of IF_6^+ at 708 cm^{-1} and 732 cm^{-1}, the first one which is more intense although of lower frequency should be assigned as ν_1 and the latter, weaker one of higher frequency as ν_2. Force constants were also calculated and claimed to support these assignments[27]. However, this extraordinary pattern was criticized[81] as being in contrast to all known MX_6 species (the BrF_6^+ cation not being known at the time). It was therefore claimed that the ratio of intensities was not significant enough, since there have also been other cases, such as $SeCl_6^{2-}$, in which the intensity of ν_2 is stronger than that of ν_1. Force field calculations using the MUBBF method were performed and claimed to support the reversed assignments $\nu_1 = 732$ and $\nu_2 = 708$ cm^{-1} in agreement with the usual pattern, in which $\nu_1 > \nu_2$ [81]. In a later study[28] it was claimed that there are many cases in which approximate force fields failed to give meaningful results and therefore should not be over-valued. Force field calculations performed on spectral data of $IF_6^+SbF_6^-$ [16] have also shown that only the use of the MUBBF method resulted in $\nu_1 < \nu_2$, whereas using UBBF, OVFF and MOVFF methods gave the opposite result of the normal pattern, namely $\nu_1 > \nu_2$.

Actually, experimental evidence helped to resolve these conflicting assignments. Polarization measurements of the Raman lines observed in $IF_6^+AsF_6^-$ solution in HF[34] have shown that the intense 711 cm^{-1} line was polarized whereas the weak 732 cm^{-1} line was depolarized. In octahedral species the totally symmetric vibration ν_1 is indeed polarized, whereas the ν_2 vibration is depolarized. This therefore provided clear support for the assignments that $\nu_1 < \nu_2$. Similar results were later also obtained in the study of $BrF_6^+AsF_6^-$ in HF solution[75], in which the intense lower frequency 661 cm^{-1} was polarized and the weaker but higher frequency at 671 cm^{-1} was depolarized, also indicating that $\nu_1 < \nu_2$.

Table 28 summarizes frequencies of ν_1 and ν_2 observed for several isoelectronic series in which it was noticed that $\Delta\nu = \nu_1 - \nu_2$ becomes smaller with increase in mass of the central atom. Similarly, $\Delta\nu$ decreases in each group of the periodic table with increase in mass of the central atom. Including the observed values for BrF_6^+ and IF_6^+ in which $\Delta\nu$ obtains even negative values indeed fits into the general pattern of this table. This trend is presumed to arise from a decrease in the non-bonded fluorine interactions with an increase in the size of the central atom[27,75].

The observed shoulder in ν_3 of ClF_6^+ at 877 cm^{-1} was explained to be due to ^{35}Cl and ^{37}Cl isotope shift, which is calculated to equal 12.5 cm^{-1}, compared to the observed split of 13 cm^{-1} [35,37]. On the other hand, the observed splitting in ν_4 of $BrF_6^+AsF_6^-$ does not seem to be due to ^{79}Br, ^{81}Br isotopes, since the observed value is larger than 2 cm^{-1}, which is the calculated one. It is therefore assumed to result from crystal field or site symmetry effects[40].

Force constants were calculated but seem to differ slightly, as listed in Table 27.

Mean amplitudes were calculated[9,10,109] and found to be at 300 °K for ClF_6^+, $\mu = 0.0426$ Å and for IF_6^+, $\mu = 0.0377$ Å.

Table 28. Vibrations ν_1 (A_{1g}), ν_2 (E_g) and $\Delta\nu = \nu_1 - \nu_2$ of isoelectronic octahedral XF_6 species and their relative intensities

	SiF_6^{2-}	PF_6^-	SF_6	ClF_6^+
ν_1	656 vs	735 vs	769 vs	679
ν_2	465 vw	563 w	640 w	630
$\Delta\nu$	191	172	129	49
	GeF_6^{2-}	AsF_6^-	SeF_6	BrF_6^+
ν_1	627 s	679 ms	708 vs	658 vs
ν_2	454 w	565 vw	662 w	668 w
$\Delta\nu$	173	114	46	– 10
	SnF_6^{2-}	SbF_6^-	TeF_6	IF_6^+
ν_1	593 s	675 s	701 s	708 vs
ν_2	465 vw	583 vw	674 w	732 w
$\Delta\nu$	128	92	27	– 24

The ClF_6^+ data from Refs.[35,37], BrF_6^+ [74,75]; all the others were listed in Ref.[28].

3.4.3 ^{19}F–NMR

All the XF_6^+ halogen-fluoride cations, show finely resolved F–NMR spectra, which are most detailed ones and are summarized in Table 29. All the observed details, splittings and intensities are in very good agreement with the details expected from spin-spin couplings as well as isotope splittings. All observations were performed in solutions of XF_6^+ compounds in HF at room temperature and in the case of ClF_6^+ and BrF_6^+, even without any additional acidification[36,75].

Two sharp sets of quartets, each of equal intensities 1:1:1:1, were observed in both ClF_6^+ and BrF_6^+ solutions. These quartets result from spin-spin coupling of six equivalent fluorines with two different isotopes of the central nuclei in natural abundance in each case, namely ^{35}Cl, ^{37}Cl and ^{79}Br, ^{81}Br, all of which have a spin of I = 3/2.

In the case of IF_6^+ solution, the F–NMR signal is split into a sextet, with approximately similar intensities[21]. This results from the fact that the only natural isotope of iodine, ^{127}I, has a spin of I = 5/2. In this case the HF solution was acidified with excess AsF_5 to suppress the broadening of IF_6^+ peaks caused by chemical exchange with the solvent according to

$$IF_6^+ + HF_2^- \rightleftharpoons IF_7 + HF$$

or

$$IF_6^+ + 2\,HF \rightleftharpoons IF_7 + H_2F^+$$

Table 29. ^{19}F–NMR data of XF_6^+ cations

Compound	Ref.	δ in ppm (ext. Ref: $CFCl_3$)	Coupling constant (Hz)	Line width (Hz)	Observed isotope coupling ratio	Gyro-magnetic ratio
$ClF_6^+PtF_6^-$	36)	– 388 (double quartet)	$J\,^{35}Cl-^{19}F$: 337 $J\,^{37}Cl-^{19}F$: 281	15	1.199	1.2014
$BrF_6^+Sb_2F_{11}^-$	75)	– 339.4 (double quartet)	$J\,^{79}Br-^{19}F$: 1575 $J\,^{81}Br-^{19}F$: 1697	35	1.077	1.0779
$BrF_6^+AsF_6^-$	75)	– 337.4 (double quartet)	$J\,^{79}Br-^{19}F$: 1587 $J\,^{81}Br-^{19}F$: 1709	49	1.077	1.0779
$IF_6^+AsF_6^-$	21)	– 70.5 (sextet)	$J\,^{127}I-^{19}F$: 2730 ± 15			
$IF_6^+AsF_6^-$	12)	– 43 ± 16				

On the other hand, no such acidification was required in case of ClF_6^+ or BrF_6^+ as both ClF_7 and BrF_7 are not known and therefore no such chemical exchanges like $ClF_6^+ + HF_2^- \rightleftharpoons ClF_7 + HF$ exist [36].

The observations of these spin-spin couplings are in contrast to all other halogen fluoride – XF_n^+ species which do not show such splittings. As already discussed (3.1.4 and 3.3.3) this results from rapid relaxation due to the interaction of the quadrupole moment of the central halogen with the asymmetric electric field gradient. However, in the case of XF_6^+ cations having a spherically symmetric electric field about the central halogen nucleus with an O_h symmetry, no such asymmetry to cause relaxation exists and these spin-spin couplings are indeed well resolved in all cases. Thus the well resolved F-NMR spectra provide strong support that the structure of XF_6^+ cations are of octahedral spherical symmetry [36,75].

The observed ratio of the peak areas of the two quartet sets in ClF_6^+ equals 3:1, which agrees well with the ratio of the natural abundance of the chlorine isotopes $^{35}Cl : ^{37}Cl = 75.4 : 24.6 = 3.06 : 1$ [36].

Similarly, the observed intensities of the two quartet sets in BrF_6^+ are about equal, also in good agreement with the ratio of the natural abundance of the bromine isotopes $^{79}Br : ^{81}Br = 50.57 : 49.43 = 1.03 : 1$.

As listed in Table 29 the coupling constants ratio of the various isotopes in ClF_6^+ and BrF_6^+ are 1.199 and 1.077 respectively. These are in very good agreement with the ratios of their gyromagnetic ratios of the respective isotopes, 1.2014 for the chlorine isotopes and 1.0779 for the bromine isotopes.

Slight changes in chemical shifts are also noticed. The center of the $^{35}ClF_6^+$ resonance is shifted by 0.15 ± 0.02 ppm to the low field of the $^{37}ClF_6^+$ one [36] and similarly that of the $^{79}BrF_6^+$ is shifted by 0.14 ± 0.02 ppm to the low field of $^{81}BrF_6^+$ [75]. This has been explained by the difference in the vibrational amplitudes of the two central halogens. The heavier atom having a smaller vibrational amplitude causing the electrons of the fluorine to be less polarized results in increased shielding of the fluorines. Both these isotopic shifts per unit mass difference, 0.07 ± 0.01 ppm for BrF_6^+ and 0.075 ± 0.01 ppm for ClF_6^+ are considerably larger than observed for the isoelectronic SeF_6 and SF_6 molecules. This seems to be according to an original suggestion[75b)] that there is a strong inverse dependence of the isotopic shift on the bond lengths, which are expected to decrease with increasing atomic number in each series of isoelectronic species.

The chemical shifts in the series ClF_2^+, ClF_4^+ and ClF_6^+ are 24, – 265 and – 388 ppm, respectively. Similarly the chemical shifts in the series BrF_2^+, BrF_4^+ and BrF_6^+ are 48, – 197 and – 338 ppm, respectively. This is as expected, that increase of oxidation state of the central halogen causes increased deshielding of the fluorine ligands. In the series IF_6^+, BrF_6^+ and ClF_6^+ the chemical shifts are – 70.5, – 338 and – 388 ppm, respectively, the increased deshielding of the fluorines as expected with increasing of the electronegativity of the central halogen. This also holds for the couples BrF_4^+, ClF_4^+ and BrF_2^+, ClF_2^+, in both cases the chemical shifts of ClF_x^+ cations being lower than those of the BrF_x^+ cations.

In the case of IF_6^+ it was noticed that cooling the solution from 34 °C to 0 °C caused the peaks to broaden. This is not a result of chemical exchange, since in this case the opposite effect of narrowing of the peaks should have resulted with decreasing temperature. The observed effect seems to be in accordance with the coupling of the nucleus and undergoing quadrupole relaxation, if the electric field gradient differs from zero. Since this is not the case in the expected O_h symmetry, it was assumed that some deformation of the octahedral symmetry happens nevertheless, as a result of collisions between molecules in solution[21)].

Broad line F–NMR have also been reported for solid $IF_6^+AsF_6^-$ [12)]. Two lines with approximately equal areas were observed and assumed to be due to the IF_6^+ and AsF_6^- ions, the chemical shift of IF_6^+ being – 43 ± 16 ppm. This is also consistent with an ionic structure, the two ions containing equal numbers of fluorines.

Another report[83)] of broad line NMR of ^{19}F, ^{127}I and ^{75}As concluded that whereas the AsF_6^- is a regular octahedron, the IF_6^+ is not.

3.4.4 Mössbauer Effect

The $IF_6^+AsF_6^-$ was studied by Mössbauer effect using ^{129}I isotope. The observed spectrum consisted of one single line, characteristic of a highly symmetric arrangement of fluorine atoms around the iodine atom[22)]. The observed isomer shift was at $\delta = -4.68$ mm sec^{-1}. The negative sign indicates removal of s electrons from the 5 s^2 shell. Assuming that removal of p-electrons from iodine takes place prior to removal of s electrons, the maximum number of removed electrons from the iodine can be calculated. Using the formula $\delta = -8.2\,h_s + 1.5\,h_p - 0.54$, in which h_p is the number of holes of p electrons and equals 6, the number of s electrons extracted from iodine h_s is calculated to be equal to 1.6. Altogether, four p-electrons and 1.6 s-electrons are extracted from iodine for all the six fluorines, or 5.6/6 = 0.93 electrons per one fluorine.

3.4.5 ClF_6^+

The ClF_6^+ cation was first reported[107] to be formed by the reaction:

$$PtF_6 + ClF_5(\text{excess}) \rightarrow ClF_6^+PtF_6^-$$

condensed into a sapphire reactor at – 196 °C and slowly warmed up to room temperature, being exposed to laboratory light for eight days. The red gas, PtF_6, disappeared, forming first a red-brown solid which slowly turned bright yellow. After pumping off the excess ClF_5, the remaining solid had no vapor pressure. It was identified as a mixture of

$$ClF_6^+PtF_6^- + ClF_4^+PtF_6^- .$$

It was later found[35,37] that on radiation of the reaction mixture at room temperature with a high pressure quartz-mercury lamp for 24 hrs a yellow-brown solid formed. Keeping the reaction mixture for an additional 24 hrs at – 20 °C without radiation allowed the color of the solid to change to yellow-orange. The solid product was identified as a mixture of ClF_2^+ and ClF_6^+ compounds of PtF_6^- having been formed according to

$$2\,ClF_5 + 2\,PtF_6 \xrightarrow[\text{UV}]{\text{unfiltered}} ClF_6^+PtF_6^- + ClF_2^+PtF_6^- + F_2$$

In a similar reaction, but allowing the UV radiation to be filtered through pyrex and keeping the reaction for 14 days, a mixture of ClF_4^+ and ClF_6^+ compounds of PtF_6^- was formed according to

$$2\,ClF_5 + 2\,PtF_6 \xrightarrow[\text{filtered UV}]{\text{pyrex}} ClF_6^+PtF_6^- + ClF_4^+PtF_6^-$$

It seems that the unfiltered UV radiation decomposes the $ClF_5 \rightarrow ClF_3 + F_2$, and this in turn forms the ClF_2^+ compound. Without this decomposition in the absence of UV radiation when filtered through pyrex, the ClF_4^+ compound was formed and not the ClF_2^+ one.

In another preparation[35,37] $FClO_2$ was condensed together with PtF_6 at – 196 °C and rapidly warmed up to – 78 °C. Keeping it for 48 hrs resulted in a mixture of ClO_2^+ and ClF_6^+ compounds formed according to

$$6\,FClO_2 + 6\,PtF_6 \rightarrow ClF_6^+PtF_6^- + 5\,ClO_2^+PtF_6^- + O_2$$

These various reported syntheses always resulted in mixtures of compounds and no pure material was obtained. The material is stable for months without decomposition when stored in FEP containers[37]. It is a very powerful oxidizer, reacting explosively with organic materials or water.

An attempt to release the parent compound ClF_7 by reacting the ClF_6^+ compound with NOF has failed[35,37]. Instead, the following reaction releases only chlorine pentafluoride

$$ClF_6^+PtF_6^- + NOF \rightarrow NO^+PtF_6^- + (ClF_7) \rightarrow ClF_5 + F_2$$

in which the expected ClF_7 decomposed to $ClF_5 + F_2$.

Similarly, thermal decomposition[107] also resulted in a mixture of $ClF_5 + F_2$ according to

$$ClF_6^+PtF_6^- \xrightarrow{140-180\,^\circ C} PtF_5 + (ClF_7) \rightarrow ClF_5 + F_2 \, .$$

3.4.6 BrF_6^+

The BrF_6^+ compounds were prepared by dissolving $KrF^+AsF_6^-$ or $Kr_2F_3^+$ salts of AsF_6^- and SbF_6^- in excess BrF_5 [74,75]. The reactions are basically according to

$$BrF_5 + KrF^+ \rightarrow BrF_6^+ + Kr$$

or

$$BrF_5 + Kr_2F_3^+ \rightarrow BrF_6^+ + KrF_2 + Kr$$

In the case of reacting the SbF_6^- salt a mixture was obtained containing

$$BrF_6^+Sb_2F_{11}^- + BrF_4^+Sb_2F_{11}^- \, .$$

Since the dimeric $As_2F_{11}^-$ is not stable, no such compound was formed when reacting an AsF_6^- salt. Similarly, no BrF_4^+ compound was formed in this case, as $BrF_4^+AsF_6^-$ has not been known owing to its instability. Therefore, a pure compound $BrF_6^+AsF_6^-$ is obtained.

A pure compound $BrF_6^+Sb_2F_{11}^-$ can be obtained from the mixture containing also $BrF_4^+Sb_2F_{11}^-$, either by pumping off the more volatile $BrF_4^+Sb_2F_{11}^-$ at 50 °C, or by vacuum sublimation, leaving behind pure $BrF_6^+Sb_2F_{11}^-$ [40]. Attempts to prepare the parent compound BrF_7 by reacting a BrF_6^+ compound with NOF have failed[75]. Instead the following reaction takes place:

$$BrF_6^+AsF_6^- + 2\,NOF \xrightarrow{-78\,°C} NO^+AsF_6^- + NO^+BrF_6^- + F_2$$

the bromine being reduced to the pentavalent state.

The BrF_6^+ compounds are very powerful oxidants[75], even more so than elemental fluorine under ambient conditions, as they oxidize xenon and oxygen gases according to

$$BrF_6^+AsF_6^- + Xe \rightarrow XeF^+AsF_6^- + BrF_5$$

or

$$BrF_6^+AsF_6^- + O_2 \rightarrow O_2^+AsF_6^- + BrF_5 + \tfrac{1}{2}F_2\ .$$

3.4.7 IF_6^+

The IF_6^+ compounds have been known for a long time and were the first ones of the XF_6^+ type to have been synthesized. As the parent compound IF_7 is known to be stable, the regular reaction by direct interaction with a proper Lewis acid takes place[114,27,82]. In an early report[114] it was assumed that the compound $IF_7 \cdot 3\,SbF_5$ should be formulated as $(IF_4)^{+3}(SbF_6^-)_3$. It was later suggested[27] to be more correctly formulated as $IF_6^+SbF_6^- \cdot xSbF_5$, with a polymeric anion such as the dimeric $Sb_2F_{11}^-$ or a higher one. This was indeed supported by the fact that various molar ratios of IF_7 and SbF_5 were found to react, such as $IF_7 \cdot 2.4\,SbF_5$. In a later study[82] it was found that reacting IF_7 with about twice the molar amount of SbF_5 formed at around 90–100 °C the $IF_6^+Sb_2F_{11}^-$, but most probably as a mixture with $IF_6^+SbF_6^-$. Only after grinding the material and reacting the solid with excess of IF_7 at 170–190 °C was the $IF_6^+SbF_6^-$ obtained. The $IF_6^+Sb_2F_{11}^-$ was also obtained[92] by reacting

$$IF_5 + KrF^+Sb_2F_{11}^- \rightarrow IF_6^+Sb_2F_{11}^- + Kr.$$

AuF_5 was found to oxidize IF_5[15] reacting according to

$$IF_5 + AuF_5 + SbF_5 \rightarrow IF_6^+SbF_6^- + AuF_3$$

Similarly, AuF_6^- also oxidizes IF_5 according to

$$IF_5 + K^+AuF_6^- + 2\,SbF_5 \rightarrow IF_6^+SbF_6^- + K^+SbF_6^- + AuF_3\ .$$

An AuF_6^- salt can be formed according to

$$3\,IF_5 + 4\,O_2^+AuF_6^- \rightarrow 3\,IF_6^+AuF_6^- + 4\,O_2 + AuF_3\ .$$

The compound $IF_6^+SbF_6^-$ was found to react and fluorinate various materials[82] such as CO, CH_4, SO_2, NO and NO_2. It also reacts with Radon gas to form a solid according to

$$Rn + IF_6^+SbF_6^- \rightarrow RnF^+SbF_6^- + IF_5 .$$

This latter reaction was suggested to be used in conjunction with a second bed of $O_2^+SbF_6^-$ reacting with Xe to form the solid $XeF^+SbF_6^-$, to trap out separately Rn and then Xe, as solid compounds. The pure gases can then be recovered by hydrolyzing the solid beds. Since no similar solid compound is known to react with Kr, it was proposed to use such a train in order to separate a mixture of Rn, Xe and Kr[82]. Similarly, it was proposed to use $IF_6^+SbF_6^-$ for analysis of Rn in air and for the purification of contaminated radioactive air.

In summary, it seems that most of the possible polyhalogen cations have indeed been prepared and identified. Their chemistry is rather diversified and various modern physical methods have been used in determining their structures and properties. Some interesting data, such as particular single-crystal X-rays are still missing. Most of all, it would be interesting to study more carefully possible chemical applications and reactions of these highly reactive compounds.

Acknowledgement. The author wishes to thank the Chemistry Department A of the Technical University of Denmark, Lyngby, Denmark, where this project started, during his stay there as guest professor.

References

1. Angenault, J., Couturier, J.C.: Rev. Chim. Minerale *9*, 701 (1972)
2. Arotsky, J., Symons, M.C.R.: Quart. Rev. *16*, 282 (1962)
3. Arotsky, J., Mishra, H.C., Symons, M.C.R.: J. Chem. Soc. *1961*, 12
4. Arotsky, J., Mishra, H,C., Symons, M.C.R.: J. Chem. Soc. *1962*, 2582
5. Aubke, F., Cady, G.H.: Inorg. Chem. *4*, 269 (1965)
6. Aynsley, E.E., Greenwood, N.N., Wharmby, D.H.W.: J. Chem. Soc. *1963*, 5369
7. Baird, H.W., Giles, H.F.: Acta Cryst. *A25*, S115 (1969)
8. Bali, A., Malhotra, K.C.: J. Inorg. Nucl. Chem. *38*, 411 (1976)
9. Baran, E.J.: Z. Chem. *14*, 204 (1974)
10. Baran, E.J.: Monatsh. Chem. *105*, 1148 (1974)
11. Baran, E.J.: Asoc. Quim. Argent. *63*, 239 (1975); C.A. *85*, 133270q
12. Barr, M.R., Dunnel, B.A.: Can. J. Chem. *48*, 895 (1970)
13. Bartlett, N., Lohmann, D.H,: J. Chem. Soc. *1962*, 5253
14. Bartlett, N., Lohmann, D.H.: ibid. *1964*, 619
15. Bartlett, N., Leary, K.: Rev. Chim. Minerale *13*, 82 (1976)
16. Basile, L.J., Hohorst, F.A., Ferraro, J.R.: Appl. Spectrosc. *29*, 260 (1975)
17. Booth, M., Morton, M.J., Gillespie, R.J.: Adv. Raman Spectrosc. *1*, 364 (1972)
18. Brown, D.H., Dixon, K.R., Sharp, D.W.A.: Chem. Comm. *1966*, 654

19. Brown, S.P., Loehr, T.M., Gard, G.H.: J. Fluor. Chem. *7*, 19 (1976)
20. Brownstein, M., Shamir, J.: Can. J. Chem. *50*, 3409 (1972)
21. Brownstein, M., Selig, H.: Inorg. Chem. *11*, 656 (1972)
22. Bukshpan, S., Soriano, J., Shamir, J.: Chem. Phys. Letters *4*, 241 (1969)
23. Carter, H.A., Aubke, F.: Can. J. Chem. *48*, 3456 (1970)
24. Charkin, O.P., Smolyar, A.E., Klimenko, N.M.: Zh. Strukt. Khim. *15*, 172 (1974) C.A. *80*, 112, 930 m
25. a) Christe, K.O., Pavlath, A.E.: Z. anorg. allg. Chem. *335*, 210 (1965)
b) Christe, K.O., Guertin, J.P.: Inorg. Chem. *4*, 905 (1965)
26. Christe, K.O., Sawodny, W.: Inorg. Chem. *6*, 313 (1967)
27. Christe, K.O., Sawodny, W.: ibid. *6*, 1783 (1967)
28. Christe, K.O., Sawodny, W.: ibid. *7*, 1685 (1968)
29. Christe, K.O., Sawodny, W.: ibid. *8*, 212 (1969)
30. Christe, K.O., Pilipovich, D.: ibid. *8*, 391 (1969)
31. Christe, K.O., Muirhead, J.S.: J. Am. Chem. Soc. *91*, 7777 (1969)
32. Christe, K.O.: J. Phys. Chem. *73*, 2792 (1969)
33. Christe, K.O., Schack, C.J.: Inorg. Chem. *9*, 2296 (1970)
34. Christe, K.O.: ibid. *9*, 2801 (1970)
35. Christe, K.O.: Inorg. Nucl. Chem. Letters *8*, 741 (1972)
36. Christe, K.O., Hon, J.F., Pilipovich, D.: Inorg. Chem. *12*, 84 (1973)
37. Christe, K.O.: ibid. *12*, 1580 (1973)
38. Christe, K.O., Sawodny, W.: ibid. *12*, 2879 (1973)
39. Christe, K.O., Wilson, R.: Inorg. Nucl. Chem. Letters *9*, 845 (1973)
40. Christe, K.O., Wilson, R.D.: Inorg. Chem. *14*, 694 (1975)
41. Christe, K.O., Curtis, E.C., Schack, C.J., Cyvin, S.J., Brunvol, J., Sawodny, W.: Spectrochim. Acta *32A*, 1142 (1976)
42. a) Cornwell, C.D., Yamasaki, J.: J. Chem. Phys. *27*, 1060 (1957)
b) Bray, P.J.: ibid. *23*, 703 (1955)
43. Chung, C., Cady, G.H.: Inorg. Chem. *11*, 2528 (1972)
44. Davies, C.G., Gillespie, R.J., Ireland, S.R., Sowa, J.M.: Can. J. Chem. *52*, 2048 (1974)
45. Dzevitskii, B.E., Sukhoverkhov, V.F.: Izv. Sib. Otd. Akad. Nauk SSSR, Ser. Khim. Nauk *2*, 54 (1968) C.A. *69*, 9251q b
46. Eachus, R.S., Symons, M.C.R.: J. Chem. Soc. Dalton Trans. *1976*, 431
47. Edwards, A.J., Jones, G.R.: Chem. Comm. *1967*, 1304
48. Edwards, A.J., Jones, G.R., Sills, R.J.C.: ibid. *1968*, 1527
49. Edwards. A.J., Jones, G.R.: J. Chem. Soc. *A1969*, 1467
50. Edwards, A.J., Sills, R.J.C.: ibid. *A1970*, 2697
51. Edwards, A.J., Jones, G.R.: ibid. *A1971*, 2318
52. Edwards, A.J., Taylor, P.: J. Chem. Soc. Dalton Trans. *1975*, 2174
53. Edwards, A.J., Christe, K.O.: ibid. *1976*, 175
54. Evans, J.C., Lo, G.Y.S.: Inorg. Chem. *6*, 836 (1967)
55. Faltens, M.O., Shirley, D.A.: J. Chem. Phys. *53*, 4249 (1970)
56. Fialkov, Y.A., Schorr, O.I.: J. Gen. Chem. SSSR *19*, a235 (1949)
57. Fialkov, Y.A., Abarbarchuk, I.L.: Ukrain. Khim. Zhur. *15*, 372 (1949) C.A. *47*, 7877c
58. Fialkov, Y.A.: Izv. Akad. SSSR, Utd. Khim. Nauk *1954*, 972 C.A. *49*, 14552 ghi
59. Fischer, J., Liimatainen, R., Bingle, J.: J. Am. Chem. Soc. *77*, 5848 (1955)
60. Forneris, R., Tavares-Forneris, Y.: J. Mol. Struct. *23*, 241 (1974)
61. Garrett, R.A., Gillespie, R.J., Senior, J.B.: Inorg. Chem. *4*, 563 (1965)
62. Gillespie, R.J., Milne, J.B.: Chem. Comm. *1966*, 158
63. Gillespie, R.J., Milne, J.B.: Inorg. Chem. *5*, 1577 (1966)
64. Gillespie, R.J., Morton, M.J.: Chem. Comm. *1968*, 1565
65. Gillespie, R.J., Milne, J.B., Morton, M.J.: Inorg. Chem. *7*, 2221 (1968)
66. Gillespie, R.J., Malhotra, K.C.: ibid. *8*, 1751 (1969)

67. Gillespie, R.J., Morton, M.J.: J. Mol. Spectrosc. *30*, 178 (1969)
68. Gillespie, R.J., Morton, M.J.: Inorg. Chem. *9*, 616 (1970)
69. Gillespie, R.J., Morton, M.J.: ibid. *9*, 811 (1970)
70. Gillespie, R.J., Morton, M.J.: Quart. Rev. *25*, 553 (1971)
71. Gillespie, R.J., Morton, M.J., Sowa, J.M.: Adv. Raman Spectrosc. *1*, 539 (1972)
72. Gillespie, R.J., Morton, M.J.: Inorg. Chem. *11*, 586 (1972)
73. Gillespie, R.J., Morton, M.J.: ibid. *11*, 591 (1972)
74. Gillespie, R.J., Ronald, J., Schrobilgen, G.J.: J.C.S. Chem. Comm. *1974*, 90
75. a) Gillespie, R.J., Schrobilgen, G.J.: Inorg. Chem. *13*, 1230 (1974)
b) Tiers, G.V.D.: J. Inorg. Nucl. Chem. *16*, 363 (1961)
76. Glemser, O., Smalc, A.: Angew. Chem. Int. Ed. *8*, 517 (1969)
77. Gortsema, F.P., Toeniskoetter, R.H.: Inorg. Chem. *5*, 1925 (1966)
78. Gottardi, W.: Monatsh. Chem. *106*, 1203 (1975)
79. Gross, U., Meinert, H., Grimmer, A.R.: Z. Chem. *10*, 441 (1970)
80. Gutman, V., Emeleus, H.J.: J. Chem. Soc. *1950*, 1046
81. Hardwick, J.L., Leroi, G.E.: Inorg. Chem. *7*, 1683 (1968)
82. Hohorst, F.A., Stein, L., Gebert, E.: ibid. *14*, 2233 (1975)
83. Hon, J.F., Christe, K.O.: J. Chem. Phys. *52*, 1960 (1970)
84. Kemmit, R.D.W., Murray, M., McRae, V.M., Peacock, R.D., Symons, M.C.R., Donnel, T.A.: J. Chem. Soc. *A1968*, 862
85. Klyagina, A.P., Klimenko, N.M., Dyatkina, M.E.: Zh. Strukt. Khim. *14*, 898 (1973)
86. Kuzmin, A.I., Shpanko, V.I., Kazakov, V.P., Sukhovezkhov, V.P., Dzevitskii, B.E., Bryukhova, E.V.: Izv. Akad. Nauk. SSSR, Ser. Fiz. *39*, 2555 (1975) C.A. *84*, 128401h
87. Lau, C., Passmore, J.: Chem. Comm. *1971*, 950
88. Lau, C., Passmore, J.: J. Chem. Soc. Dalton Trans. *1973*, 2528
89. Lind, M.D., Christe, K.O.: Inorg. Chem. *11*, 608 (1972)
90. Lynton, H., Passmore, J.: Can. J. Chem. *49*, 2539 (1971)
91. Masson, I.: J. Chem. Soc. *1938*, 1708
92. McKee, C., Adams, C.J., Zalkin, A., Bartlett, N.: J. Chem. Soc. Chem. Comm. *1973*, 26
93. Meinert, H., Gross, U., Grimmer, A.R.: Z. Chem. *10*, 226 (1970)
94. Merryman, D.J., Edwards, P.A., Corbett, J.D., McCarley, R.E.: J. Chem. Soc. Chem. Comm. *1972*, 779
95. Merryman, D.J., Corbett, J.D.: Inorg. Chem. *13*, 1258 (1974)
96. Merryman, D.J., Edwards, P.A., Corbett, J.D., McCarley, R.E.: Inorg. Chem. *13*, 1471 (1974)
97. Merryman, D.J., Corbett, J.D., Edwards, P.A.: ibid. *14*, 428 (1975)
98. Olah, G.A., Comisarow, M.B.: J. Am. Chem. Soc. *90*, 5033 (1968)
99. Olah. G.A., Comisarow, M.B.: J. Am. Chem. Soc. *91*, 2172 (1969)
100. Passmore, J., Taylor, P.: J. Chem. Soc. Dalton Trans. *1976*, 804
101. Paul, R.C., Arora, C.L., Malhotra, K.C.: J. Inorg. Nucl. Chem. *33*, 991 (1971)
102. Paul, R.C., Arora, C.L., Malhotra, K.C.: Indian J. Chem. *9*, 473 (1971)
103. Paul, R.C., Puri, J.K., Arora, C.L., Malhotra, K.C.: ibid. *9*, 1384 (1971)
104. Richards, G.W., Woolf, A.A.: J. Fluor. Chem. *1*, 129 (1971)
105. Roberto, F.Q., Mamantov, G.: Inorg. Chim. Acta *2*, 173 (1968)
106. Roberto, F.Q., Mamantov, G.: ibid. *2*, 317 (1968)
107. Roberto, F.Q.: Inorg. Nucl. Chem. Letters *8*, 737 (1972)
108. Ruff, O.: Ber. Deut. Chem. Ges. *39*, 4310 (1906); ibid. *48*, 2068 (1915)
109. Sanyal, D.N., Verma, D.N., Dixit, L.: Acta Cienca Indica *1*, 47 (1974) C.A. *82*, 131208q
110. Schmeisser, M., Pammer, E.: Angew. Chem. *69*, 781 (1957)
111. Schmeisser, M., Ludovici, W.: Z. Naturforsch. *20b*, 602 (1965)
112. Schmeisser, M., Ludovici, W., Naumann, D., Sartori, P., Scharf, E.: Chem. Ber. *101*, 4214 (1968)
113. Seel, F., Detmer, O.: Angew. Chem. *70*, 163 (1958)
114. Seel, F., Detmer, O.: Z. anorg. allg. Chem. *301*, 113 (1959)

115. Selig, H., Shamir, J.: Inorg. Chem. *3*, 294 (1964)
116. Senior, J. B., Grover, J. L.: Can. J. Chem. *49*, 2688 (1971)
117. a) Shamir, J.: Isr. J. Chem. *17*, 37 (1978)
b) Shamir, J., Yaroslavsky, I.: Isr. J. Chem. *7*, 495 (1969)
118. Shamir, J., Lustig, M.: Inorg. Nucl. Chem. Letters *8*, 958 (1972)
119. Shamir, J., Rafaeloff, R.: Spectrochim. Acta *29*, 873 (1973)
120. Shamir, J., Lustig, M.: Inorg. Chem. *12*, 1108 (1973)
121. Sharpe, A. G.: J. Chem. Soc. *1949*, 2901
122. Sharpe, A. G.: ibid. *1950*, 3444
123. Sheft, I., Hyman, H. H., Katz, J. J.: J. Am. Chem. Soc. *75*, 5221 (1953)
124. Sheft, I., Martin, A. F., Katz, J. J.: ibid. *78*, 1557 (1956)
125. Stein, L.: Science *175*, 1463 (1972)
126. Sufonov, V. V., Abramova, E. A., Korshunov, B. G.: Zh. Neogr. Khim. *18*, 568 (1973) C.A. *78*, 16863v
127. Sukhoverkhov, V. F., Dzevitskii, D. E.: Dokl. Akad. Nauk. SSSR *170*, 1099 (1966) C.A. *67*, 59300j
128. Sukhoverkhov, V. F., Dzevitskii, B. Z.: ibid. *177*, 611 (1967) C.A. *68*, 55243r
129. Sukhoverkhov, V. F., Shpanko, V. I.: Zh. Neorg. Khim. *20*, 3083 (1975) C.A. *84*, 50451c
130. Sukhoverkhov, V. F., Shpanko, V. I.: ibid. *21*, 1109 (1976) C.A. *84*, 185532a
133. Surles, T., Hyman, H. H., Quarterman, L. A., Popov, A. I.: ibid. *10*, 913 (1971)
134. Surles, T., Perkins, A., Quarterman, L. A., Hyman, H. H., Popov, A. I.: J. Inorg. Nucl. Chem. *34*, 3561 (1972)
135. Surles, T., Quarterman, L. A., Hyman, H. H.: J. Fluor. Chem. *3*, 293 (1973)
136. Surles, T., Quarterman, L. A., Hyman, H. H.: ibid. *3*, 453 (1973)
137. Toy, M. S., Cannon, W. A.: J. Phys. Chem. *70*, 2241 (1966)
138. Ungemach, S. R., Schaefer, H. F.: Chem. Phys. Letters *38*, 407 (1976)
139. Ungemach, S. R., Schaefer, H. F.: J. Am. Chem. Soc. *98*, 1658 (1976)
140. Vonk, C. G., Wiebenga, E. H.: Rec. Trav. Chim. *78*, 913 (1959)
141. Vonk, C. G., Wiebenga, E. H.: Acta Cryst. *12*, 859 (1959)
142. Wilson, W. W., Aubke, F.: Inorg. Chem. *13*, 326 (1974)
143. Wilson, W. W., Dalziel, J. R., Aubke, F.: J. Inorg. Nucl. Chem. *37*, 665 (1975)
144. Wilson, W. W., Winfield, J. M., Aubke, F.: J. Fluor. Chem. *7*, 245 (1976)
145. Woolf, A. A., Emeleus, H. J.: J. Chem. Soc. *1949*, 2865
146. Woolf, A. A., Greenwood, N. N.: ibid. *1950*, 2200
147. Woolf, A. A.: ibid. *1950*, 3678
148. Yeats. P. A., Wilson, W. W., Aubke, F.: Inorg. Nucl. Chem. Letters *9*, 209 (1973)

Author-Index Volumes 1—37

Ahrland, S.: Factors Contributing to (b)-behaviour in Acceptors. Vol. 1, pp. 207—220

Ahrland, S.: Thermodynamics of Complex Formation between Hard and Soft Acceptore and Donors. Vol. 5, pp. 118—149.

Ahrland, S.: Thermodynamics of the Stepwise Formation of Metal-Ion Complexes in Aqueous Solution. Vol. 15, pp. 167—188.

Allen, G. C., Warren, K. D.: The Electronic Spectra of the Hexafluoro Complexes of the First Transition Series. Vol. 9, pp. 49—138.

Allen, G. C., Warren, K. D.: The Electronic Spectra of the Hexafluoro Complexes of the Second and Third Transition Series. Vol. 19, pp. 105—165.

Babel, D.: Structural Chemistry of Octahedral Fluorocomplexes of the Transition Elements. Vol. 3, pp. 1—87.

Baker, E. C., Halstead, G. W., Raymond, K. N.: The Structure and Bonding of 4*f* and 5*f* Series Organometallic Compounds. Vol. 25, pp. 21—66.

Baughan, E. C.: Structural Radii, Electron-cloud Radii, Ionic Radii and Solvation. Vol. 15, pp. 53—71.

Bayer, E., Schretzmann, P.: Reversible Oxygenierung von Metallkomplexen. Vol. 2, pp. 181—250.

Bearden, A. J., Dunham, W. R.: Iron Electronic Configurations in Proteins: Studies by Mössbauer Spectroscopy. Vol. 8, pp. 1—52.

Blasse, G.: The Influence of Charge-Transfer and Rydberg States on the Luminescence Properties of Lanthanides and Actinides. Vol. 26, pp. 43—79.

Blauer, G.: Optical Activity of Conjugated Proteins. Vol. 18, pp. 69—129.

Bonnelle, C.: Band and Localized States in Metallic Thorium, Uranium and Plutonium, and in Some Compounds, Studied by X-Ray Spectroscopy. Vol. 31, pp. 23—48.

Bradshaw, A. M., Cederbaum, L. S., Domcke, W.: Ultraviolet Photoelectron Spectroscopy of Gases Adsorbed on Metal Surfaces. Vol. 24, pp. 133—170.

Braterman, P. S.: Spectra and Bonding in Metal Carbanyls. Part A: Bonding. Vol. 10, pp. 57—86.

Braterman, P. S.: Spectra and Bonding in Metal Carbonyls. Part B: Spectra and Their Interpretation. Vol. 26, pp. 1—42.

Bray, R. C., Swann, J. C.: Molybdenum-Containing Enzymes. Vol. 11, pp. 107—144.

van Bronswyk, W.: The Application of Nuclear Quadrupole Resonance Spectroscopy to the Study of Transition Metal Compounds. Vol. 7, pp. 87—113.

Buchanan, B. B.: The Chemistry and Function of Ferredoxin. Vol. 1, pp. 109—148.

Buchler, J. W., Kokisch, W., Smith, P. D.: *Cis*, *Trans*, and *Metal* Effects in Transition Metal Porphyrins. Vol. 34, pp. 79—134.

Bulman, R. A.: Chemistry of Plutonium and the Transuranics in the Biosphere. Vol. 34, pp. 39—77.

Burdett, J. K.: The Shapes of Main-Group Molecules; A Simple Semi-Quantitative Molecular Orbital Approach. Vol. 31, pp. 67—105.

Campagna, M., Wertheim, G. K., Bucher, E.: Spectroscopy of Homogeneous Mixed Valence Rare Earth Compounds. Vol. 30, pp. 99—140.

Cheh, A. M., Neilands, J. P.: The δ-Aminolevulinate Dehydratases: Molecular and Environmental Properties. Vol. 29, pp. 123—169.

Ciampolini, M.: Spectra of 3d Five-Coordinate Complexes. Vol. 6, pp. 52—93.

Clark, R. J. H., Stewart, B.: The Resonance Raman Effect. Review of the Theory and of Applications in Inorganic Chemistry. Vol. 36, pp. 1—80.

Cook, D. B.: The Approximate Calculation of Molecular Electronic Structures as a Theory of Valence. Vol. 35, pp. 37—86.

Cox, P.A.: Fractional Parentage Methods for Ionisation of Open Shells of *d* and *f* Electrons. Vol. 24, pp. 59—81.

Crichton, R.R.: Ferritin. Vol. 17, pp. 67—134.

Daul, C., Schläpfer, C.W., von Zelewsky, A.: The Electronic Structure of Cobalt (II) Complexes with Schiff Bases and Related Ligands. Vol. 36, pp. 129—171.

Dehnicke, K., Shihada, A.-F.: Structural and Bonding Aspects in Phosphorus Chemistry-Inorganic Derivates of Oxohalogeno Phosphoric Acids. Vol. 28, pp. 51—82.

Drago, R.S.: Quantitative Evaluation and Prediction of Donor-Acceptor Interactions. Vol. 15, pp. 73—139.

Duffy, J.A.: Optical Electronegativity and Nephelauxetic Effect in Oxide Systems. Vol. 32, pp. 147—166.

Dunn, M.F.: Mechanisms of Zinc Ion Catalysis in Small Molecules and Enzymes. Vol. 23, pp. 61—122.

Ermer, O.: Calculations of Molecular Properties Using Force Fields. Applications in Organic Chemistry. Vol. 27, pp. 161—211.

Erskine, R.W., Field, B.O.: Reversible Oxygenation. Vol. 28, pp. 1—50.

Fajans, K.: Degrees of Polarity and Mutual Polarization of Ions in the Molecules of Alkali Fluorides, SrO, and BaO. Vol. 3, pp. 88—105.

Fee, J.A.: Copper Proteins — Systems Containing the "Blue" Copper Center. Vol. 23, pp. 1—60.

Feeney, R.E., Komatsu, S.K.: The Transferrins. Vol. 1, pp. 149—206.

Felsche, J.: The Crystal Chemistry of the Rare-Earth Silicates. Vol. 13, pp. 99—197.

Ferreira, R.: Paradoxical Violations of Koopmans' Theorem, with Special Reference to the 3d Transition Elements and the Lanthanides. Vol. 31, pp. 1—21.

Fraga, S., Valdemoro, C.: Quantum Chemical Studies on the Submolecular Structure of the Nucleic Acids. Vol. 4, pp. 1—62.

Fraústo da Silva, J.J.R., Williams, R.J.P.: The Uptake of Elements by Biological Systems. Vol. 29, pp. 67—121.

Fricke, B.: Superheavy Elements. Vol. 21, pp. 89—144.

Fuhrhop, J.-H.: The Oxidation States and Reversible Redox Reactions of Metalloporphyrins. Vol. 18, pp. 1—67.

Furlani, C., Cauletti, C.: He(I) Photoelectron Spectra of *d*-Metal Compounds. Vol. 35, pp. 119—169.

Gillard, R.D., Mitchell, P.R.: The Absolute Configuration of Transition Metal Complexes. Vol. 7, pp. 46—86.

Griffith, J.S.: On the General Theory of Magnetic Susceptibilities of Polynuclear Transition-metal Compounds. Vol. 10, pp. 87—126.

Gutmann, V., Mayer, U.: Thermochemistry of the Chemical Bond. Vol. 10, pp. 127—151.

Gutmann, V., Mayer, U.: Redox Properties: Changes Effected by Coordination. Vol. 15, pp. 141—166.

Gutmann, V., Mayer, H.: Application of the Functional Approach to Bond Variations under Pressure. Vol. 31, pp. 49—66.

Hall, D.I., Ling, J.H., Nyholm, R.S.: Metal Complexes of Chelating Olefin-Group V Ligands. Vol. 15, pp. 3—51.

Harnung, S.E., Schäffer, C.E.: Phase-fixed 3-Γ Symbols and Coupling Coefficients for the Point Groups. Vol. 12, pp. 201—255.

Harnung, S.E., Schäffer, C.E.: Real Irreducible Tensorial Sets and their Application to the Ligand-Field Theory. Vol. 12, pp. 257—295.

Hathaway, B.J.: The Evidence for "Out-of-the-Plane" Bonding in Axial Complexes of the Copper(II) Ion. Vol. 14, pp. 49—67.

Hellner, E. E.: The Frameworks (Bauverbände) of the Cubic Structure Types. Vol. 37, pp. 61—140.

von Herigonte, P.: Electron Correlation in the Seventies. Vol. 12, pp. 1—47.

Hill, H. A. O., Röder, A., Williams, R. J. P.: The Chemical Nature and Reactivity of Cytochrome P-450. Vol. 8, pp. 123—151.

Hogenkamp, H. P. C., Sando, G. N.: The Enzymatic Reduction of Ribonucleotides. Vol. 20, pp. 23—58.

Hoffman, D. K., Ruedenberg, K., Verkade, J. G.: Molecular Orbital Bonding Concepts in Polyatomic Molecules — A Novel Pictorial Approach. Vol. 33, pp. 57—96.

Hubert, S., Hussonnois, M., Guillaumont, R.: Measurement of Complexing Constants by Radiochemical Methods. Vol. 34, pp. 1—18.

Hudson, R. F.: Displacement Reactions and the Concept of Soft and Hard Acids and Bases. Vol. 1, pp. 221—223.

Hulliger, F.: Crystal Chemistry of Chalcogenides and Pnictides of the Transition Elements. Vol. 4, pp. 83—229.

Iqbal, Z.: Intra- und Inter-Molecular Bonding and Structure of Inorganic Pseudohalides with Triatomic Groupings. Vol. 10, pp. 25—55.

Izatt, R. M., Eatough, D. J., Christensen, J. J.: Thermodynamics of Cation-Macrocyclic Compound Interaction. Vol. 16, pp. 161—189.

Jerome-Lerutte, S.: Vibrational Spectra and Structural Properties of Complex Tetracyanides of Platinum, Palladium and Nickel. Vol. 10, pp. 153—166.

Jørgensen, C. K.: Electric Polarizability. Innocent Ligands and Spectroscopic Oxidation States. Vol. 1, pp. 234—248.

Jørgensen, C. K.: Recent Progress in Ligand Field Theory. Vol. 1, pp. 3—31.

Jørgensen, C. K.: Relations between Softness, Covalent Bonding, Ionicity and Electric Polarizability. Vol. 3, pp. 106—115.

Jørgensen, C. K.: Valence-Shell Expansion Studied by Ultra-violet Spectroscopy. Vol. 6, pp. 94—115.

Jørgensen, C. K.: The Inner Mechanism of Rare Earths Elucidated by Photo-Electron Spectra. Vol. 13, pp. 199—253.

Jørgensen, C. K.: Partly Filled Shells Constituting Anti-bonding Orbitals with Higher Ioniza tion Energy than their Bonding Counterparts. Vol. 22, pp. 49—81.

Jørgensen, C. K.: Photo-electron Spectra of Non-metallic Solids and Consequences for Quantum Chemistry. Vol. 24, pp. 1—58.

Jørgensen, C. K.: Narrow Band Thermoluminescence (Candoluminescence) of Rare Earths in Auer Mantles. Vol. 25, pp. 1—20.

Jørgensen, C. K.: Deep-lying Valence Orbitals and Problems of Degeneracy and Intensities in Photo-electron Spectra. Vol. 30, pp. 141—192.

Jørgensen, C. K.: Predictable Quarkonium Chemistry. Vol. 34, pp. 19—38.

Kimura, T.: Biochemical Aspects of Iron Sulfur Linkage in None-Heme Iron Protein, with Special Reference to "Adrenodoxin". Vol. 5, pp. 1—40.

Kjekshus, A., Rakke, T.: Considerations on the Valence Concept. Vol. 19, pp. 45—83.

Kjekshus, A., Rakke, T.: Geometrical Considerations on the Marcasite Type Structure. Vol. 19, pp. 85—104.

König, E.: The Nephelauxetic Effect. Calculation and Accuracy of the Interelectronic Repulsion Parameters I. Cubic High-Spin d^2, d^3, d^7 and d^8 Systems. Vol. 9, pp. 175—212.

Koppikar, D. K., Sivapullaiah, P. V., Ramakrishnan, L., Soundararajan, S.: Complexes of the Lanthanides with Neutral Oxygen Donor Ligands. Vol. 34, pp. 135—213.

Krumholz, P.: Iron(II) Diimine and Related Complexes. Vol. 9, pp. 139—174.

Labarre, J. F.: Conformational Analysis in Inorganic Chemistry: Semi-Empirical Quantum Calculation vs. Experiment. Vol. 35, pp. 1—35.

Lehn, J.-M.: Design of Organic Complexing Agents. Strategies towards Properties. Vol. 16, pp. 1—69.

Linarès, C., Louat, A., Blanchard, M.: Rare-Earth Oxygen Bonding in the $LnMO_4$ Xenotime Structure. Vol. 33, pp. 179—207.

Lindskog, S.: Cobalt(II) in Metalloenzymes. A Reporter of Structure-Function Relations. Vol. 8, pp. 153—196.

Llinás, M.: Metal-Polypeptide Interactions: The Confirmational State of Iron Proteins. Vol. 17, pp. 135—220.

Lucken, E.A.C.: Valence-Shell Expansion Studied by Radio-Frequency Spectroscopy. Vol. 6, pp. 1—29.

Ludi, A., Güdel, H. U.: Structural Chemistry of Polynuclear Transition Metal Cyanides. Vol. 14, pp. 1—21.

Maggiora, G. M., Ingraham, L. L.: Chlorophyll Triplet States. Vol. 2, pp. 126—159.

Magyar, B.: Salzebullioskopie III. Vol. 14, pp. 111—140.

Mayer, U., Gutmann, V.: Phenomenological Approach to Cation-Solvent Interactions. Vol. 12, pp. 113—140.

Mildvan, A. S., Grisham, C. M.: The Role of Divalent Cations in the Mechanism of Enzyme Catalyzed Phosphoryl and Nucleotidyl. Vol. 20, pp. 1—21.

Moreau-Colin, M. L.: Electronic Spectra and Structural Properties of Complex Tetracyanides of Platinum, Palladium and Nickel. Vol. 10, pp. 167—190.

Morris, D. F. C.: Ionic Radii and Enthalpies of Hydration of Ions. Vol. 4, pp. 63—82.

Mooris, D. F. C.: An Appendix to Structure and Bonding. Vol. 4 (1968). Vol. 6, pp. 157—159.

Müller, A., Baran, E. J., Carter, R. O.: Vibrational Spectra of Oxo-, Thio-, and Selenometallates of Transition Elements in the Solid State. Vol. 26, pp. 81—139.

Müller, A., Diemann, E., Jørgensen, C. K.: Electronic Spectra of Tetrahedral Oxo, Thio and Seleno Complexes. Formed by Elements of the Beginning of the Transition Groups. Vol. 14, pp. 23—47.

Müller, U.: Strukturchemie der Azide. Vol. 14, pp. 141—172.

Murrell, J. N.: The Potential Energy Surfaces of Polyatomic Molecules. Vol. 32, pp. 93—146.

Neilands, J. B.: Naturally Occurring Non-porphyrin Iron Compounds. Vol. 1, pp. 59—108.

Neilands, J. B.: Evolution of Biological Iron Binding Centers. Vol. 11, pp. 145—170.

Nieboer, E.: The Lanthanide Ions as Structural Probes in Biological and Model Systems. Vol. 22, pp. 1—47.

Novack, A.: Hydrogen Bonding in Solids. Correlation of Spectroscopic and Cristallographic Data. Vol. 18, pp. 177—216.

Oelkrug, D.: Absorption Spectra and Ligand Field Parameters of Tetragonal 3*d*-Transition Metal Fluorides. Vol. 9, pp. 1—26.

Oosterhuis, W. T.: The Electronic State of Iron in Some Natural Iron Compounds: Determination by Mössbauer and ESR Spectroscopy. Vol. 20, pp. 59—99.

Orchin, M., Bollinger, D. M.: Hydrogen-Deuterium Exchange in Aromatic Compounds. Vol. 23, pp. 167—193.

Peacock, R. D.: The Intensities of Lanthanide $f \leftrightarrow f$ Transitions. Vol. 22, pp. 83—122.

Penneman, R. A., Ryan, R. R., Rosenzweig, A.: Structural Systematics in Actinide Fluoride Complexes. Vol. 13, pp. 1—52.

Reinen, D.: Ligand-Field Spectroscopy and Chemical Bonding in Cr^{3+}-Containing Oxidic Solids. Vol. 6, pp. 30—51.

Reinen, D.: Kationenverteilung zweiwertiger $3d^n$-Ionen in oxidischen Spinell-, Granat- und anderen Strukturen. Vol. 7, pp. 114—154.

Reinen, D., Friebel, C.: Local and Cooperative Jahn-Teller Interactions in Model Structures. Spectroscopic and Structural Evidence. Vol. 37, pp. 1—60.

Reisfeld, R.: Spectra and Energy Transfer of Rare Earths in Inorganic Glasses. Vol. 13, pp. 53—98.

Reisfeld, R.: Radiative and Non-Radiative Transitions of Rare Earth Ions in Glasses. Vol. 22, pp. 123—175.

Reisfeld, R.: Excited States and Energy Transfer from Donor Cations to Rare Earths in the Condensed Phase. Vol. 30, pp. 65—97.

Sadler, P. J.: The Biological Chemistry of Gold: A Metallo-Drug and Heavy-Atom Label with Variable Valency, Vol. 29, pp. 171—214.

Schäffer, C. E.: A Perturbation Representation of Weak Covalent Bonding. Vol. 5, pp. 68—95.

Schäffer, C. E.: Two Symmetry Parameterizations of the Angular-Overlap Model of the Ligand-Field. Relation to the Crystal-Field Model. Vol. 14, pp. 69—110.

Schneider, W.: Kinetics and Mechanism of Metalloporphyrin Formation. Vol. 23, pp. 123—166.

Schubert, K.: The Two-Correlations Model, a Valence Model for Metallic Phases. Vol. 33, pp. 139—177.

Schutte, C. J. H.: The Ab-Initio Calculation of Molecular Vibrational Frequencies and Force Constants. Vol. 9, pp. 213—263.

Shamir, J.: Polyhalogen Cations. Vol. 37, pp. 141—210.

Shannon, R. D., Vincent, H.: Relationship between Covalency, Interatomatic Distances, and Magnetic Properties in Halides and Chalcogenides. Vol. 19, pp. 1—43.

Shriver, D. F.: The Ambident Nature of Cyanide. Vol. 1, pp. 32—58.

Siegel, F. L.: Calcium-Binding Proteins. Vol. 17, pp. 221—268.

Simon, A.: Structure and Bonding with Alkali Metal Suboxides. Vol. 36, pp. 81—127.

Simon, W., Morf, W. E., Meier, P. Ch.: Specificity for Alkali and Alkaline Earth Cations of Synthetic and Natural Organic Complexing Agents in Membranes. Vol. 16, pp. 113—160.

Simonetta, M., Gavezzotti, A.: Extended Hückel Investigation of Reaction Mechanisms. Vol. 27, pp. 1—43.

Sinha, S. P.: Structure and Bonding in Highly Coordinated Lanthanide Complexes. Vol. 25, pp. 67—147.

Sinha, S. P.: A Systematic Correlation of the Properties of the f-Transition Metal Ions. Vol. 30, pp. 1—64.

Smith, D. W.: Ligand Field Splittings in Copper(II) Compounds. Vol. 12, pp. 49—112.

Smith, D. W., Williams, R. J. P.: The Spectra of Ferric Haems and Haemoproteins. Vol. 7, pp. 1—45.

Smith, D. W.: Applications of the Angular Overlap Model. Vol. 35, pp. 87—118.

Speakman, J. C.: Acid Salts of Carboxylic Acids, Crystals with some "Very Short" Hydrogen Bonds. Vol. 12, pp. 141—199.

Spiro, G., Saltman, P.: Polynuclear Complexes of Iron and their Biological Implications. Vol. 6, pp. 116—156.

Strohmeier, W.: Problem und Modell der homogenen Katalyse. Vol. 5, pp. 96—117.

Thompson, D. W.: Structure and Bonding in Inorganic Derivatives of β-Diketones. Vol. 9 pp. 27—47.

Thomson, A. J., Williams, R. J. P., Reslova, S.: The Chemistry of Complexes Related to *cis*-$Pt(NH_3)_2Cl_2$. An Anti-Tumour Drug. Vol. 11, pp. 1—46.

Tofield, B. C.: The Study of Covalency by Magnetic Neutron Scattering. Vol. 21, pp. 1—87.

Trautwein, A.: Mössbauer-Spectroscopy on Heme Proteins. Vol. 20, pp. 101—167.

Truter, M. R.: Structures of Organic Complexes with Alkali Metal Ions. Vol. 16, pp. 71—111.

Vahrenkamp, H.: Recent Results in the Chemistry of Transition Metal Clusters with Organic Ligands. Vol. 32, pp. 1—56.

Wallace, W. E., Sankar, S. G., Rao, V. U. S.: Field Effects in Rare-Earth Intermetallic Compounds. Vol. 33, pp. 1—55.

Warren, K. D.: Ligand Field Theory of Metal Sandwich Complexes. Vol. 27, pp. 45—159.

Warren, K. D.: Ligand Field Theory of f-Orbital Sandwich Complexes. Vol. 33, pp. 97—137.

Watson, R. E., Perlman, M. L.: X-Ray Photoelectron Spectroscopy. Application to Metals and Alloys. Vol. 24, pp. 83—132.

Weakley, T. J. R.: Some Aspects of the Heteropolymolybdates and Heteropolytungstates. Vol. 18, pp. 131—176.

Weissbluth, M.: The Physics of Hemoglobin. Vol. 2, pp. 1—125.

Weser, U.: Chemistry and Structure of some Borate Polyol Compounds. Vol. 2, pp. 160—180.

Weser, U.: Reaction of some Transition Metals with Nucleic Acids and their Constituents. Vol. 5, pp. 41—67.

Weser, U.: Structural Aspects and Biochemical Function of Erythrocuprein. Vol. 17, pp. 1—65.

Willemse, J., Cras, J. A., Steggerda, J. J., Keijzers, C. P.: Dithiocarbamates of Transition Group Elements in "Unusual" Oxidation State. Vol. 28, pp. 83—126.

Williams, R. J. P., Hale, J. D.: The Classification of Acceptors and Donors in Inorganic Reactions. Vol. 1, pp. 249—281.

Williams, R. J. P., Hale, J. D.: Professor Sir Ronald Nyholm. Vol. 15, p. 1 and 2.

Wilson, J. A.: A Generalized Configuration-Dependent Band Model for Lanthanide Compounds and Conditions for Interconfiguration Fluctuations. Vol. 32, pp. 57—91.

Winkler, R.: Kinetics and Mechanism of Alkali Ion Complex Formation in Solution. Vol. 10, pp. 1—24.

Wood, J. M., Brown, D. G.: The Chemistry of Vitamin B_{12}-Enzymes. Vol. 11, pp. 47—105.

Wüthrich, K.: Structural Studies of Hemes and Hemoproteins by Nuclear Magnetic Resonance Spectroscopy. Vol. 8, pp. 53—121.

Zumft, W. G.: The Molecular Basis of Biological Dinitrogen Fixation. Vol. 29, pp. 1—65.